VOLUME ONE HUNDRED AND ELEVEN

INTERNATIONAL REVIEW OF NEUROBIOLOGY

Neurobiology of Acupuncture

INTERNATIONAL REVIEW OF NEUROBIOLOGY

VOLUME 111

SERIES EDITORS

R. ADRON HARRIS

Waggoner Center for Alcohol and Drug Addiction Research
The University of Texas at Austin
Austin, Texas, USA

PETER JENNER

Division of Pharmacology and Therapeutics
GKT School of Biomedical Sciences
King's College, London, UK

VOLUME ONE HUNDRED AND ELEVEN

INTERNATIONAL REVIEW OF NEUROBIOLOGY

Neurobiology of Acupuncture

Edited by

BAI-YUN ZENG

Neurodegenerative Disease Research Group, Institute of Pharmaceutical Science, School of Biomedical Sciences King's College London, London, United Kingdom

KAICUN ZHAO

Department of Natural Sciences, Middlesex University London, The Burroughs, Hendon, London, United Kingdom

FAN-RONG LIANG

Acupuncture and Tuina School, Chengdu University of Traditional Chinese Medicine, Chengdu, China

ELSEVIER

AMSTERDAM • BOSTON • HEIDELBERG • LONDON
NEW YORK • OXFORD • PARIS • SAN DIEGO
SAN FRANCISCO • SINGAPORE • SYDNEY • TOKYO
Academic Press is an imprint of Elsevier

AP

Academic Press is an imprint of Elsevier
32 Jamestown Road, London NW1 7BY, UK
Radarweg 29, PO Box 211, 1000 AE Amsterdam, The Netherlands
The Boulevard, Langford Lane, Kidlington, Oxford, OX5 1GB, UK
225 Wyman Street, Waltham, MA 02451, USA
525 B Street, Suite 1800, San Diego, CA 92101-4495, USA

First edition 2013

ISBN: 978-0-12-411545-3
ISSN: 0074-7742

For information on all Academic Press publications
visit our website at store.elsevier.com

Printed and bound by CPI Group (UK) Ltd, Croydon, CR0 4YY

CONTENTS

CONTRIBUTORS

Kwang Seok Ahn
Department of Pathology, College of Korean Medicine, Kyung Hee University, Seoul, Republic of Korea

Seung-Hoon Choi
Department of Pathology, College of Korean Medicine, Kyung Hee University, Seoul, and Korea Institute of Oriental Medicine, Daejeon, Republic of Korea

Cai-Lian Cui
Neuroscience Research Institute & Department of Neurobiology, School of Basic Medical Sciences, and Key Laboratory for Neuroscience, Ministry of Education /National Health and Family Planning Commission, Peking University, Beijing, PR China

Fulvio Florenzano
Confocal Microscopy Unit, European Brain Research Institute (EBRI)—Institute of Cellular Biology and Neurobiology, National Research Council (CNR), Rome, Italy

Ching-Liang Hsieh
Graduate Institute of Integrated Medicine, College of Chinese Medicine; Acupuncture Research Center, and Department of Chinese Medicine, China Medical University Hospital, Taichung, Taiwan, ROC

Peter Jenner
Neurodegenerative Disease Research Group, Institute of Pharmaceutical Science, School of Biomedical Sciences, King's College London, London, United Kingdom

Juan Li
Acupuncture and Tuina School, Chengdu University of Traditional Chinese Medicine, Chengdu, China

Xin Li
Department of Anesthesiology, Xijing Hospital, Fourth Military Medical University, Xi'an, China

Yi-jing Li
Neuroscience Research Institute & Department of Neurobiology, School of Basic Medical Sciences, and Key Laboratory for Neuroscience, Ministry of Education /National Health and Family Planning Commission, Peking University, Beijing, PR China

Fan-rong Liang
Acupuncture and Tuina School, Chengdu University of Traditional Chinese Medicine, Chengdu, China

John C. Longhurst
Department of Medicine, Physiology and Biophysics, and Pharmacology, School of Medicine, University of California, Medical Sciences, Irvine, California, USA

Luigi Manni
Institute of Translational Pharmacology, National Research Council (CNR), Rome, Italy

Min-Ho Nam
Department of Pathology, College of Korean Medicine, Kyung Hee University, Seoul, Republic of Korea

Hisao Nishijo
System Emotional Science, University of Toyama, Toyama, Japan

Stefania Lucia Nori
Department of Pharmaceutical and Biomedical Sciences (FARMABIOMED) Nanomates, University of Salerno, Fisciano, Italy

Taketoshi Ono
Department of Judo Neurophysiotherapy, Graduate School of Medicine and Pharmaceutical Sciences, University of Toyama, Toyama, Japan

Karen Pilkington
Faculty of Science and Technology, University of Westminster, London, United Kingdom

Virginia Protto
Institute of Translational Pharmacology, National Research Council (CNR), Rome, Italy

Kazushige Sakai
Department of Judo Neurophysiotherapy, Graduate School of Medicine and Pharmaceutical Sciences, University of Toyama, Toyama, Japan

Sarah Salvage
Neurodegenerative Disease Research Group, Institute of Pharmaceutical Science, School of Biomedical Sciences, King's College London, London, United Kingdom

Marzia Soligo
Institute of Translational Pharmacology, National Research Council (CNR), Rome, Italy

Toku Takahashi
Department of Neurology and Department of Surgery, Medical College of Wisconsin and Zablocki VA Medical Center, Milwaukee, Wisconsin, USA

Kouich Takamoto
System Emotional Science, and Department of Judo Neurophysiotherapy, Graduate School of Medicine and Pharmaceutical Sciences, University of Toyama, Toyama, Japan

Stephanie Tjen-A-Looi
Department of Medicine, School of Medicine, University of California, Irvine, California, USA

Susumu Urakawa
System Emotional Science, and Department of Judo Neurophysiotherapy, Graduate School of Medicine and Pharmaceutical Sciences, University of Toyama, Toyama, Japan

Qiang Wang
Department of Anesthesiology, Xijing Hospital, Fourth Military Medical University, Xi'an, China

Liu-Zhen Wu
Neuroscience Research Institute & Department of Neurobiology, School of Basic Medical Sciences, and Key Laboratory for Neuroscience, Ministry of Education /National Health and Family Planning Commission, Peking University, Beijing, PR China

Jing-jing Xing
Acupuncture and Tuina School, Chengdu University of Traditional Chinese Medicine, Chengdu, China

Jung-Sheng Yu
Department of Chinese Medicine, Chi Mei Medical Center, and Department of Cosmetic Science and Institute of Cosmetic Science, College of Pharmacy and Science, Chia Nan University of Pharmacy & Science, Tainan, Taiwan, ROC

Bai-Yun Zeng
Neurodegenerative Disease Research Group, Institute of Pharmaceutical Science, School of Biomedical Sciences, King's College London, London, United Kingdom

Kaicun Zhao
Department of Natural Sciences, Middlesex University London, The Burroughs, Hendon, London, United Kingdom

Yi Zhuang
Acupuncture and Tuina School, Chengdu University of Traditional Chinese Medicine, Chengdu, China

PREFACE

Acupuncture is a fundamental part of traditional Chinese medicine. Over its 3000 years of development, a wealth of experience has been accumulated in the practice of acupuncture, attesting to a wide range of diseases and conditions that can be effectively treated by this approach. Modern acupuncture can be defined as insertion of fine thin needles into the specific points on the body with mechanical, electrical, or other physical manipulations, which stimulate nerve receptors both directly or indirectly mechanical coupling with connective tissue surrounding the needles. In general, the acupuncture stimulation, through the local reflex and central nervous system, induces endocrine, neuroendocrine, autonomic, and systemic behavioral responses. This suggests that acupuncture therapy may beneficially affect a whole body even if it, using a few fine needles, stimulates only limited sites of the body.

Within past three decades, there has been a rapid progress in acupuncture research including neuroscience due to the development of modern technology and ever increasing demand for a more efficient and cost-effective health-care system. We believe that there is a need to compile a book to summarize the recent developments in acupuncture research and, in particular, the neurobiology of acupuncture, and that this will promote understanding and enhance acupuncture research in the future. In this volume, a talented group of scientists and/or clinicians review the major breakthrough and capture the status of the fields, from clinic issues to basic science and feedback to the clinic, from body systems to tissues, to cells and molecules; from past to present days and future challenges. We have tried to cover as many topics as possible although it is ultimately impossible to get every topic in one volume regarding vast amount of information in the literature.

Recently, efficacy in acupuncture has been intensively debated due to the failure of some large scales of clinical trials in chronic pain. This is a critical and urgent issue for future acupuncture research. "Acupuncture point specificity" by J-.J. Xing reviews recent development in basic science and clinical studies on the role of acupoint specificity. Evidence cumulated from brain image and biological studies showed that acupoint specificity does exist although acupuncture needling, acupoint location, and more importantly acupoint specificity-related issues, such as sham acupuncture, and placebo phenomena should be seriously considered before analyzing the efficacy of acupuncture.

Bench science provides fundamental insight and mechanisms of many subjects including acupuncture. "Effects of acupuncture needling with specific sensation on cerebral hemodynamics and autonomic nervous activity in humans" by K. Takamoto focuses on the cerebral hemodynamic responses and autonomic nervous activity changes induced by acupuncture stimulation with or without de-qi sensation. Interestingly, the authors argued that acupuncture-induced specific de-qi sensation is more important than specific acupoint stimulation for inducing cerebral hemodynamic responses and changes in autonomic nervous activity. Then, "Acupuncture stimulation induces neurogenesis in adult brain" by M-.H. Nam reviews recent studies on the effect of stimulation at ST36 and GV20 for adult neurogenesis. They showed that acupuncture-induced upregulation of neurotrophic factors such GDNF, BDNF, and activation of primo vascular system in the brain all contribute to the enhanced neurogenesis in the brain. "Acupuncture stimulation and neuroendocrine regulation" by J-.S. Yu provides evidence from literature review that somatic stimulation by acupuncture has specific pathways and central sites for its action in the spinal cord and different regions of brain, and needling-specific acupoint triggers the release of neurotransmitters, neurohormones, and opioids which have wide range of effects on different body systems. "Acupuncture and neurotrophin modulation" by M. Soligo reviews evidence from animal models and human subjects that acupuncture stimulation increases neurotrophin expression/action not only in peripheral system and target organs but also in spinal cord and brain although further studies are needed to investigate the underlying mechanisms of acupuncture on neurotrophin modulation.

"Acupuncture regulation of blood pressure: Two decades of research" by J.C. Longhurst reviews the development in acupuncture regulation of blood pressure in their laboratory. Their studies showed that acupoint stimulation activated somatic input to different brain regions, including ventral hypothalamic arcuate nucleus and midbrain vlPAG, leading to prolonged release of opioids and GABA and nociceptin, in addition to other neurotransmitter, acting postsynaptically to directly or indirectly modulate autonomic outflow, guiding to a long-lasting reduction of blood pressure. "Effect and mechanism of acupuncture on gastrointestinal diseases" by T. Takahashi focuses on effect of acupuncture stimulation on gastrointestinal system and finds that acupuncture stimulation of somato-sensory neurons activates various nuclei at the CNS, leading to modulating the imbalance between sympathetic and parasympathetic activity, releasing opioids and

OXT. These mechanisms underlie acupuncture treatment of GI disorders such as dysmotility, visceral hypersensitivity, and visceral pain.

"Acupuncture therapy for psychiatric illness" by K. Pilkington summarizes acupuncture-induced central mediators such as serotonin, norepinephrine, dopamine, and glutamate and their roles in specific psychological disorders such as anxiety disorders, depression, bipolar disorder, and schizophrenia. "Acupuncture for the treatment of insomnia" by K. Zhao provides a comprehensive understanding of the efficacy of acupuncture therapy for insomnia and a list of acupoints used in clinical trials between 2007 and 2013; and very useful suggestions that integrated approaches to explore the efficacy of acupuncture are applicable not only to insomnia but also to other conditions. "Acupuncture for the treatment of drug addiction" by C-.L. Cui retrospects three important steps in the acupuncture treatment of substance abuse, including pioneer work by Dr. Wen in Hong Kong, in 1972, that acupuncture at four body points and two ear points combined with electrical stimulation can relieve opioid withdrawal signs in the addicts. Then, Dr. Smith, in New York, finalized a protocol (1985), using only ear points without electrical stimulation for the treatment of cocaine dependence. Recently, Dr. Han from Beijing characterized a protocol (2005), using electrical stimulation of identified frequencies on body points to ameliorate heroin withdrawal signs and reduce relapse of heroin use.

Recently, the use of acupuncture in neurological disorders has been expanded from stroke to neurodegenerative diseases such as Parkinson's and Alzheimer's diseases. "Acupuncture therapy for stroke patients" by X. Li reviews literature of stroke and provides evidence from both animal models and clinic studies that acupuncture therapy is not only effective in improving stroke function recovery but can also be used as a preventive strategy which could lead to ischemic tolerance when combined with electrotherapy. "Current development of acupuncture research in Parkinson's disease" by B-.Y. Zeng focuses on the underlying mechanisms of acupuncture on Parkinson's models. Evidence has showed that acupuncture has neuroprotective potential by increasing release of neurotrophic factors in basal ganglia area, it can slow down cell death by exerting antioxidant and antiinflammatory effects, and further, it modulated neuronal activity in the basal ganglia output structure. "Effect and Mechanism of Acupuncture on Alzheimer's Disease" by B.Y. Zeng showed that acupuncture stimulation at Alzheimer's models enhanced cholinergic neurotransmission, trophic factor releasing, improve synaptic plasticity, and decrease the levels of Aβ proteins in the hippocampus and relevant brain regions. fMRI image of patients with AD demonstrated that acupuncture increased in the activity in the temporal lobe and prefrontal lobe, paralleling to the improving memory and cognitive function.

"History of acupuncture research" by Y. Zhuang not only retrospects the past acupuncture research but also reviews the current achievements and discusses the challenges faced by acupuncture researchers in future, in particular how to improve efficacy of acupuncture and minimize the impact of sham acupuncture and placebo phenomena, how to improve methodology in design and evaluation of clinical studies, and furthermore, how to reconcile the individualized treatment in TCM with blinding and standardized controlling clinical trials. It seems that the need for unified standards and evaluation index particularly in acupuncture clinical studies may be one of the rationales.

Finally, we would like to thank the excellent group of scientists who contributed a brilliant set of chapters to make up this volume, and dedicated editorial staff to accomplish this book.

Bai-Yun Zeng
Kaicun Zhao
Fan-rong Liang

CHAPTER ONE

History of Acupuncture Research

Yi Zhuang*, Jing-jing Xing*, Juan Li*, Bai-Yun Zeng†, Fan-rong Liang*,1

*Acupuncture and Tuina School, Chengdu University of Traditional Chinese Medicine, Chengdu, China
†Neurodegenerative Disease Research Group, Institute of Pharmaceutical Science, School of Biomedical Sciences, King's College London, London, United Kingdom
[1]Corresponding author: e-mail address: acuresearch@126.com

Contents

Abstract

The acupuncture has been practiced in China for more than 3000 years and was spread to Europe and American from the sixteenth to the nineteenth century. The history of acupuncture research was initiated in the eighteenth century and developed rapidly since then. In the past, physicians tried hard to apply acupuncture into clinical practice, while scientists were focused on the possible characteristics of acupoints and meridians. In the modern time, scientists have strived hard to evaluate the real effectiveness of acupuncture and the underlying physiological and biological mechanisms of acupuncture. Reviewing research history from past to present, we are delighted to witness this wonderful development. Accumulated evidences that acupuncture is beneficial in various conditions significantly enhanced our understanding the mechanisms of acupuncture treatment. However, there is still no conclusive evidence in acupuncture clinical studies. The clinical research still needs great improving, while the basic research results need to be appropriately transformed into clinical outcomes. Based on current achievements, we believe that although the challenges and difficulties exist, a more

International Review of Neurobiology, Volume 111
ISSN 0074-7742
http://dx.doi.org/10.1016/B978-0-12-411545-3.00001-8

collaborative, innovative, and integrated approach will help us to achieve further progress in future acupuncture research.

1. INTRODUCTION

Acupuncture has been practiced in China for more than 3000 years and is a natural healing therapy, which has gained increasing popularity and acceptance between public and heathcare professionals worldwide. It was introduced to Korea and Japan in the sixth century AD and spread to the Europe and North America in the sixteenth to nineteenth century AD. In 1971, a report by James Reston in the New York Times about his experience of acupuncture treatment in China exposed countless American for the first time to acupuncture. Since then, acupuncture research has been extensively conducted worldwide. In this chapter, we retrospect the past acupuncture research, review the current achievement, and discuss the challenges faced by acupuncture research in the future.

2. ACUPUNCTURE RESEARCH IN THE PAST

Acupuncture research in the past began in the 1800s and peaked in 1987, when the World Federation of Acupuncture and Moxibustion was formally founded. It comprised of two aspects: clinical and basic research. The studies in this period mainly focused on the pain-related disorders and the nature of acupoint and meridians, providing a solid underpinning for the development and promotion of acupuncture.

2.1. Clinical research

The first literature introducing acupuncture to Europe was published in 1683, by a Dutch doctor Wilhelm Ten Rhijne, who observed the practice of therapeutic needling used in patients with arthritis when he visited Japan (Berman et al., 2004; Bivins, 2001). Then, a German physician Englebert Kaempfer described the acupoints and equipments associated with the acupuncture needles, medical cases, needle holders, and assisted hammers in his book "History of Japan" in 1728 following his tour of Southeast Asia and Japan in the late sixteenth century (Bivins, 2001). However, it took almost 100 years to persuade the European physician to accept the clinical application of acupuncture. The first physician who applied acupuncture into clinical practice was Dr. Louis Joseph Berlioz in France. He published his case reports in 1816, which contained acupuncture therapy for rheumatism, arthritis, and stiff muscles and joints. The first English monograph on "acupuncturation," as it was generally called at that time, was written by a young surgeon named James Morss Churchill in 1823. He advised needle insertion on the site of pain in the myofascial disorders (Campbell, 2002). At the same time, a number of articles appeared in the scientific literatures such as a Lancet editorial article entitled "Acupuncturation" in 1823 (White, 2004). In 1825, an American, Benjamin Bache Franklin, firstly tested acupuncture analgesia on the prisoners in 1825 (Cassedy, 1974). The following year, he published the first clinical study of the acupuncture modality, a report of his observations on acupuncture for lumbago (low back pain), which may be the first article of acupuncture in the United States (Meng, Xu, & Lao, 2011). Probably the first published British case series, presenting the success of acupuncture in 1000 patients with severe sciatica, was in 1893 (Ernst, 2001).

The development of acupuncture has progressed unceasingly in the time of the twentieth century. The Frenchmen George Soulie de Morant (1878–1955), who was convinced of the importance of acupuncture when he witnessed the effect of acupuncture treatment during an epidemic of cholera in Beijing, got in touch with acupuncture therapy and learned by heart. He published many articles and works on acupuncture and his book "l'Acuponcture chinoise," which systematically introduced the acupoints and meridians, is still regarded as a classic work on acupuncture (De Morant, 1994). Another contributor was Roger de la Fuye, a student of de Morant, who combined the trigger points with acupuncture and drug injection, which greatly improved the clinical effect. Further, he wrote many textbooks on acupuncture and homeopathy (Stollberg, 2006). Due to the increasing interest in acupuncture, the French Acupuncture and Moxibustion Society was established in Paris in the 1940s, the first academic association of acupuncture and moxibustion in the world. In addition, the French physician P. Nogier firstly published the diagram of auricular points in 1957, which described auricular acupuncture treatments for a wide variety of illnesses (Ceniceros & Brown, 1998). These works contributed to spreading acupuncture therapy worldwide.

The acupuncture researches in the past were mostly clinical observation, individual case report, and personal experience introduction, not the systematic clinical trials. In the 1960s, the conception of biostatistics and methodologies, such as randomization that are essential to controlled clinical trials in Western medicine, has been widely accepted by acupuncture researchers in Asian countries. In the same period, the first controlled clinical trial was conducted in Japan (Shichido, 1996).

In the 1970s, the acupuncture research has step into a modern time. It was all enlightened by the famous report from James Reston in the New York Times in 1971 (Reston, 1971), which described his personal experience that the acupuncture therapy alleviated his abdominal distension after surgery. One year later, US President Richard Nixon visited China and watched the acupuncture performance. The *Journal of the American Medical Association* (JAMA) has firstly reported two articles in 1971, which separately introduced the medical care in China (Dimond, 1971a) and acupuncture anesthesia (Dimond, 1971b). Then, more scientists and medical doctors paid great attentions on the acupuncture phenomena and discussed its real clinical effectiveness.

In 1972, the National Institutes of Health (NIH) in the United States gave its first grant to acupuncture research (Ulett, Han, & Han, 1998). In 1973, the Food and Drug Administration (FDA) in United States has labeled acupuncture needles as "investigational" medical devices (Lytle, 1996). Simultaneously, some researchers discussed about the possibility of double-blind method and efficacy in acupuncture research (Chein & Shapito, 1973; Mark, 1973).

The first influential randomized controlled trial (RCT) on acupuncture was published in NEJM in 1975. Gaw et al. randomly assigned 40 patients with osteoarthritis pain into experimental or control group to assess the pain reduction by acupuncture. However, both experimental and control groups showed a reduction in pain after the treatments (Gaw, Chang, & Shaw, 1975). Another study was conducted to evaluate the chronic shoulder pain relief by acupuncture in 1976. The result showed no statistically significant differences between the classic and placebo acupuncture (the needles did not penetrate the skin) (MOORE & BERK, 1976). Then, Cahn, Carayon, Hill, and Flamant (1978) evaluated the analgesic effect of acupuncture by a double-blind controlled trial in 90 patients undergoing gastroscopy, which revealed real acupuncture could perform easier and better tolerance in endoscopy than control group. In the 1980s, acupuncture research expanded from pain-related conditions to other medical disorders. Pui Fung, Kit Wun Chow, and Yeung So (1986) showed verum significantly reduced the exercise-induced asthma compared to sham acupuncture (SA). At the same time, Jobst et al. (1986) applied acupuncture into patients with chronic obstructive pulmonary disease. Three-week experiment indicated the traditional acupuncture is more beneficial than placebo acupuncture in terms of subjective scores of breathlessness and six-minute walking distance; however, objective measures of lung function were unchanged in either group.

Take together, acupuncture has been gradually accepted by Western societies. In 1979, FDA classified acupuncture needles as class III (investigational) medical device but allows their clinical use by licensed practitioners (Hammerschlag, 2000). At the same year, WHO listed 43 kinds of diseases and conditions that can be cured by acupuncture and moxibustion, such as nausea and vomiting, pain, addictive diseases, asthma and bronchitis, and rehabilitation of stroke. The acupuncture research has stepped into a new era.

2.2. Basic research

The acupuncture basic research was also originated in the eighteenth century and has gathered increasing interest and acceptance worldwide. In 1755, the Vienna physician Gerard van Swieten wrote down his observation on the physiological communication involved in alleviation of pain using acupuncture and moxibustion (Bo-Ying & Grant, 2001). After that, Roughment combined physiological knowledge with acupuncture and deduced that acupuncture was a kind of counterirritation therapy in 1798 (Birch & Lewith, 2007). This proposal may roughly sketch the model of mechanism underlying acupuncture analgesia. Another important innovation during this time was the electrical stimulation employed in acupuncture basic research. Sarlandiere employed electrical stimulation to the inserted needles to see the altered treatment effects in 1825 (Lu & Needham, 1980). Moreover, other physicians, such as Trousseau, Pidoux, and Duchenne, reintroduced the electrical stimulation into the treatment of chronic pain (Willer, Roby, & Le Bars, 1984).

The acupuncture studies in Japan were initiated during the Meiji Restoration in the latter part of the nineteenth century, which were conducted in a way different from the Western countries. The Japanese scientists and doctors tried to analyze the meridians, acupoints, and potential mechanism by Western-style medical theory. Ohkubo Tekisai in 1894 held a view that acupuncture was a kind of stimulation on the nervous system. Therefore, his style of acupuncture was puncturing on the sympathetic ganglion (Lu & Needham, 1980). On the other hand, Hidetsurumaru et al. were concentrated on the influence of acupuncture and moxibustion on the autonomic nerve and blood (Chen, 2006). The phenomenon of propagated sensations along meridians was first observed by Japanese physician Yoshio Nagahama in 1946 (Zhu, 1998). Later, Yoshio Nakatani used an electrodermal measurement technology to discover that there were many galvanic points scattered on the skin in 1950. Similar to the classic acupoints, these galvanic points could be matched into lines, named Ryodoraku, which could not only reflect and diagnose the diseases but also balance the body's disorders through some appropriate points (Zhu, 1998). Till now, many Japanese acupuncturists still believe in the correlation between Ryodoraku and the state of autonomic nervous system.

Korean physicians held a totally different view on meridians and acupoints. In the early 1960s, Kim Bongham claimed that he had found the anatomical and physiological basis of meridians, which caught quite a stir over the world. In his studies, he stained out some special ducts and nodes in the subcutaneous tissue, organ surface, and nerve tissues, entitled as Bongham ducts and Bongham corpuscle (Soh, 2012a). However, due to lack of details in the methodology and nonreproducible results, his studies were ignored for almost 40 years. Recently, Korean scientist Soh Kwangsup began to continue Kim's theory and renamed as primo-vascular system to the further acupuncture investigation (Soh, 2012b).

Other scientists utilized modern techniques to analyze the characteristics of meridians and acupoints. In 1961, the French physician Niboyet found that the acupoints have a lower electrical resistance than the surrounding skin (Zhu, 1981). Furthermore, Voll investigated the electrodermal properties of the meridians and acupoints, known as "Electro-acupuncture according to Voll" or EAV and found that almost two-thirds of the EVA points were classic acupoints (Voll, 1975, 1980). These studies significantly enhanced our understanding of biophysical characteristics of the meridians and acupoints.

One important acupuncture research that occurred during the 1970–1980s was the findings of biological and physiological mechanism of acupuncture analgesia. In 1965, the physiologist Melzack and Wall proposed the "gate control theory," which was believed to be related with the plausible mechanism of acupuncture anesthesia (Man & Chen, 1972). Inspired by this theory, Chinese scholar, Professor Zhang Xiangtong, explained acupuncture analgesia mechanism by electrophysiology (Chang, 1978). He suggested that acupuncture analgesia was a result of acupoint afferent impulses and painful areas afferent impulses interacting in the brain. This hypothesis has greatly motivated the other scientists pursuing the acupuncture research in this direction. In 1972, Professor Han Jisheng and his colleagues collected the cerebrospinal fluid from rabbits that received acupuncture stimulation and infused into others, providing the first evidence that acupuncture alleviates pain by releasing neuromodulatory substances in the brain (Wang, Kain, & White, 2008). A few years later, the scientists have discovered two polypeptides released from the brain with potent opiate agonist activity, called endogenous morphine-like factor or endogenous opiate-like substance (Hughes et al., 1975; Terenius & Wahlström, 1975). Subsequently, Mayer, Price, and Rafii (1977) demonstrated that the naloxone, an antagonist of opiate, could block the acupuncture antinociceptive effect. Furthermore, it has been reported that that opiate receptor knockout CXBK mice had poor electroacupuncture (EA) analgesia effect (Peets & Pomeranz, 1978), suggesting that the endogenous opioids play an important role in acupuncture analgesia. Other researches indicated the involvement of nonopiate mechanisms in acupuncture analgesia as well. A report published in the journal Science in 1982 demonstrated both neural and hormonal pathways and both opiate and nonopiate substances play important roles in the complex modulation of pain transmission (Watkins & Mayer, 1982). In addition, scientists also observed acupuncture tolerance similar to morphine tolerance (Han, Tang, Huang, Liang, & Zhang, 1979). This phenomenon may attribute to the antiopiate substances accompanied with the generation of endogenous opiate substances during the acupuncture stimulation. Later, both Faris, Komisaruk, Watkins, and Mayer (1983) and Han, Ding, and Fan (1985) reported the cholecystokinin octapeptide (CCK-8) had the antagonism function to acupuncture and morphine analgesia, which results in tolerance in the rats (Faris et al., 1983; Han et al., 1985).

3. CURRENT RESEARCH IN ACUPUNCTURE

Since the establishment of the World Federation of Acupuncture and Moxibustion in 1987, the acupuncture research has stepped into a blossom era. In 1995, the Western Pacific Region of the World Health Organization (WHO) announced Acupuncture Clinical Research Specification. In 1996, the US FDA has redefined the conception of acupuncture and moxibustion and admitted them as therapeutic methods. Another revolutionary issue was the NIH Consensus Development Conference on Acupuncture in 1997, which evaluated available scientific information and efficacy of acupuncture in many disorders. In 2007, WHO drafted evidence-based acupuncture and moxibustion clinical practice guideline, which covered five diseases or symptoms, like depression, migraine, Bell's palsy, herpes zoster, and dysphagia after stroke. All of these events greatly promoted the development of acupuncture. Moreover, the acupuncture development was greatly enhanced in many countries, for example, the German government has funded many acupuncture researches in recent years.

3.1. Clinical research

The clinical research in the present is much different from before. More organizations and institutions have supported clinical researches with various targets. The clinical trial designation becomes more rigorous. The conception of evidence-based medicine provides a great impact on the designations of clinical trials. Most clinical researches currently are large-scaled RCTs, some providing convincing evidences for acupuncture application. Further, the covered diseases and symptoms are more diverse and multisystem. Since scientists and physicians gradually understand that acupuncture can be utilized to treat various kinds of diseases, the emphasis has been transferred from analgesia-related disorders to other diseases, like cancers, digestive and cardiovascular disorders, obstetrics and gynecology problems and withdrawal symptoms, and neurodegenerative diseases. In this part, we will discuss the clinical trials of some diseases and symptoms.

3.1.1 Acupuncture analgesia

Acupuncture analgesia has been applied in diverse kinds of pain-related disorders, especially in musculoskeletal and connective tissue diseases. Acupuncture is commonly used to treat headache and migraine. Many clinical trials were launched to assess the efficacy of acupuncture. Vickers et al. reported that the acupuncture effect in chronic headache is prolonged and beneficial to patients, deserving being incorporated by National Health Service in the United Kingdom (Vickers, 2004). However, two acupuncture clinical trials in patients with chronic pain sponsored by the German government found that although the number of days with headache was reduced by acupuncture compared to baseline, there was no significant difference between the real acupuncture and placebo acupuncture (Linde et al., 2005; Melchart, 2005). There was no suggestion about the cause of conflict results. Two trials in China on the migraine prophylaxis revealed that acupuncture was more effective than flunarizine in decreasing days of migraine attacks (Wang et al., 2011) but is only a minor effect when compared with SA (Li et al., 2012). These controversial results led to the doubt of effectiveness of verum acupuncture (VA) treatment (Diener et al., 2006).

The efficacy of acupuncture for lower back pain remains a controversy. Studies of acupuncture analgesia effect showed no difference in patients with lower back pain between VA and sham treatment groups (Brinkhaus et al., 2006; Cherkin, 2009). However, a clinical trial found weak evidence of effect of acupuncture at 12 months but stronger evidence of a small benefit at 24 months in patients with persistent nonspecific low back pain (Thomas, 2006).

Although osteoarthritis has been managed by acupuncture for prolonged time, the efficacy of acupuncture is still unclear. Many clinical trials demonstrated acupuncture plus diclofenac was more effective than placebo acupuncture plus diclofenac for osteoarthritis symptoms such as pain relief and joint function improvement (Berman et al., 2004; Vas, 2004). Compared with minimal acupuncture or no acupuncture, some researchers claimed that acupuncture was more effective in clinical scores but faded over time (Witt et al., 2005). However, others claimed the addition of acupuncture to a course of physiotherapy provided no additional improvement in pain relief, and only small benefits were observed in true acupuncture group and nonpenetrating acupuncture group (Foster et al., 2007; Witt et al., 2006). Studies about the influence of expectations on clinical outcome, pooled from four RCT of acupuncture in patients with pain, found a close link between better improvement and higher outcome expectations (Linde et al., 2007).

3.1.2 Other disorders

Abundant clinical trials investigated acupuncture for cancer treatment-related side effects. Acupuncture was used to treat breast cancer chemotherapy-induced emesis, showing that adjunct EA was more effective than minimal needling or antiemetic pharmacotherapy alone, although the effect had limited duration (Shen et al., 2000). In breast cancer patients with vasomotor symptoms, acupuncture treatment not only showed parallel therapeutic effect to venlafaxine, the chosen drug for the symptoms, but also did not induce adverse side effect and improved the energy in those patients (Walker et al., 2010). Further, acupuncture significantly improved joint pain, stiffness, and fatigue in patients with breast cancer (Crew et al., 2010; Molassiotis et al., 2012). Recent studies showed that radiation-induced dry mouth in cancer patients, both measured by subjective scores and objective salivary flow rates, could be remarkably improved by acupuncture (Meng et al., 2012). All the results suggested acupuncture could be a safe, effective, and durable therapy for cancer treatment-related side effects.

Acupuncture management for gastroenterological problems has always been a hot spot. Early literature has implied acupuncture could induce more tolerance to gastroscopy in patients than placebo acupuncture (Cahn et al., 1978). This efficacy of acupuncture was verified in another study in 2003 (Fanti et al., 2003). In the 1990s, scientists paid more attention to gastric physiology and pathology. EA was suggested to reduce the severity of symptoms in motion sickness (Hu, Stern, & Koch, 1992). Another study showed acupuncture significantly reduced in gastric acid secretion (Lux et al., 1994). A latest survey showed that EA not only exert analgesic for postoperative ileus after laparoscopic surgery but also reduce the duration and hospital stay compared with SA (Ng et al., 2013).

A clinical trial of cardiovascular conditions showed acupuncture has an additional beneficial effect in patients with severe, intensively treated angina pectoris, including the number of angina attacks per week, the performance before onset of pain during exercises, intensity of pain at maximal workload, ST-segment depressions, and quality of life (Richter, Herlitz, & Hjalmarson, 1991). However, when another clinical trial compared the differences between the individualized acupuncture, standardized acupuncture, and invasive SA for hypertension, the outcomes indicated that active acupuncture (individualized and standardized) provided no greater benefit than invasive SA in reducing systolic or diastolic BP (Macklin et al., 2006). A further randomized trial for lowering BP implied that mean 24 h ambulatory blood pressures could be lowered by traditional acupuncture, but the effect disappeared after cessation of acupuncture treatment (Flachskampf et al., 2007). In summary, more evidence is required to testify the effectiveness of acupuncture in cardiovascular system.

Acupuncture has been used in allergic diseases and symptoms for a long time. Clinical trials in patients with eczema showed that acupuncture is effective in improving type I hypersensitivity itch and is equivalent to oral antihistamine drugs (Pfab et al., 2009, 2012). The latest trial published in 2013 provided scientific evidence that acupuncture could significantly improve the quality of life for the patients with seasonal allergic rhinitis and reduce antihistamine use measures compared with SA and cetirizine alone (Brinkhaus et al., 2013).

3.2. Basic research

3.2.1 The nature of acupoints and meridian

The physiological and biological nature of acupoint and meridian remains elusive for years. The recent studies focusing on the connective tissue reported that loose connective tissue may account for the needle-grasp effect shown at acupoints (Langevin, Churchill, & Cipolla, 2001; Langevin, Churchill, Fox, et al., 2001; Langevin & Yandow, 2002). Furthermore, some clinical studies indicated the mast cells in the connective tissue played an important role in acupuncture analgesia (Zhang et al., 2007). Studies on meridians, using nuclear tracer, demonstrated that migration speed and pattern of a radioactive tracer along pathways matching with meridian in patients had neither a vascular nor a lymphatic origin (Darras, de Vernejoul, & Albarede, 1992). The electrical characteristics of acupoints and meridians suggested that acupoints have local electrical resistance/impedance minima with diameters of approximately 1–4 mm, while the meridians have lower electrical impedance and higher capacitance compared to adjacent controls (Darras et al., 1992). Although primo-vascular system (PVS), originated from Kim Bonghan' study, was regarded as an extension of acupuncture meridians (Soh, 2009), there is no sufficient evidences to support the correlation between PVS and traditional Chinese meridians.

3.2.2 The underlying mechanism of acupuncture effect

The mechanisms of acupuncture stimulation have gained a lot of interest and have been extensively investigated. It was reported that individual differences of acupuncture analgesia are associated with inherited genetic factors and the density of CCK receptors (Chae, Park, Hahm, Yi, & Lee, 2006; Lee et al., 2002; Wan, Wilson, Han, & Mogil, 2001). Other studies explored brain regional activity associated with acupuncture analgesia by means of functional magnetic resonance imaging (fMRI) and positron emission tomography (PET) and confirmed many brain structures indeed involved in the modulation of acupuncture analgesia (Fang, Krings, Weidemann, Meister, & Thron, 2004; Hsieh et al., 2001; Hui et al., 2000; Kong et al., 2002). In addition, abundant evidences demonstrated that frequency-dependent EA analgesia is mediated by the different opioid receptor subtypes (Han, 2003; Kim, Min, Na, & Park, 2004; Wang, Zhang, Wang, Cao, & Han, 2005; Zhang et al., 2004). While many studies concentrating on central molecular mechanisms of acupuncture reported that ERK1/2 signal pathway and the downstream events NF-kB, c-fos, and c-jun play important roles in acupuncture effect (Guo et al., 1996; Park et al., 2002; Song et al., 2006), acupuncture could trigger an increased level of interstitial adenosine at the local acupoint, which reduced severity of chronic pain in mice by activating adenosine A1 receptors, while the adenosine A1 receptor gene knockout mice did not show the same responses (Goldman et al., 2010). The similar results were confirmed in human subjects (Takano et al., 2012). This suggests that acupuncture analgesia could be mediated synergistically through the central agents and local agents.

Studies of effect of acupuncture on cardiovascular system demonstrated that acupuncture could diminish regional myocardial ischemia by reduction in cardiac oxygen demand and decrease pressor response (Li, Pitsillides, Rendig, Pan, & Longhurst, 1998). The further investigations indicated that acupuncture could reduce the heightened sympathetic tone through regulation on group III and IV somatic afferents and endogenous opiate system, which lowered myocardial oxygen demand (Chao et al., 1999; Zhou, Fu, Tjen-A-Looi, Li, & Longhurst, 2005). Meanwhile, other clinical trials implied that acupuncture also mediate specific opioid receptors in the rostral ventrolateral medulla (rVLM) (Li, 2001). Recently, other brain regions have been identified to be related with acupuncture regulation on cardiovascular system, like midbrain vlPAG and arcuate nucleus (Li, Tjen-A-Looi, & Longhurst, 2006; Tjen-A-Looi, Li, & Longhurst, 2006).

A series of studies concerning effect of acupuncture on immune system demonstrated that successive EA stimulation at ST36 could enhance splenic NK cell activity in normal rats and mice through increased releasing of IFN-γ and β-endorphin (Hisamitsu, Kasahara, Umezawa, Ishino, & Hisamitsu, 2002; Sato, Yu, Guo, Kasahara, & Hisamitsu, 1996). Other studies demonstrated EA stimulation significantly reduced the elevated serum levels of IgE by suppressing the increase of Th2 cytokines (Park et al., 2004).

Endocrine and metabolic disorders, including obesity and ovulatory dysfunction, have been treated with acupuncture. Recent studies showed that acupuncture could reduce the high levels of ovarian nerve growth factor (NGF), corticotrophin-releasing factor, and endothelin-1 concentrations—all markers for sympathetic activity—and increase low hypothalamic β-endorphin concentrations and immune function (Stener-Victorin & Lindholm, 2004). Moreover, low-frequency EA modulated the sympathetic activity by expressing mRNA and proteins of α_{1a}-, α_{1b}-, α_{1d}-, and β_2-adrenoceptors and the NGF receptor p75NTR and immunohistochemical expression of tyrosine hydroxylase (Manni, Lundeberg, Holmang, Aloe, & Stener-Victorin, 2005).

4. UTILIZATION OF MODERN TECHNOLOGY IN ACUPUNCTURE RESEARCH

In the past decade, advanced modern technology has been introduced into the acupuncture research for helping understand the central and peripheral mechanism of acupuncture.

Noninvasive neuroimaging techniques have been extensively used in acupuncture studies in the past decade, showing therapeutic effect of acupuncture could modulate neural activity in many cortical and subcortical (i.e., somatosensory, brainstem, limbic, and cerebellum) brain areas both in healthy subjects and in many conditions. Hui et al., by using the BOLD fMRI, observed a concerted attenuation of signal intensity in the limbic system when subjects experienced deqi sensation, indicating a close link between neural activity and deqi sensation (Hui et al., 2005). Positron emission tomography studies showed VA and SA have dramatically different effects on central the opioid receptor-binding ability. VA caused a long-term increase in μ-opioid receptor (MOR), whereas SA led to a decreased MOR (Harris et al., 2009). Another example of good correlation between acupuncture stimulation and brain function is a report from Napadow et al. (2007a, 2007b). FMRI showed that the patients with carpal tunnel syndrome (CTS) had sensorimotor hyperactivation and an overlapping or blurred representation of adjacent fingers within the primary somatosensory cortex. Acupuncture treatment for 5 weeks significantly improved all symptoms of CTS and reduced hyperactivation and overlapping in the sensorimotor cortex (Napadow et al., 2007a, 2007b). PET-CT, studying the brain responses on acupuncture clinical effect for patients with functional dyspepsia (FD), showed that the potential mechanism of acupuncture treating FD is to modulate the homeostatic afferent processing network, providing support for use of acupuncture in clinical practice (Zeng et al., 2012). In addition to these visualized tools, omics technologies have been employed into acupuncture research, including genomics, proteomics, and metabolomics. cDNA microarray analysis demonstrated that 68 genes, which were differentially expressed more than twofold in a neuropathic pain model, could be restored to normal after EA treatment (Lee et al., 2003). EA treatment could affect 10% of the genes of rhesus monkeys with cerebral ischemia, which were involved in signal transduction, cell-cycle control, metabolism, the stress response, and DNA repair (Guo et al., 2004). Sung et al. reported 36 different proteins expressed between the EA group and nontreated group rats with neuropathic pain (Sung et al., 2004). The proteins associated with inflammation, enzyme metabolism, and signal transduction were restored to the normal levels in rats with neuropathic pain after EA treatment (Sung et al., 2004). Metabolomics is a new technology for studying networks of metabolites in the body and its interaction with the environment. Nuclear magnetic resonance-based metabonomic technique demonstrated that acupuncture stimulation significantly changes the levels of leucine/isoleucine, lactate and glucose, and lipids towards those of healthy control, although this is only a proof-of-principle study due to the limited number of recruited subjects (Wu et al., 2010).

5. FUTURE OF ACUPUNCTURE RESEARCH

5.1. Challenges facing acupuncture research

Although a lot of progress has been made in acupuncture research during the past few decades, there still remain many challenges in the future of acupuncture development. For example, treatments based on syndrome differentiation and individualized treatment protocol are regarded to be fundamental parts of TCM. On the other hand, blinding and standardized controlling are also technically difficult in acupuncture practice. Moreover, there are various kinds of placebo control groups, like SA, nonacupuncture, and minimal acupuncture, making clinical trial studies more complicated and reducing the efficacy of acupuncture. How to reconcile those factors in the acupuncture research in particular clinical acupuncture studies are the big challenges faced by the acupuncture research community. The need for the unified standard and evaluation indexes particularly in acupuncture clinical study might be one of the rationales.

Although basic research, performed mostly in animal models and health human subjects, reported very encouraging physiological effects of acupuncture, the results from clinical trials are less encouraging or minimal at the best. It is very likely, based on our present knowledge, that the animal models employed for mechanistic studies of acupuncture are not within the explanatory framework of fundamentals of acupuncture treatment such as "qi," meridians, and acupoints and there is an urgent need to develop better animal models in which all those essential parts can be illustrated.

5.2. Directions of future acupuncture research

There is mounting evidence proving acupuncture treatment is superior to usual care. However, acupuncture treatment is, at most, only marginally more effective than SA, which, when compared to no treatment, is associated with larger effect sizes than when conventional placebos are compared to no treatment. These lead to a hypothesis that acupuncture may be a placebo intervention, which is just needling into the skin at any positions of the body. However, the truth is that acupuncture is a complex intervention with placebo effect, nonspecific physiological effects to needle insertion, and specific effect. The future acupuncture research not only needs to emphasize acupoint effective specificity but also needs to identify key physiological and psychological nonneedle components of acupuncture treatment to minimize specific nonneedle effects in sham treatment.

The translational medicine in acupuncture research means the future researches need to make a bridge connecting the laboratory mechanism with the clinical outcomes. The various needling parameters, such as depth, frequency, angle, and twirling, have a great impact on the acupuncture effect. The acupuncture also could influence numerous biomarkers testified in the laboratory. Together, all these could improve the correlations between the needling parameters, biomarker changes, and clinical outcomes.

5.3. Methodology, design, and evaluation of clinical studies

The high-quality RCTs indicate that large-scale, well-designed trials such as real randomization, rational controlling, and rigorous blinding are necessary in the future of acupuncture clinical studies. The real randomization includes two aspects: a method of getting randomly assigned sequence and hiding procedure during the implementation allocation. It could prevent psychological bias from the acupuncturists and patients. Rational controlling is one of the important methods to remove the placebo effect of acupuncture. The most controlled methods in China are different acupuncture techniques, like placebo needle, nonspecific acupoint, SA, and minimal acupuncture. The clinical researches in the future need more scientifically advanced controlled methods, providing a more objective evaluation of acupuncture effect. It is also very important to perform blinding control in acupuncture practice. Most trials about acupuncture are single-blind, while the double-blind mentioned usually refer to the subjects and evaluators being blinded instead of acupuncturists. Even though, rigorous and explicit blinding remains necessary in the practice.

Referred to the clinical outcome evaluation, the research in the future should focus on the more objective and reliable measurements, like symptom and function scales and life quality evaluation. Some biomarkers tested from the laboratory could also be brought into TCM outcome criteria, which could better reflect the acupuncture effect. Furthermore, the evaluation in the future should put more emphasis on the efficacy, effectiveness, and cost-effectiveness. The effectiveness is comparing to the waiting list control, while the efficacy is comparing to the nonpenetrating and penetrating SA. The cost-effectiveness in the healthcare insurance should also be seriously considered in order to offer a sustainable and affordable approach to millions of peoples with different conditions.

6. SUMMARY

Acupuncture research has a very long history. It initiated in the eighteenth century and grew rapidly up to now. Scientists and clinicians have strived hard to understand the physiological and biological mechanism of acupuncture effect and evaluate the real effectiveness of acupuncture in clinical practice. Reviewing research history from past to present, we are delighted to witness this rapid and wonderful development. So far, cumulative evidences demonstrated that acupuncture is beneficial in various conditions. Moreover, the underlying mechanism can partly be interpreted by modern advanced science and technology. However, there is still a shortage of conclusive evidence in clinical trials. Further, there are many very important issues surrounding acupuncture research, such as underlying mechanisms of acupuncture needling, acupoint specificity, and understanding how individual factors of acupuncture treatment interact and translate into physiological and clinical outcome, that need to be solved before maximizing clinical beneficial of acupuncture treatment.

REFERENCES

Berman, B. M., Lao, L., Langenberg, P., Lee, W. L., Gilpin, A. M., & Hochberg, M. C. (2004). Effectiveness of acupuncture as adjunctive therapy in osteoarthritis of the knee: A randomized, controlled trial. *Annals of Internal Medicine*, *141*(12), 901–910.

Birch, S., & Lewith, G. (2007). Acupuncture research: The story so far. In H. MacPherson, R. Hammerschlag, G. Lewith, & R. Schnyer (Eds.), *Acupuncture research*. London: Elsevier.

Bivins, R. (2001). The needle and the lancet: Acupuncture in Britain, 1683–2000. *Acupuncture in Medicine*, *19*(1), 2–14.

Bo-Ying, M., & Grant, A. (2001). The transmission of Traditional Chinese Medicine (TCM) to England (outline). In A. K. L. Chan, G. K. Clancey, & H. C. Loy (Eds.), *Historical Perspectives on East Asian Science, Technology, and Medicine* (p. 214). Singapore: Singapore University Press.

Brinkhaus, B., Ortiz, M., Witt, C. M., Roll, S., Linde, K., Pfab, F., et al. (2013). Acupuncture in patients with seasonal allergic rhinitis. A randomized trial. *Annals of Internal Medicine, 158*(4), 225–234.

Brinkhaus, B., Witt, C. M., Jena, S., Linde, K., Streng, A., Wagenpfeil, S., et al. (2006). Acupuncture in patients with chronic low back pain: A randomized controlled trial. *Archives of Internal Medicine, 166*(4), 450.

Cahn, A., Carayon, P., Hill, C., & Flamant, R. (1978). Acupuncture in gastroscopy. *The Lancet, 311*(8057), 182–183.

Campbell, A. (2002). Acupuncture, expertise and cross-cultural medicine. *Acupuncture in Medicine, 20*(1), 49–50.

Cassedy, J. H. (1974). Early uses of acupuncture in the United States, with an addendum (1826) by Franklin Bache, MD. *Bulletin of the New York Academy of Medicine, 50*(8), 892.

Ceniceros, S., & Brown, G. R. (1998). Acupuncture: A review of its history, theories, and indications. *Southern Medical Journal, 91*(12), 1121–1125.

Chae, Y., Park, H. J., Hahm, D. H., Yi, S. H., & Lee, H. (2006). Individual differences of acupuncture analgesia in humans using cDNA microarray. *The Journal of Physiological Sciences, 56*(6), 425–431.

Chang, H. T. (1978). The neurophysiological basis of acupuncture analgesia. *Science in Chinese, 4*, 010.

Chao, D. M., Shen, L. L., Tjen-A-Looi, S., Pitsillides, K. F., Li, P., & Longhurst, J. C. (1999). Naloxone reverses inhibitory effect of electroacupuncture on sympathetic cardiovascular reflex responses. *American Journal of Physiology-Heart and Circulatory Physiology, 276*(6), H2127–H2134.

Chein, E. Y., & Shapito, A. K. (1973). Evaluation of acupuncture. *JAMA: The Journal of the American Medical Association, 224*(11), 1533–1534.

Chen, C. H. (2006). An investigation of developmental history of Japanese acupuncture-moxibustion. MA degree Thesis, Jinan: Jinan University.

Cherkin, D. C. (2009). A randomized trial comparing acupuncture simulated acupuncture, and usual care for chronic low back pain. *Archives of Internal Medicine, 169*(9), 858. http://dx.doi.org/10.1001/archinternmed.2009.65.

Crew, K. D., Capodice, J. L., Greenlee, H., Brafman, L., Fuentes, D., Awad, D., et al. (2010). Randomized, blinded, sham-controlled trial of acupuncture for the management of aromatase inhibitor-associated joint symptoms in women with early-stage breast cancer. *Journal of Clinical Oncology, 28*(7), 1154–1160. http://dx.doi.org/10.1200/jco.2009.23.4708.

Darras, J.-C., de Vernejoul, P., & Albarede, P. (1992). Nuclear medicine and acupuncture: A study on the migration of radioactive tracers after injection at acupoints. *American Journal of Acupuncture, 20*(3), 245–256.

De Morant, G. S. (1994). *Chinese acupuncture.* Paradigm Publications.

Diener, H. C., Kronfeld, K., Boewing, G., Lungenhausen, M., Maier, C., Molsberger, A., et al. (2006). Efficacy of acupuncture for the prophylaxis of migraine: A multicentre randomised controlled clinical trial. *The Lancet Neurology, 5*(4), 310–316. http://dx.doi.org/10.1016/s1474-4422(06)70382-9.

Dimond, E. G. (1971a). Medical education and care in People's Republic of China. *JAMA: The Journal of the American Medical Association, 218*(10), 1552–1557.

Dimond, E. G. (1971b). Acupuncture anesthesia. *JAMA, 218*(10), 1558–1563.

Ernst, E. (2001). Pricking the skin: A history of acupuncture. *The Lancet, 358*(9279), 427–428. http://dx.doi.org/10.1016/s0140-6736(01)05538-6.

Fang, J., Krings, T., Weidemann, J., Meister, I., & Thron, A. (2004). Functional MRI in healthy subjects during acupuncture: Different effects of needle rotation in real and false acupoints. *Neuroradiology*, *46*(5), 359–362.
Fanti, L., Gemma, M., Passaretti, S., Guslandi, M., Testoni, P. A., Casati, A., et al. (2003). Electroacupuncture analgesia for colonoscopy: A prospective, randomized, placebo-controlled study. *The American Journal of Gastroenterology*, *98*(2), 312–316.
Faris, P. L., Komisaruk, B. R., Watkins, L. R., & Mayer, D. J. (1983). Evidence for the neuropeptide cholecystokinin as an antagonist of opiate analgesia. *Science*, *219*(4582), 310–312.
Flachskampf, F. A., Gallasch, J., Gefeller, O., Gan, J., Mao, J., Pfahlberg, A. B., et al. (2007). Randomized trial of acupuncture to lower blood pressure. *Circulation*, *115*(24), 3121–3129. http://dx.doi.org/10.1161/circulationaha.106.661140.
Foster, N. E., Thomas, E., Barlas, P., Hill, J. C., Young, J., Mason, E., et al. (2007). Acupuncture as an adjunct to exercise based physiotherapy for osteoarthritis of the knee: Randomised controlled trial. *BMJ: British Medical Journal*, *335*(7617), 436.
Gaw, A. C., Chang, L. W., & Shaw, L.-C. (1975). Efficacy of acupuncture on osteoarthritic pain. A controlled, double-blind study. *The New England Journal of Medicine*, *293*(8), 375.
Goldman, N., Chen, M., Fujita, T., Xu, Q., Peng, W., Liu, W., et al. (2010). Adenosine A1 receptors mediate local anti-nociceptive effects of acupuncture. *Nature Neuroscience*, *13*(7), 883–888. http://dx.doi.org/10.1038/nn.2562.
Guo, J., Gao, H., Chen, J., Zhao, P., Cao, X., Li, Y., et al. (2004). Modulation of the gene expression in the protective effects of electroacupuncture against cerebral ischemia: A cDNA microarray study. *Acupuncture & Electro-Therapeutics Research*, *29*(3–4), 173.
Guo, H. F., Tian, J., Wang, X., Fang, Y., Hou, Y., & Han, J. (1996). Brain substrates activated by electroacupuncture (EA) of different frequencies (II): Role of Fos/Jun proteins in EA-induced transcription of preproenkephalin and preprodynorphin genes. *Molecular Brain Research*, *43*(1), 167–173.
Hammerschlag, R. (2000). Funding of acupuncture research by the National Institutes of Health: A brief history. *Clinical Acupuncture and Oriental Medicine*, *1*(3), 133–138.
Han, J. S. (2003). Acupuncture: Neuropeptide release produced by electrical stimulation of different frequencies. *Trends in Neurosciences*, *26*(1), 17–22.
Han, J. S., Ding, X. Z., & Fan, S. G. (1985). Is cholecystokinin octapeptide (CCK-8) a candidate for endogenous antiopioid substrates? *Neuropeptides*, *5*(4), 399–402.
Han, J., Tang, J., Huang, B., Liang, X., & Zhang, N. (1979). Acupuncture tolerance in rats: Anti-opiate substrates implicated. *Chinese Medical Journal*, *92*(9), 625.
Harris, R. E., Zubieta, J. K., Scott, D. J., Napadow, V., Gracely, R. H., & Clauw, D. J. (2009). Traditional Chinese acupuncture and placebo (sham) acupuncture are differentiated by their effects on μ-opioid receptors (MORs). *NeuroImage*, *47*(3), 1077–1085. http://dx.doi.org/10.1016/j.neuroimage.2009.05.083.
Hisamitsu, T., Kasahara, T., Umezawa, T., Ishino, T., & Hisamitsu, N. (2002). *The effect of acupuncture on natural killer cell activity. Paper presented at the international congress series.*
Hsieh, J. C., Tu, C. H., Chen, F. P., Chen, M. C., Yeh, T. C., Cheng, H. C., et al. (2001). Activation of the hypothalamus characterizes the acupuncture stimulation at the analgesic point in human: A positron emission tomography study. *Neuroscience Letters*, *307*(2), 105–108.
Hu, S., Stern, R., & Koch, K. (1992). Electrical acustimulation relieves vection-induced motion sickness. *Gastroenterology*, *102*(6), 1854.
Hughes, J., Smith, T., Kosterlitz, H., Fothergill, L. A., Morgan, B., & Morris, H. (1975). Identification of two related pentapeptides from the brain with potent opiate agonist activity. *Nature*, *258*(5536), 577.

Hui, K. K., Liu, J., Makris, N., Gollub, R. L., Chen, A. J., Moore, C. I., et al. (2000). Acupuncture modulates the limbic system and subcortical gray structures of the human brain: Evidence from fMRI studies in normal subjects. *Human Brain Mapping, 9*(1), 13–25.

Hui, K. K. S., Liu, J., Marina, O., Napadow, V., Haselgrove, C., Kwong, K. K., et al. (2005). The integrated response of the human cerebro-cerebellar and limbic systems to acupuncture stimulation at ST 36 as evidenced by fMRI. *NeuroImage, 27*(3), 479–496. http://dx.doi.org/10.1016/j.neuroimage.2005.04.037.

Jobst, K., Mcpherson, K., Brown, V., Fletcher, H., Mole, P., Hua Chen, J., et al. (1986). Controlled trial of acupuncture for disabling breathlessness. *The Lancet, 328*(8521), 1416–1419.

Kim, J. H., Min, B.-I., Na, H. S., & Park, D. S. (2004). Relieving effects of electroacupuncture on mechanical allodynia in neuropathic pain model of inferior caudal trunk injury in rat: Mediation by spinal opioid receptors. *Brain Research, 998*(2), 230–236.

Kong, J., Ma, L., Gollub, R. L., Wei, J., Yang, X., Li, D., et al. (2002). A pilot study of functional magnetic resonance imaging of the brain during manual and electroacupuncture stimulation of acupuncture point (LI-4 Hegu) in normal subjects reveals differential brain activation between methods. *The Journal of Alternative & Complementary Medicine, 8*(4), 411–419.

Langevin, H. M., Churchill, D. L., & Cipolla, M. J. (2001). Mechanical signaling through connective tissue: A mechanism for the therapeutic effect of acupuncture. *The FASEB Journal, 15*(12), 2275–2282.

Langevin, H. M., Churchill, D. L., Fox, J. R., Badger, G. J., Garra, B. S., & Krag, M. H. (2001). Biomechanical response to acupuncture needling in humans. *Journal of Applied Physiology, 91*(6), 2471–2478.

Langevin, H. M., & Yandow, J. A. (2002). Relationship of acupoints and meridians to connective tissue planes. *The Anatomical Record, 269*(6), 257–265.

Lee, G. S., Han, J. B., Shin, M. K., Hong, M. C., Kim, S. W., Min, B. I., et al. (2003). Enhancement of electroacupuncture-induced analgesic effect in cholecystokinin-A receptor deficient rats. *Brain Research Bulletin, 62*(2), 161–164.

Lee, G., Rho, S., Shin, M., Hong, M., Min, B. I., & Bae, H. (2002). The association of cholecystokinin-A receptor expression with the responsiveness of electroacupuncture analgesic effects in rat. *Neuroscience Letters, 325*(1), 17–20.

Li, P. (2001). Rostral ventrolateral medullary opioid receptor subtypes in the inhibitory effect of electroacupuncture on reflex autonomic response in cats. *Autonomic Neuroscience, 89*(1), 38–47.

Li, P., Pitsillides, K. F., Rendig, S. V., Pan, H. L., & Longhurst, J. C. (1998). Reversal of reflex-induced myocardial ischemia by median nerve stimulation: A feline model of electroacupuncture. *Circulation, 97*(12), 1186–1194. http://dx.doi.org/10.1161/01.cir.97.12.1186.

Li, P., Tjen-A-Looi, S. C., & Longhurst, J. C. (2006). Excitatory projections from arcuate nucleus to ventrolateral periaqueductal gray in electroacupuncture inhibition of cardiovascular reflexes. *American Journal of Physiology-Heart and Circulatory Physiology, 290*(6), H2535–H2542.

Li, Y., Zheng, H., Witt, C. M., Roll, S., Yu, S.-G., Yan, J., et al. (2012). Acupuncture for migraine prophylaxis: A randomized controlled trial. *Canadian Medical Association Journal, 184*(4), 401–410.

Linde, K., Streng, A., Jürgens, S., Hoppe, A., Brinkhaus, B., Witt, C., et al. (2005). Acupuncture for patients with migraine. *JAMA: The Journal of the American Medical Association, 293*(17), 2118–2125.

Linde, K., Witt, C. M., Streng, A., Weidenhammer, W., Wagenpfeil, S., Brinkhaus, B., et al. (2007). The impact of patient expectations on outcomes in four randomized controlled trials of acupuncture in patients with chronic pain. *Pain, 128*(3), 264–271. http://dx.doi.org/10.1016/j.pain.2006.12.006.

Lu, G.-D., & Needham, J. (1980). *Celestial lancets: A history and rationale of acupuncture and moxa*. Routledge.

Lux, G., Hagel, J., Bäcker, P., Bäcker, G., Vogl, R., Ruppin, H., et al. (1994). Acupuncture inhibits vagal gastric acid secretion stimulated by sham feeding in healthy subjects. *Gut, 35*(8), 1026–1029.

Lytle, C. D. (1996). History of the food and drug administration's regulation of acupuncture devices. *The Journal of Alternative and Complementary Medicine, 2*(1), 253–256.

Macklin, E. A., Wayne, P. M., Kalish, L. A., Valaskatgis, P., Thompson, J., Pian-Smith, M. C. M., et al. (2006). Stop Hypertension with the Acupuncture Research Program (SHARP): Results of a randomized, controlled clinical trial. *Hypertension, 48*(5), 838–845. http://dx.doi.org/10.1161/01.HYP.0000241090.28070.4c.

Man, P., & Chen, C. (1972). Mechanism of acupunctural anesthesia. The two-gate control theory. *Diseases of the Nervous System, 33*(11), 730.

Manni, L., Lundeberg, T., Holmang, A., Aloe, L., & Stener-Victorin, E. (2005). Effect of electro-acupuncture on ovarian expression of alpha (1)-and beta (2)-adrenoceptors, and p75 neurotrophin receptors in rats with steroid-induced polycystic ovaries. *Reproductive Biology and Endocrinology, 3*, 21.

Mark, L. C. (1973). Double-blind studies of acupuncture. *JAMA: The Journal of the American Medical Association, 225*(12), 1532.

Mayer, D. J., Price, D. D., & Rafii, A. (1977). Antagonism of acupuncture analgesia in man by the narcotic antagonist naloxone. *Brain Research, 121*, 368–372.

Melchart, D. (2005). Acupuncture in patients with tension-type headache: Randomised controlled trial. *BMJ: British Medical Journal, 331*(7513), 376–382. http://dx.doi.org/10.1136/bmj.38512.405440.8F.

Melzack, R., & Wall, P. D. (1965). Pain mechanism: A new theory. *Science, 150*(3699), 971–979.

Meng, Z., Garcia, M. K., Hu, C., Chiang, J., Chambers, M., Rosenthal, D. I., et al. (2012). Randomized controlled trial of acupuncture for prevention of radiation-induced xerostomia among patients with nasopharyngeal carcinoma. *Cancer, 118*(13), 3337–3344. http://dx.doi.org/10.1002/cncr.26550.

Meng, X., Xu, S., & Lao, L. (2011). Clinical acupuncture research in the West. *Frontiers of Medicine, 5*(2), 134–140. http://dx.doi.org/10.1007/s11684-011-0135-9.

Molassiotis, A., Bardy, J., Finnegan-John, J., Mackereth, P., Ryder, D. W., Filshie, J., et al. (2012). Acupuncture for cancer-related fatigue in patients with breast cancer: A pragmatic randomized controlled trial. *Journal of Clinical Oncology, 30*(36), 4470–4476.

Moore, M. E., & Berk, S. N. (1976). Acupuncture for chronic shoulder pain. An experimental study with attention to the role of placebo and hypnotic susceptibility. *Annals of Internal Medicine, 84*(4), 381–384.

Napadow, V., Kettner, N., Liu, J., Li, M., Kwong, K., Vangel, M., et al. (2007a). Hypothalamus and amygdala response to acupuncture stimuli in carpal tunnel syndrome. *Pain, 130*(3), 254–266.

Napadow, V., Liu, J., Li, M., Kettner, N., Ryan, A., Kwong, K. K., et al. (2007b). Somatosensory cortical plasticity in carpal tunnel syndrome treated by acupuncture. *Human Brain Mapping, 28*(3), 159–171.

Ng, S. S. M., Leung, W. W., Mak, T. W. C., Hon, S. S. F., Li, J. C. M., Wong, C. Y. N., et al. (2013). Electroacupuncture reduces duration of postoperative ileus after laparoscopic surgery for colorectal cancer. *Gastroenterology, 144*(2), 307–313. http://dx.doi.org/10.1053/j.gastro.2012.10.050, e301.

Park, M. B., Ko, E., Ahn, C., Choi, H., Rho, S., Shin, M. K., et al. (2004). Suppression of IgE production and modulation of Th1/Th2 cell response by electroacupuncture in DNP-KLH immunized mice. *Journal of Neuroimmunology, 151*(1), 40–44.

Park, H. J., Lee, H. S., Lee, H. J., Yoo, Y. M., Lee, H. J., Kim, S., et al. (2002). Decrease of the electroacupuncture-induced analgesic effects in nuclear factor-kappa B1 knockout mice. *Neuroscience Letters, 319*(3), 141–144.

Peets, J., & Pomeranz, B. (1978). CXBK mice deficient in opiate receptors show poor electroacupuncture analgesia. *Nature*, *273*, 675–676.
Pfab, F., Huss-Marp, J., Gatti, A., Fuqin, J., Athanasiadis, G. I., Irnich, D., et al. (2009). Influence of acupuncture on type I hypersensitivity itch and the wheal and flare response in adults with atopic eczema: A blinded, randomized, placebo-controlled, crossover trial. *Allergy*, *65*(7), 903–910. http://dx.doi.org/10.1111/j.1398-9995.2009.02284.x.
Pfab, F., Kirchner, M. T., Huss-Marp, J., Schuster, T., Schalock, P. C., Fuqin, J., et al. (2012). Acupuncture compared with oral antihistamine for type I hypersensitivity itch and skin response in adults with atopic dermatitis: A patient- and examiner-blinded, randomized, placebo-controlled, crossover trial. *Allergy*, *67*(4), 566–573. http://dx.doi.org/10.1111/j.1398-9995.2012.02789.x.
Pui Fung, K., Kit Wun Chow, O., & Yeung So, S. (1986). Attenuation of exercise-induced asthma by acupuncture. *The Lancet*, *328*(8521), 1419–1422.
Reston, J. (1971). Now, let me tell you about my appendectomy in Peking. *New York Times*, *1*(6).
Richter, A., Herlitz, J., & Hjalmarson, A. (1991). Effect of acupuncture in patients with angina pectoris. *European Heart Journal*, *12*(2), 175–178.
Sato, T., Yu, Y., Guo, S., Kasahara, T., & Hisamitsu, T. (1996). Acupuncture stimulation enhances splenic natural killer cell cytotoxicity in rats. *The Japanese Journal of Physiology*, *46*(2), 131–136.
Shen, J., Wenger, N., Glaspy, J., Hays, R. D., Albert, P. S., Choi, C., et al. (2000). Electroacupuncture for control of myeloablative chemotherapy-induced emesis. *JAMA: The Journal of the American Medical Association*, *284*(21), 2755–2761.
Shichido, T. (1996). Clinical evaluation of acupuncture and moxibustion. *Ido No Nippon Journal*, *623*(8), 102.
Soh, K. S. (2009). Bonghan circulatory system as an extension of acupuncture meridians. *Journal of Acupuncture and Meridian Studies*, *2*(2), 93–106.
Soh, K. S. (2012a). A brief history of the Bong-Han theory and the primo vascular system. In *The primo vascular system* (pp. 3–5). Springer.
Soh, K. S. (2012b). Current state of research on the primo vascular system. In *The primo vascular system* (pp. 25–39). Springer.
Song, L., Zhu, Z., Duan, X., Liu, X., Fan, J., Ju, G., et al. (2006). [Effects of electroacupuncture at "Zusanli" (ST 36) on ERK1/2 phosphorylation in the dorsal horn of spinal cord of the rat]. *Zhongguo Zhen Jiu= Chinese Acupuncture & Moxibustion*, *26*(5), 362.
Stener-Victorin, E., & Lindholm, C. (2004). Immunity and β-endorphin concentrations in hypothalamus and plasma in rats with steroid-induced polycystic ovaries: Effect of low-frequency electroacupuncture. *Biology of Reproduction*, *70*(2), 329–333.
Stollberg, G. (2006). Acupuncture in Western Europe. In D. Schirmer et al. (Ed.), *Hybridising East and West* (pp. 259–284). Berlin: LIT inc.
Sung, H. J., Kim, Y. S., Kim, I. S., Jang, S. W., Kim, Y. R., Na, D. S., et al. (2004). Proteomic analysis of differential protein expression in neuropathic pain and electroacupuncture treatment models. *Proteomics*, *4*(9), 2805–2813.
Takano, T., Chen, X., Luo, F., Fujita, T., Ren, Z., Goldman, N., et al. (2012). Traditional acupuncture triggers a local increase in adenosine in human subjects. *The Journal of Pain*, *13*(12), 1215–1223. http://dx.doi.org/10.1016/j.jpain.2012.09.012.
Terenius, L., & Wahlström, A. (1975). Search for an endogenous ligand for the opiate receptor. *Acta Physiologica Scandinavica*, *94*(1), 74–81.
Thomas, K. J. (2006). Randomised controlled trial of a short course of traditional acupuncture compared with usual care for persistent non-specific low back pain. *BMJ: British Medical Journal*, *333*(7569), 620–623. http://dx.doi.org/10.1136/bmj.38878.907361.7C.

Tjen-A-Looi, S. C., Li, P., & Longhurst, J. C. (2006). Midbrain vlPAG inhibits rVLM cardiovascular sympathoexcitatory responses during electroacupuncture. *American Journal of Physiology-Heart and Circulatory Physiology, 290*(6), H2543–H2553.

Ulett, G. A., Han, J., & Han, S. (1998). Traditional and evidence-based acupuncture: History, mechanisms, and present status. *Southern Medical journal, 91*(12), 1115–1120.

Vas, J. (2004). Acupuncture as a complementary therapy to the pharmacological treatment of osteoarthritis of the knee: Randomised controlled trial. *BMJ: British Medical Journal, 329*(7476), 1210–1216. http://dx.doi.org/10.1136/bmj.38238.601447.3A.

Vickers, A. J. (2004). Acupuncture for chronic headache in primary care: Large, pragmatic, randomised trial. *BMJ: British Medical Journal, 328*(7442), 740–744. http://dx.doi.org/10.1136/bmj.38029.421863.EB.

Voll, R. (1975). Twenty years of electroacupuncture diagnosis in Germany. A progress report. *American Journal of Acupuncture, 3*(1), 7–17.

Voll, R. (1980). The phenomenon of meridian testing in electroacupuncture according to Voll. *American Journal of Acupuncture, 8*(1), 97–104.

Walker, E. M., Rodriguez, A. I., Kohn, B., Ball, R. M., Pegg, J., Pocock, J. R., et al. (2010). Acupuncture versus venlafaxine for the management of vasomotor symptoms in patients with hormone receptor–positive breast cancer: A randomized controlled trial. *Journal of Clinical Oncology, 28*(4), 634–640.

Wan, Y., Wilson, S. G., Han, J.-S., & Mogil, J. S. (2001). The effect of genotype on sensitivity to electroacupuncture analgesia. *Pain, 91*(1), 5–13.

Wang, S.-M., Kain, Z. N., & White, P. (2008). Acupuncture analgesia: I. The scientific basis. *Anesthesia & Analgesia, 106*(2), 602–610.

Wang, L. P., Zhang, X. Z., Guo, J., Liu, H. L., Zhang, Y., Liu, C. Z., et al. (2011). Efficacy of acupuncture for migraine prophylaxis: A single-blinded, double-dummy, randomized controlled trial. *Pain, 152*(8), 1864–1871. http://dx.doi.org/10.1016/j.pain.2011.04.006.

Wang, Y., Zhang, Y., Wang, W., Cao, Y., & Han, J. S. (2005). Effects of synchronous or asynchronous electroacupuncture stimulation with low versus high frequency on spinal opioid release and tail flick nociception. *Experimental Neurology, 192*(1), 156–162.

Watkins, L. R., & Mayer, D. J. (1982). Organization of endogenous opiate and nonopiate pain control systems. *Science, 216*(4551), 1185–1192.

White, A. (2004). A brief history of acupuncture. *Rheumatology, 43*(5), 662–663. http://dx.doi.org/10.1093/rheumatology/keg005.

Willer, J., Roby, A., & Le Bars, D. (1984). Psychophysical and electrophysiological approaches to the pain-relieving effects of heterotopic nociceptive stimuli. *Brain, 107*(4), 1095–1112.

Witt, C., Brinkhaus, B., Jena, S., Linde, K., Streng, A., Wagenpfeil, S., et al. (2005). Acupuncture in patients with osteoarthritis of the knee: A randomised trial. *The Lancet, 366*(9480), 136–143. http://dx.doi.org/10.1016/s0140-6736(05)66871-7.

Witt, C. M., Jena, S., Brinkhaus, B., Liecker, B., Wegscheider, K., & Willich, S. N. (2006). Acupuncture in patients with osteoarthritis of the knee or hip: A randomized, controlled trial with an additional nonrandomized arm. *Arthritis & Rheumatism, 54*(11), 3485–3493. http://dx.doi.org/10.1002/art.22154.

Wu, Q., Zhang, Q., Sun, B., Yan, X., Tang, Y., Qiao, X., et al. (2010). 1H NMR-based metabonomic study on the metabolic changes in the plasma of patients with functional dyspepsia and the effect of acupuncture. *Journal of Pharmaceutical and Biomedical Analysis, 51*(3), 698–704.

Zeng, F., Qin, W., Ma, T., Sun, J., Tang, Y., Yuan, K., et al. (2012). Influence of acupuncture treatment on cerebral activity in functional dyspepsia patients and its relationship with efficacy. *The American Journal of Gastroenterology, 107*(8), 1236–1247. http://dx.doi.org/10.1038/ajg.2012.53.

Zhang, D., Ding, G., Shen, X., Yao, W., Zhang, Z., Zhang, Y., et al. (2007). [Influence of mast cell function on the analgesic effect of acupuncture of' Zusanli"(ST 36) in rats]. *Zhen ci yan jiu= Acupuncture research/[Zhongguo yi xue ke xue yuan Yi xue qing bao yan jiu suo bian ji*, *32*(3), 147.

Zhang, R. X., Lao, L., Wang, L., Liu, B., Wang, X., Ren, K., et al. (2004). Involvement of opioid receptors in electroacupuncture-produced anti-hyperalgesia in rats with peripheral inflammation. *Brain Research*, *1020*(1–2), 12.

Zhou, W., Fu, L. W., Tjen-A-Looi, S. C., Li, P., & Longhurst, J. C. (2005). Afferent mechanisms underlying stimulation modality-related modulation of acupuncture-related cardiovascular responses. *Journal of Applied Physiology*, *98*(3), 872–880.

Zhu, Z. (1981). Research advances in the electrical specificity of meridians and acupoints. *American Journal of Acupuncture*, *9*, 203–216.

Zhu, B. (1998). *Scientific foundations of acupuncture-moxibustion*. Qingdao: Qingdao Publishing House, 446–450.

CHAPTER TWO

Effects of Acupuncture Needling with Specific Sensation on Cerebral Hemodynamics and Autonomic Nervous Activity in Humans

Kouich Takamoto[*,†], Susumu Urakawa[*,†], Kazushige Sakai[†], Taketoshi Ono[†], Hisao Nishijo[*,1]

[*]System Emotional Science, University of Toyama, Toyama, Japan
[†]Department of Judo Neurophysiotherapy, Graduate School of Medicine and Pharmaceutical Sciences, University of Toyama, Toyama, Japan
[1]Corresponding author: e-mail address: nishijo@med.u-toyama.ac.jp

Contents

Abstract

Effective therapeutic factors in acupuncture therapy include specific stimulation points, called acupoints, and specific sensations, called de-qi, that are induced by needling manipulation. Human neuroimaging studies have reported that acupuncture stimulation with de-qi sensations induced specific activity patterns in the brain and modulated autonomic nervous activity. However, acupoints and nonacupoints have been reported to induce de-qi sensations. Thus, it remains unclear whether these physiological responses induced by acupuncture and associated with therapeutic efficacy are related to specific stimulation sites (acupoints) or unique de-qi sensations. This review focuses on the cerebral hemodynamic responses and autonomic nervous activity changes induced by acupuncture stimulation at acupoints and nonacupoints with and without

International Review of Neurobiology, Volume 111
ISSN 0074-7742
http://dx.doi.org/10.1016/B978-0-12-411545-3.00002-X

de-qi sensations. We argue that the specific sensations induced by acupuncture are more important than the specific stimulation sites for inducing cerebral hemodynamic and autonomic responses and that autonomic responses during acupuncture, which might be important for therapeutic efficacy, might be mediated though the brain activity changes exemplified by the cerebral hemodynamic responses during acupuncture.

1. BACKGROUND

Acupuncture is a traditional medicine that can be traced back more than 3000 years in China (Han & Ho, 2011). Acupuncture has been used to effectively treat a wide variety of disorders, such as psychoneurotic disorders, cardiovascular disorders, and, especially, pain, by inserting needles into specific acupuncture points, which are called acupoints, on the patient's body (Esch, Guarna, Bianchi, Zhu, & Stefano, 2004). Some of the effective therapeutic factors in acupuncture therapy that have been suggested include specific stimulation points (acupoints) and specific sensations, which are called de-qi, induced by needling manipulation (Shi, Yang, Liu, & Wang, 2012). The specific points are carefully defined for acupuncture therapy in traditional Chinese medicine (TCM) (Low, 2001). In TCM theory, the 12 meridians are defined as the channels through which energy, which is known as qi, flows. The abnormal flow of qi in the meridians is associated with various diseases in the principal Chinese organs (Filshie & White, 1998). It is believed that stimulation at the acupoints along the meridians associated with specific diseases restores its natural flow (Longhurst, 2010). However, recent studies have reported inconsistent results. Several randomized controlled trials conducted in patients with fibromyalgia (Assefi et al., 2005) and chronic epicondylitis (Fink, Wolkenstein, Karst, & Gehrke, 2002) have reported that acupuncture stimulation at the acupoints defined by TCM theory is not significantly more effective than sham acupuncture, which is defined as stimulation at points away from the treatment acupoints (i.e., nonacupoints); however, other studies have demonstrated acupoint specificity for therapeutic efficacy (Zhang, Bian, & Lin, 2010). Thus, findings of acupoint specificity in acupuncture efficacy have been controversial.

Specific sensations called de-qi that are induced by needling have also been considered important in acupuncture therapy since the early classical texts (Kaptchuk, 2002; Kong et al., 2007). de-qi sensations have included many sensations, such as aching, soreness, pressure, heaviness, fullness, warmth, cooling, numbness, tingling, and dull pain, in previous questionnaires (Kong et al., 2007). The most common sensations that were specified in several questionnaires that evaluated de-qi sensations were heaviness and numbness (Park et al., 2013). In a study that investigated subjective sensations and patients' attitudes in acupuncture, 82% of the patients considered the needle sensation to be an important factor of acupuncture treatment (White et al., 2008). In addition, 68% of them thought that stronger needle sensations led to better treatment results. Furthermore, 47 of 86 acupuncture practitioners in China and United States agree that dull pain is de-qi, and over half agree that de-qi is beneficial, while sharp pain is harmful (Hui et al., 2011). A total of 73% of them believed that there was a positive correlation between de-qi and treatment efficacy. A clinical trial that investigated the effects of real and sham acupuncture on osteoarthritis has reported that the experience of de-qi can be used as a predictor of significant improvement (Takeda & Wessel, 1994). Thus, clinical studies have reported that the specific sensation that is called de-qi is one of the important factors in acupuncture therapy. However, there is a lack of adequate experimental data, indicating significant relationships among de-qi sensations and other therapeutic factors, such as acupuncture points, needling techniques, and afferent fiber types (e.g., Aδ- or C-fibers) (Hui et al., 2007).

Furthermore, the relationships between acupoints and de-qi are controversial. Several studies have reported significant relationships between de-qi and acupoints. Roth, Maret-Maric, and Adler (1997) have reported that needle sensations are significantly stronger with traditional needling at acupoints than with sham acupuncture at nonacupoints (Roth et al., 1997). However, another study has reported that there are no significant differences in needle sensations between acupoints and nonacupoints in any body location, including the hands, feet, and legs (Vincent, Richardson, Black, & Pither, 1989). Taken together, although some clinical data have reported relationships among acupoints, specific sensations (de-qi), and therapeutic efficacy, the clinical data are controversial.

In western medicine, myofascial trigger points (MTrPs) have been used instead of acupoints when referring to the targeted points in acupuncture therapy. MTrPs have been defined as highly localized and exquisitely tender spots in a palpable taut band. When the points are pressed, a characteristic pattern of referred pain and a flexion withdrawal reflex (jump sign) are elicited (Travell & Simons, 1983). Melzack, Stillwell, and Fox (1977) have reported that 71% of identified MTrPs correlate anatomically with acupoints. A recent review paper has also suggested that the clinical correspondence between MTrPs and classical acupoints in the treatment of pain disorders is likely 95% or higher (Dorsher, 2008).

Acupuncture stimulation at MTrPs is widely used in the treatment of myofascial pain syndrome, fibromyalgia, chronic fatigue, and secondary muscular strain due to other diseases. Clinical trial studies have reported that acupuncture at MTrPs in patients with chronic neck pain and low back pain significantly reduces pain compared to stimulation at non-MTrPs and classical acupoints (Itoh, Katsumi, Hirota, & Kitakoji, 2007; Itoh, Katsumi, & Kitakoji, 2004). Sensations similar to de-qi are obtained by acupuncture stimulation at MTrPs (Kung et al., 2001). However, MTrP specificity in evoking acupuncture sensations and physiological responses has not been challenged as it has in the acupuncture studies on classical acupoints.

2. NEUROIMAGING STUDIES OF ACUPUNCTURE THERAPY IN HUMANS

The number of human neuroimaging studies that have used positron emission tomography (PET), single-photon emission computed tomography, and functional magnetic resonance imaging (fMRI) to investigate acupuncture mechanisms has dramatically increased since 2000 (Dhond, Kettner, & Napadow, 2007; Huang et al., 2012). Several studies have reported that the cerebral hemodynamic responses occur mainly in the somatosensory areas, motor areas, basal ganglia, cerebellum, and limbic system and that areas for higher cognitive functions are more prominent (increased or decreased) during acupuncture stimulation at acupoints compared to stimulation at nonacupoints (Huang et al., 2012). However, Fang et al. (2009) have shown that hemodynamic responses are decreased in multiple brain regions, including the limbic–paralimbic–neocortical system, such as the medial prefrontal cortex (mPFC), the amygdala, hippocampus, and precuneus, regardless of the different acupoints (Liv3, Lv2, ST44) and regardless of the stimulation of acupoints and nonacupoints. However, Cho et al. (1998) have reported that acupuncture stimulation at acupoints in the foot that are related to vision, such as BL67, activates the visual cortex, while the visual cortex is not activated by stimulation at nonacupoints that were located 2–5 cm away from these vision-related acupoints. However, these results were not replicated in two subsequent studies (Li, Cheung, Ma, & Yang, 2003; Siedentopf et al., 2002). Furthermore, a recent study has reported that activities in the occipital cortex are not significantly different when stimulation is conducted at two vision-related acupoints and one nonacupoint (Kong et al., 2009). These studies have consistently indicated that acupuncture affects brain activity. However, the studies have reported quite controversial results with respect to the efficacy of specific acupoints and the acupoint specificity to specific organs.

Recent neuroimaging studies have also investigated brain activity changes induced by acupuncture in terms of de-qi. The areas commonly activated by acupuncture with de-qi are SI, SII, the thalamus, MI, the cerebellum, the insula, and the inferior parietal lobe, while the brain areas commonly deactivated by acupuncture with de-qi are the anterior cingulate cortex (ACC), the amygdala, the hippocampus, the parahippocampus, the hypothalamus, the frontal cortex, and the posterior cingulate cortex (Sun et al., 2013). However, stimulation at acupoints and at nonacupoints has been shown to induce the de-qi sensation (Fang et al., 2009; Vincent et al., 1989). Thus, it remains unclear whether specific brain activities induced by acupuncture are related to specific stimulation sites or unique de-qi sensations.

3. CEREBRAL HEMODYNAMIC RESPONSES ASSOCIATED WITH de-qi SENSATIONS

Although acupuncture has been applied to MTrPs in western medicine, it is unknown whether acupuncture at MTrPs can induce unique de-qi sensations and specific brain activity as has been shown with acupuncture at classical acupoints. Therefore, we investigated the effects of acupuncture with and without specific de-qi sensations on the cerebral hemodynamic responses in human subjects with functional near-infrared spectroscopy (fNIRS) (Takamoto et al., 2010). In that study, we inserted acupuncture needles into both MTrPs and non-MTrPs in the right extensor muscle in the forearm (Takamoto et al., 2010). MTrPs are defined as areas that have a palpable taut band where tenderness and specific referred pain are induced by pressing these areas. The non-MTrPs were placed 2 cm away and proximal to a given MTrPs of the right extensor muscle in the forearm. Acupuncture needles were inserted vertically 1.0 cm into the skin. fNIRS recordings were initiated 2 min after the insertion of the needle. The acupuncture needles were manipulated with a vertical back-and-forth movement for 15 s, which was followed by a resting period of 60 s. The needle remained there during the resting period. The procedure was repeated eight times. The intensities, frequencies, and durations of the acupuncture stimulation were similar in both MTrPs and non-MTrPs. The subjects were instructed to press a button with their left hand whenever they felt a de-qi sensation (aching, soreness, pressure, heaviness, fullness, warmth, cooling, numbness, tingling, or dull pain) in the given MTrPs or non-MTrPs. After the recording, the de-qi sensations were scored by self-evaluation according to a previous method (Wu et al., 2002). The subjects were asked to comprehensively evaluate their de-qi sensations on a scale of 1–10 (0, no sensation; 5, moderate de-qi sensation; 10, maximum de-qi sensation). Cerebral hemodynamic responses were recorded with fNIRS during acupuncture stimulation at the MTrPs and non-MTrPs. The changes in hemoglobin (Hb) concentrations (oxygenated-Hb (oxy-Hb), deoxygenated-Hb (deoxy-Hb), and total Hb (oxy-Hb + deoxy-Hb)) from baseline were estimated based on the modified Lambert–Beer law (Seiyama, Hazeki, & Tamura, 1988; Wray, Cope, Delpy, Wyatt, & Reynolds, 1988).

The behavioral results indicated that both the frequencies and intensities of the de-qi sensations after MTrP stimulation were greater than those after non-MTrP stimulation (Fig. 2.1). Almost all subjects mainly felt a heavy/dull sensation among all of the de-qi sensation categories, as has been shown in previous studies with needling at classic acupoints (Hui et al., 2007; Kong et al., 2007). This indicated that the acupuncture

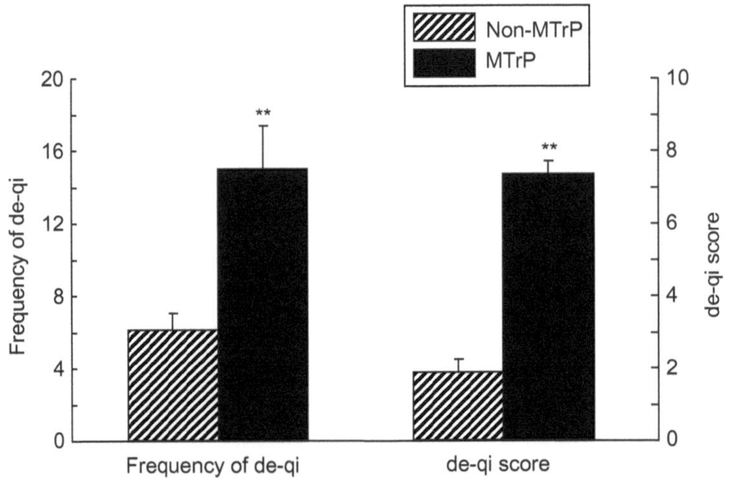

Figure 2.1 Comparison of de-qi frequencies and de-qi scores (sensation intensity) between the myofascial trigger points (MTrPs) and non-MTrPs. The error bars indicate the standard error of the mean (SEM). Acupuncture stimulation at MTrPs more frequently induced strong de-qi sensations compared to stimulation at non-MTrPs, and this difference was significant (**, $P < 0.01$).

sensations at the MTrPs were quite similar to those at classic acupoints. Figure 2.2 shows a typical example of the cerebral hemodynamic responses (topographical maps of *T*-values) to acupuncture stimulation at non-MTrPs without de-qi sensation (A) and at MTrPs with de-qi sensation (B). The hemodynamic responses were statistically analyzed with a general linear model and a boxcar function that approximated the hemodynamic responses (Takeuchi et al., 2009). In this analysis, the start of the boxcar function was gradually delayed up to 20 s. Negative *T*-values indicated that the oxy-Hb concentrations were significantly decreased from baseline levels. Changes in cerebral hemodynamic responses (*T*-values) were less evident for needle manipulation at the non-MTrPs without de-qi sensation (A), while the *T*-values decreased in the bilateral supplementary motor area (SMA) after needle manipulation at the MTrPs (B) when a 0-s delay was applied. When the delay was gradually increased up to 20 s, we observed that the *T*-values in the more anterior regions, including the presupplementary motor area (pre-SMA) and mPFC, were also decreased with MTrP needling. These results have suggested that deactivation in the SMA gradually spread to more anterior regions (pre-SMA, mPFC) with longer latencies.

Figure 2.3 shows the group-averaged changes in hemodynamic responses in the mPFC, pre-SMA, and SMA 20 s after the start of needle manipulation. The data were separately averaged in terms of their de-qi sensation and stimulation sites. The results showed that acupuncture stimulation with de-qi sensation significantly induced a decrement in hemodynamic activity (deactivation) in these three regions regardless of the stimulation sites (MTrPs or non-MTrPs) (Fig. 2.3). Furthermore, we analyzed the relationships between the latencies of the hemodynamic responses and the latencies when the subjects felt a de-qi sensation. The latencies of the hemodynamic responses were analyzed by a general linear model (Takeuchi et al., 2009). In the SMA, the onset latencies of the hemodynamic responses were significantly correlated with those of subjective de-qi sensations (Takamoto et al., 2010). This temporal relationship between de-qi sensations and cerebral hemodynamic responses strongly suggests that the specific de-qi sensation is important for inducing cerebral hemodynamic responses.

Consistent with the present results, a previous fMRI study has also reported deactivation in the mPFC with acupuncture stimulation at classical acupoints with de-qi sensation (Sun et al., 2013). Generally, deactivation is defined as a decrease in fMRI signal intensity with a decrease in regional cerebral blood flow in PET studies (Shimada, Hiraki, Matsuda, & Oda, 2004).

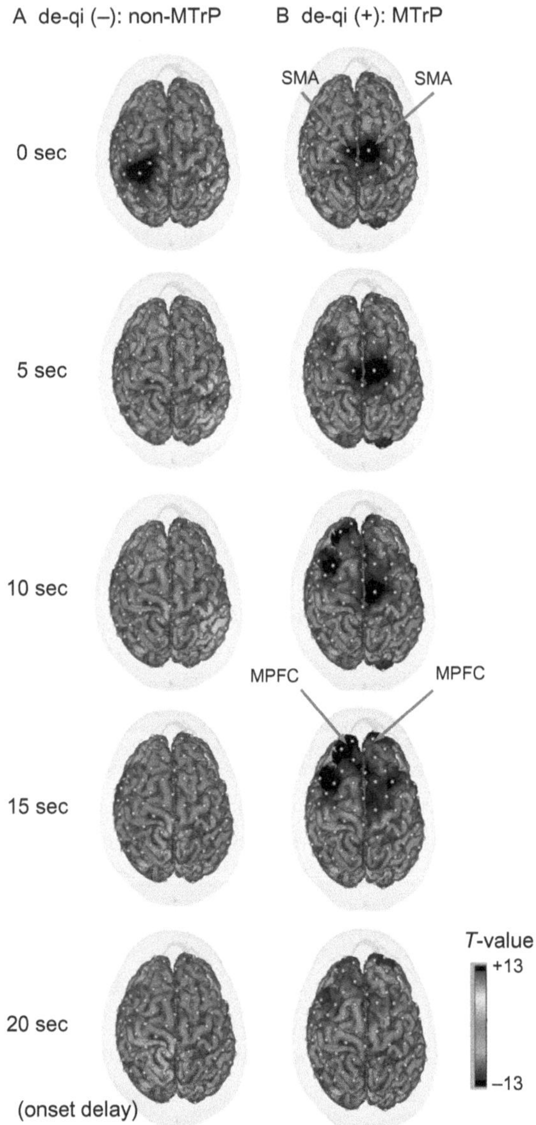

Figure 2.2 Examples of topographical *T*-value maps of the responses to acupuncture stimulation at non-MTrPs without the de-qi sensation (A) and MTrPs with de-qi sensation (B) in a typical subject. Note that the *T*-values decreased prominently in the supplementary motor area (SMA) on both sides and that the brain regions with lower negative *T*-values spread from the SMA to more anterior parts of the prefrontal cortex (PFC), including the pre-SMA and the medial PFC (mPFC). (For color version of this figure, the reader is referred to the online version of this chapter.)

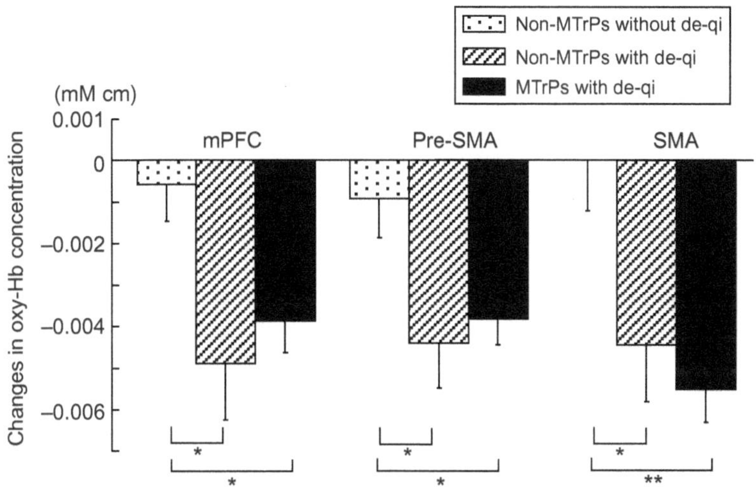

Figure 2.3 Comparison of the group-averaged changes in oxygenated-hemoglobin (oxy-Hb) concentrations in the mPFC, pre-SMA, and SMA regions among the three conditions (non-MTrP stimulation without de-qi sensation, non-MTrP stimulation with de-qi sensation, and MTrP stimulation with de-qi sensation). Note that the oxy-Hb concentrations in the mPFC, pre-SMA, and SMA are decreased significantly during stimulation at non-MTrP with de-qi sensation and at MTrP with de-qi sensation compared to non-MTrP stimulation without de-qi sensation ($P < 0.05$). The error bars indicate the SEM. $^{*}P < 0.05$. $^{**}P < 0.01$.

Hemodynamic responses (oxy-Hb concentration changes) measured by NIRS are strongly correlated with signal intensity changes in fMRI (Strangman, Culver, Thompson, & Boas, 2002; Yamamoto & Kato, 2002). These findings have suggested that a decrease in oxy-Hb concentrations in NIRS studies is comparable to the deactivation that has been reported in fMRI and PET studies (Shimada et al., 2004). These findings suggest that the brain effects of de-qi sensations induced by needling might be similar regardless of the stimulation of MTrPs or classic acupoints. Furthermore, similar hemodynamic responses were elicited regardless of the stimulation of MTrPs or non-MTrPs when de-qi sensations were induced. Thus, we suggest that the specific sensations induced by acupuncture stimulation rather than the specific points stimulated (MTrPs/acupoints or non-MTrPs/nonacupoints) are more important for inducing acupuncture effects on the brain.

In the next sections, we discuss the possibility that the de-qi sensations induced by acupuncture stimulation might be beneficial to several disorders, including chronic pain, psychiatric diseases, and diseases related to autonomic dysfunction, because of its suppressive effects on the SMA and mPFC.

4. PRE-SMA AND SMA RESPONSES TO ACUPUNCTURE STIMULATION

The pre-SMA and SMA, which are located in Brodmann area 6, are believed to be involved in general motor functions (Luppino, Matelli, Camarda, & Rizzolatti, 1993; Picard & Strick, 2001). However, previous studies have reported the following: (1) SMA activity is increased by noxious stimulation, such as heat stimulation (Farrell, Laird, & Eqan, 2005; Hsieh et al., 1995; Kwan, Crawley, Mikulis, & Davis, 2000); (2) SMA activity is correlated with the intensity of pain (Coghill, Sang, Maisog, & Iadarola, 1999) or an unpleasant sensation (Drzezga et al., 2001); (3) SMA activity is increased in patients experiencing phantom pain (Dettmers et al., 2001; Willoch et al., 2000) and allodynia (Peyron et al., 2004); and (4) the SMA receives afferent inputs from the ACC, which is involved in the emotional evaluation of pain (Morecraft & Van-Hoesen, 1992, 1993; Wang, Shima, Sawamura, & Tanji, 2001). Our results have shown that de-qi sensations were temporally correlated with hemodynamic responses in the SMA. Furthermore, an fMRI study has reported that activity in the pre-SMA and SMA is significantly increased in response to finger movements of the affected extremity in patients with complex regional pain syndrome (CRPS) compared to healthy controls (Maihöfner et al., 2007). These findings suggest that activity in the pre-SMA and SMA is increased in subjects or patients with acute or chronic pain and that acupuncture stimulation with de-qi sensations may ameliorate the pain through its suppressive effects on the pre-SMA and SMA regions.

However, volumetric MRI studies have reported that brain gray matter in the SMA is reduced in patients with fibromyalgia (Puri et al., 2010) and unilateral coxarthrosis (Rodriguez-Raecke, Niemeier, Ihle, Ruether, & May, 2013). In addition, the brain gray matter in the SMA has been shown to be increased after hip joint endoprosthetic surgery in these patients (Rodriguez-Raecke et al., 2013). Currently, the exact mechanisms of these changes in gray matter volume due to pain have not yet been elucidated, although some researchers have suggested that changes in motor functions might induce such volume changes (Rodriguez-Raecke et al., 2013). The relationship between gray matter volume changes and brain activity changes in fMRI studies also remains unclear. Further studies are required to elucidate any possible relationships among subjective pain, hemodynamic responses, and volumetric changes in these regions.

5. PREFRONTAL CORTICAL RESPONSES TO ACUPUNCTURE STIMULATION

The prefrontal cortex (PFC) has been shown to contribute to various higher brain functions, such as working memory, learning, attention, and emotional behavior (Birrell & Brown, 2000; Ragozzino, 2000; Ridderinkhof, Ullsperger, Crone, & Nieuwenhuis, 2004). The PFC has been included in the brain regions of the so-called pain matrix (Schweinhardt & Bushnell, 2010). In patients with chronic low back pain, spontaneous pain intensity has been shown to correlate with brain activity in the mPFC (Baliki et al., 2006). Lidocaine treatment of these patients has been shown to significantly decrease brain activity in the mPFC (Baliki, Geha, Apkarian, & Chialvo, 2008). Furthermore, neuropsychological studies have reported that activity in specific parts of the brain, including the mPFC, is increased in the resting period, and this resting-state activity has been termed the default-mode of brain activity (Raichle et al., 2001; Sonuga-Barke & Castellanos, 2007). In patients with chronic low back pain, deactivation in several key default-mode network regions, including the mPFC, is significantly suppressed compared to healthy subjects (Baliki, Geha, Apkarian, & Chialvo, 2008). A recent arterial spin labeling study has reported that the default-mode network encodes clinical pain (Loggia et al., 2013). These previous studies have suggested that pathological chronic pain is associated with hyperactivity of the mPFC. Therefore, acupuncture needling might suppress chronic pain through its inhibitory effects on activity in the mPFC. Consistently, fMRI studies have reported that activity of the default-mode network regions, including the mPFC, is decreased during acupuncture stimulation with de-qi sensations (Hui et al., 2005, 2009). Furthermore, when de-qi is mixed with sharp pain, deactivation is attenuated or these regions are even activated (Hui et al., 2005, 2009).

Several previous studies have reported that mPFC activity is increased in various psychiatric diseases, such as schizophrenia (Taylor, Welsh, Chen, Velander, & Liberzon, 2007), social phobia (Blair et al., 2008), and panic disorder (Sakai et al., 2006). Excessive default-mode activity has also been reported in patients with attention-deficit/hyperactivity disorder (Sonuga-Barke & Castellanos, 2007) and schizophrenia (Zhou et al., 2007). Acupuncture has been applied to some of these psychiatric disorders (Pilkington, Kirkwood, Rampes, Cummings, & Richardson, 2007; Rathbone & Xia, 2005). These findings suggest that the therapeutic effects of acupuncture in patients with psychiatric diseases might be mediated through its inhibitory effects on mPFC activity.

6. de-qi SENSATION AND AUTONOMIC FUNCTIONS

Several previous studies have reported that acupuncture modulated autonomic nervous activity. Acupuncture stimulation has been shown to change the sympathovagal balance toward vagal predominance (Huang et al., 2005; Nishijo, Mori, Yosikawa, & Yazawa, 1997). Acupuncture has been applied to various pathological conditions involving autonomic dysfunction, such as nausea and vomiting (Streitberger, Ezzo, & Schneider, 2006). Abnormalities in the autonomic nervous system may generate and sustain chronic pain (Passatore & Roatta, 2006; Schott, 1999). Previous studies have reported that sympathetic overactivation and/or parasympathetic hypoactivation are observed in patients with chronic low back pain (Gockel, Lindholm, Niemist, & Hurri, 2008). Thus, acupuncture may treat diseases associated with autonomic dysfunction by modulating autonomic nervous activity. However, a recent systemic review has indicated that there is no convincing evidence that acupuncture can modify autonomic functions based on heart rate (HR) variability (HRV) compared to sham acupuncture with minimal penetration into nonacupoints (Lee et al., 2010).

We hypothesized that de-qi sensations induced by acupuncture may modulate autonomic nervous activity and investigated the relationships among de-qi sensations induced by acupuncture manipulation, the effects on sympathetic and parasympathetic autonomic functions, and electroencephalographic (EEG) changes (Sakai et al., 2007). An acupuncture needle was inserted into the right trapezius muscle of the subjects, and acupuncture manipulation was repeated in order to induce specific acupuncture sensations repeatedly while the needle was left in the muscle. The results indicated that acupuncture manipulation significantly decreased HR. A spectral analysis of the HRV indicated that acupuncture manipulation significantly decreased the low-frequency (LF) components of both HRV and systolic blood pressure (SBP) variability (SBPV), which is an index of sympathetic activity, and significantly reduced the ratio of LF to the high-frequency (HF) component of HRV (LF/HF, which is an index of sympathetic activity). Furthermore, there was a significant negative correlation between the changes in the LF/HF ratio of HRV and the number of specific acupuncture sensations that were reported and a significant positive correlation between the HF, which is an index of parasympathetic activity, of the HRV and the number of acupuncture sensations (Fig. 2.4). Analyses of the EEG data indicated that acupuncture manipulation nonspecifically increased the power of all of the spectral bands, especially the theta and alpha bands of the EEGs. However, the changes in the HF and total power, which is the overall activity of the autonomic nervous system, of the HRV were positively correlated with changes in the theta, alpha, and gamma powers, while the changes in the LF of the SBPV and the LF/HF of the HRV were negatively correlated with the changes in the power of all of the spectral bands (Fig. 2.5). These results were consistent with the suggestion that the autonomic changes induced by acupuncture manipulation with de-qi sensations might be

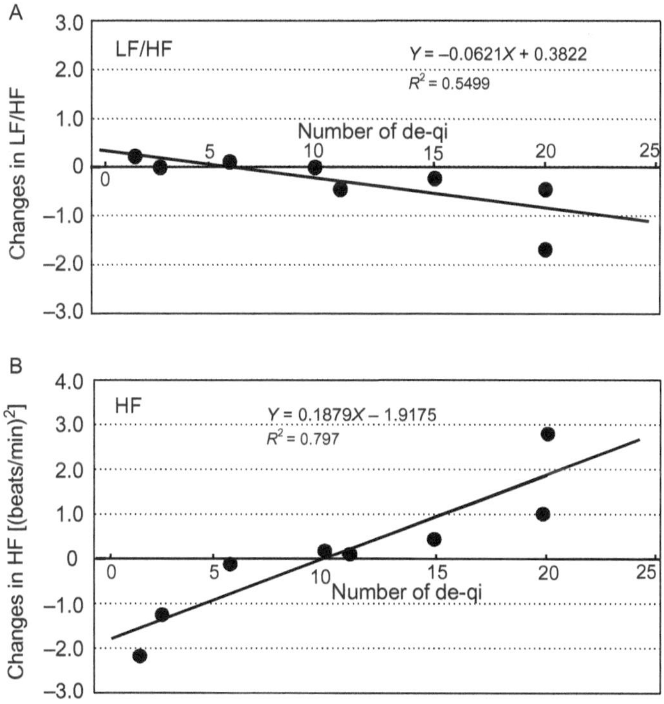

Figure 2.4 Correlation between the number of de-qi sensations and the low-frequency/high-frequency (LF/HF) ratio of heart rate variability (HRV) (A) and between the number of de-qi sensations and the HF component of HRV (B). Note the significant negative and positive correlations between the number of de-qi sensations and the changes in the LF/HF ratio and HF component, respectively. Ordinates: changes in the LF/HF ratio of HRV between the preacupuncture control and postacupuncture period (A) and changes in the HF component of HRV between the preacupuncture control and postacupuncture period (B); abscissa: number of button presses indicating de-qi. *Modified from Sakai et al. (2007).*

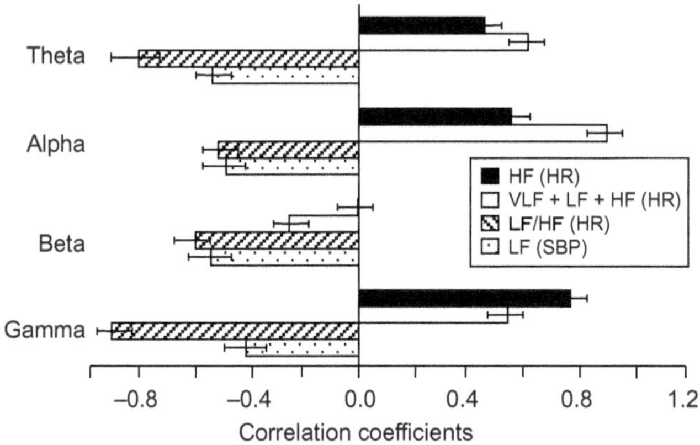

Figure 2.5 Relationships between the changes in electroencephalographic spectral powers and those in autonomic functions compared to the preacupuncture control period. Note that the HF component of HRV and total power (very low frequency component (VLF) + LF + HF) were positively correlated with theta, alpha, and gamma power, while the LF/HF ratio of HRV and the LF component of the systolic blood pressure (SBP) variability were negatively correlated with theta, alpha, beta, and gamma power. Abscissa, Pearson's correlation coefficient.

mediated through the central nervous system and especially through the forebrain, as shown in the EEG changes.

Previous fMRI studies have also reported significant relationships among de-qi sensations induced by acupuncture manipulation, the effects on autonomic functions, and fMRI signal changes. Beissner, Deichmann, Henke, and Bär (2012) have investigated the relationships among de-qi sensations, brain activity, and autonomic responses during acupuncture stimulation at PC6. The intensity of the de-qi sensations induced by acupuncture stimulation at PC6 is negatively correlated with the individual subject's HR. Furthermore, the de-qi sensations are associated with activity increases in the cortical and brainstem regions, including the primary and secondary somatosensory cortices, the ACC, the middle and superior frontal gyrus, the insula, the dorsolateral prefrontal cortex, and the locus coeruleus, while this sensation is associated with activity decreases (deactivation) in the ventromedial prefrontal cortex, the orbitofrontal cortex, and the perigenual and subgenual cingulate cortex (vmPFC/pgACC). In addition, activity in the vmPFC, the dorsolateral prefrontal cortex, the pgACC, the hypothalamus, the periaqueductal gray, the rostral ventromedial medulla, and the ventrolateral medulla is associated with both de-qi sensations and HR responses. Napadow et al. (2012) have investigated the relationships between brain activities and different autonomic responses during acupuncture stimulation at different acupoints. Acupuncture stimulation at ST36 and SP9 induces higher de-qi scores compared to stimulation at nonacupoints. In particular, stimulation at SP9 induces greater sharp pain and skin conductance responses (SCR), which are associated with activity increase in the insula, while the intensity of the de-qi sensations is greater with acupuncture stimulation at ST36 and correlated with HR reduction, which is associated with activity decreases in the default-mode network regions, including the mPFC. These studies suggest that the mPFC might play an important role in inducing autonomic responses.

Previous neuroanatomical and noninvasive imaging studies have reported that the mPFC sends efferent fibers directly to the hypothalamus and brainstem, which is involved in autonomic and behavioral control under stressful conditions (Hadjipavlou, Dunckley, Behrens, & Tracey, 2006; Ongür, An, & Price, 1998; Rempel-Clower & Barbas, 1998). Activities in the mPFC have been shown to be positively correlated with blood adrenocorticotropic hormone levels (Liberzon et al., 2007) and the SCR reflecting sympathetic activity (Critchley, Elliott, Mathias, & Dolan, 2000). Furthermore, our previous studies have reported that hemodynamic activity in the dorsal part of the mPFC, which is reduced (deactivated) by needling with de-qi sensations (Takamoto et al., 2010), is significantly positively correlated with the LF/HF and negatively correlated with the HF (Yasui et al., 2010). These results suggest that de-qi sensations might induce deactivation in the dorsal part of the mPFC and vmPFC/pgACC, which in turn might increase parasympathetic nervous activity and decrease sympathetic nervous activity.

These changes in autonomic nervous activities induced by acupuncture might be beneficial to pain control. Abnormalities of autonomic functions have been implicated in the chronic pain in certain diseases. For example, sympathetic overactivation has been reported in patients with fibromyalgia with diffuse tenderness and musculoskeletal pain (Furlan et al., 2005), causalgia (Baron, Levine, & Fields, 1999), and whiplash-associated disorders (Passatore & Roatta, 2006). This abnormality in the sympathetic nervous system might generate and sustain chronic pain (Janig, 1992; Passatore & Roatta, 2006; Schott, 1999). In chronic sympathetically mediated pain, which is a subcategory of reflex sympathetic dystrophy (CRPS type 1), the blockade of sympathetic outflow relieved the pain without interfering with the sensory afferent fibers (Apkarian, Thomas, Krauss, & Szeverenyi, 2001). Furthermore, morphological studies have reported that the sympathetic nervous system, which controls the blood vessels within muscles, simultaneously controls both the extrafusal and intrafusal muscle fibers by the branching of the collaterals (Bombardi et al., 2006; Selkowitz, 1992). Accordingly, skeletal muscle tension depends on sympathetic activity, and musculoskeletal pain is caused by increased tension due to an excessive activation of the sympathetic nervous system by stress (Barker & Saito, 1981; Passatore, Filippi, & Grassi, 1985). Thus, these results suggest that acupuncture with specific sensations, which suppresses sympathetic nervous activity, is beneficial to pain relief.

7. WHAT ARE ACUPOINTS?

Previous studies have reported that both the de-qi sensation and acupuncture analgesia are blocked by intramuscular, but not subcutaneous, procaine injections at acupuncture points (Chiang, Chang, Chu, & Yang, 1973). Furthermore, there were no significant differences in the sensations evoked by needling between acupoints and nonacupoints at the epidermis, dermis, and fascia level, although there were differences at the intramuscular level (Park et al., 2011). Needle penetration into the superficial layer (epidermis, dermis, and fascia level) induces pricking and sharp sensations, while needle penetration into the deep layer (intramuscular level) induces a high degree of deep, dull, heavy, and spreading sensations. In addition, sensations induced by saline injections into the muscle are similar to de-qi sensations (Henderson, Bandler, Gandevia, & Macefield, 2006). Thus, these results suggest that de-qi sensations are associated with certain receptors confined to the intramuscular level. In the deep tissue, including the muscles, a large population of polymodal nociceptors has been found in thin afferent fibers, such as the Aδ- and C-fibers. Previous studies have suggested that these afferent fibers are involved in the de-qi sensations induced by acupuncture stimulation at acupoints (Lu, 1983; Wang, Yao, Xian, & Hou, 1985). A close correlation between a particular de-qi sensation and the excitation of thin afferent fibers responsive to bradykinin has been reported (Kawakita & Gotoh, 1996; Kawakita et al., 2006). The discharge of polymodal nociceptors in the deep tissues and de-qi sensations was provoked simultaneously by manual acupuncture, and the frequencies of discharge were well correlated with the intensity of the de-qi sensations (Kawakita & Gotoh, 1996). We have reported that both the frequencies and intensities of de-qi sensations are greater following stimulation at MTrPs compared to stimulation at non-MTrPs (Takamoto et al., 2010). Furthermore, previous studies have proposed that acupoints are specifically located in the areas densely innervated by the somatic sensory nerves compared to other regions (nonacupoints) (Chan, 1984; Li, Zhang, & Xie, 2004). These results suggest that both acupoints and MTrPs are not anatomically distinct regions but, rather, regions in the muscles with dense sensory innervation where de-qi sensations can be relatively easily and frequently obtained.

8. LIMITATIONS

Several controversial issues in acupuncture studies have been discussed. Two clinical trials have reported that there were no significant differences in pain relief between those who felt de-qi and those who did not (Foster et al., 2007; White, Prescott, & Lewith, 2010). Furthermore, except for acupuncture analgesia and anesthesia to acute pain, evidence indicating a significant relationship between de-qi and therapeutic effects has been rare (Kong et al., 2007). Thus, the relationship between de-qi and treatment efficacy is controversial. Second, analgesia and the modulation of the autonomic nervous system during acupuncture might be attributed to placebo effects. In clinical trials, several studies have reported that acupuncture is no more effective than sham acupuncture, which is defined as needle insertion into the superficial level and no manipulation at nonacupoints (Diener et al., 2006; Haake et al., 2007; Scharf et al., 2006). Furthermore, nonpenetrating sham acupuncture has been shown to induce significant placebo effects in the treatment of subjective pain complaints (Kaptchuk et al., 2006). Third, almost all previous neuroimaging studies on acupuncture have used healthy subjects, and the therapeutic effects of the de-qi sensation on brain activity have not been analyzed in patients. Fourth, the relationships between the individual sensations evoked during acupuncture and de-qi are unclear. Chiang et al. (1973) have reported that numbness, fullness, and soreness sensations are associated with acupuncture analgesia, suggesting that the de-qi sensation is associated with these sensations. However, another study has reported that burning, intense, pulsating, and stinging sensations are significantly correlated with acupuncture analgesia (Park et al., 2013). Thus, there is no consensus about the sensory characteristics of de-qi. Further studies are required to address these issues in order to clarify the mechanisms of acupuncture.

9. CONCLUSIONS

Acupuncture stimulation-induced de-qi sensations might be effective in the treatment of various disorders, including chronic pain, psychiatric diseases, and diseases associated with autonomic nervous dysfunction. These effects might be mediated through the central nervous system, including the mPFC, the pre-SMA, and the SMA. Deactivation of these areas during acupuncture increases parasympathetic nervous activity and decreases sympathetic nervous activity, which might be beneficial to various disorders, including chronic pain with autonomic dysfunction. Furthermore, de-qi sensations, rather than specific stimulation sites, might be more important in the induction of acupuncture effects.

ACKNOWLEDGMENTS

This work was supported partly by JSPS (Japan Society for the Promotion of Science) Asian Core Program, and the Ministry of Education, Science, Sports and Culture, Grant-in-Aid for Scientific Research (B) (22240051).

REFERENCES

Apkarian, A. V., Thomas, P. S., Krauss, B. R., & Szeverenyi, N. M. (2001). Prefrontal cortical hyperactivity in patients with sympathetically mediated chronic pain. *Neuroscience Letters*, *311*, 193–197.

Assefi, N. P., Sherman, K. J., Jacobsen, C., Goldberg, J., Smith, W. R., & Buchwald, D. (2005). A randomized clinical trial of acupuncture compared with sham acupuncture in fibromyalgia. *Annals of Internal Medicine*, *143*, 10–19.

Baliki, M. N., Chialvo, D. R., Geha, P. Y., Levy, R. N., Harden, T. B., Parrish, T. B., et al. (2006). Chronic pain and the emotional brain: Specific brain activity associated with spontaneous fluctuations of intensity of chronic back pain. *The Journal of Neuroscience*, *26*, 12165–12173.

Baliki, M. N., Geha, P. Y., Apkarian, A. V., & Chialvo, D. R. (2008). Beyond feeling: Chronic pain hurts the brain, disrupting the default-mode network dynamics. *The Journal of Neuroscience*, *28*, 1398–1403.

Baliki, M. N., Geha, P. Y., Jabakhanji, R., Harden, N., Schnitzer, T. J., & Apkarian, A. V. (2008). A preliminary fMRI study of analgesic treatment in chronic back pain and knee osteoarthritis. *Molecular Pain*, *4*, 47.

Barker, D., & Saito, M. (1981). Autonomic innervation of receptors and muscle fibres in cat skeletal muscle. *Proceedings of the Royal Society of London—Series B: Biological Sciences*, *212*, 317–332.

Baron, R., Levine, J. D., & Fields, H. L. (1999). Causalgia and reflex sympathetic dystrophy: Does the sympathetic nervous system contribute to the generation of pain? *Muscle & Nerve*, *22*, 678–695.

Beissner, F., Deichmann, R., Henke, C., & Bär, K. J. (2012). Acupuncture—Deep pain with an autonomic dimension? *NeuroImage*, *60*, 653–660.

Birrell, J. M., & Brown, V. J. (2000). Medial frontal cortex mediates perceptual attentional set shifting in the rat. *The Journal of Neuroscience*, *20*, 4320–4324.

Blair, K., Geraci, M., Devido, J., McCaffrey, D., Chen, G., Vythilingam, M., et al. (2008). Neural response to self- and other referential praise and criticism in generalized social phobia. *Archives of General Psychiatry*, *65*, 1176–1184.

Bombardi, C., Grandis, A., Chiocchetti, R., Bortolami, R., Johansson, H., & Lucchi, M. L. (2006). Immunohistochemical localization of alpha(1a)-adrenoreceptors in muscle spindles of rabbit masseter muscle. *Tissue & Cell*, *38*, 121–125.

Chan, S. H. (1984). What is being stimulated in acupuncture: Evaluation of the existence of a specific substrate. *Neuroscience and Biobehavioral Reviews, 8*, 25–33.
Chiang, C. Y., Chang, C. T., Chu, H. L., & Yang, L. F. (1973). Peripheral afferent pathway for acupuncture analgesia. *Scientia Sinica, 16*, 210–217.
Cho, Z. H., Chung, S. C., Jones, J. P., Park, J. B., Park, H. J., Lee, H. J., et al. (1998). New findings of the correlation between acupoints and corresponding brain cortices using functional MRI. *Proceedings of the National Academy of Sciences of the United States of America, 95*, 2670–2673.
Coghill, R. C., Sang, C. N., Maisog, J. M., & Iadarola, M. J. (1999). Pain intensity processing within the human brain: A bilateral, distributed mechanism. *The Journal of Neurophysiology, 82*, 1934–1943.
Critchley, H. D., Elliott, R., Mathias, C. J., & Dolan, R. J. (2000). Neural activity relating to generation and representation of galvanic skin conductance responses: A functional magnetic resonance imaging study. *The Journal of Neuroscience, 20*, 3033–3040.
Dettmers, C., Adler, T., Rzanny, R., van-Schayck, R., Gaser, C., Weiss, T., et al. (2001). Increased excitability in the primary motor cortex and supplementary motor area in patients with phantom limb pain after upper limb amputation. *Neuroscience Letters, 307*, 109–112.
Dhond, R. P., Kettner, N., & Napadow, V. (2007). Neuroimaging acupuncture effects in the human brain. *The Journal of Alternative and Complementary Medicine, 13*, 603–616.
Diener, H. C., Kronfeld, K., Boewing, G., Lungenhausen, M., Maier, C., Molsberger, A., et al. (2006). Efficacy of acupuncture for the prophylaxis of migraine: A multicentre randomised controlled clinical trial. *Lancet Neurology, 5*, 310–316.
Dorsher, P. T. (2008). Can classical acupuncture points and trigger points be compared in the treatment of pain disorders? Birch's analysis revisited. *The Journal of Alternative and Complementary Medicine, 14*, 353–359.
Drzezga, A., Darsow, U., Treede, R. D., Siebner, H., Frisch, M., Munz, F., et al. (2001). Central activation by histamine-induced itch: Analogies to pain processing: A correlational analysis of O-15 H2O positron emission tomography studies. *Pain, 92*, 295–305.
Esch, T., Guarna, M., Bianchi, E., Zhu, W., & Stefano, G. B. (2004). Commonalities in the central nervous system's involvement with complementary medical therapies: Limbic morphinergic processes. *Medical Science Monitor, 10*, MS6–MS17.
Fang, J., Jin, Z., Wang, Y., Li, K., Kong, J., Nixon, E. E., et al. (2009). The salient characteristics of the central effects of acupuncture needling: Limbic-paralimbic-neocortical network modulation. *Human Brain Mapping, 30*, 1196–1206.
Farrell, M. J., Laird, A. R., & Eqan, G. F. (2005). Brain activity associated with painfully hot stimuli applied to the upper limb: A meta-analysis. *Human Brain Mapping, 25*, 129–139.
Filshie, J., & White, A. (1998). *Medical acupuncture: A western scientific approach*. Philadelphia: Elsevier Limited.
Fink, M., Wolkenstein, E., Karst, M., & Gehrke, A. (2002). Acupuncture in chronic epicondylitis: A randomized controlled trial. *Rheumatology (Oxford, England), 41*, 205–209.
Foster, N. E., Thomas, E., Barlas, P., Hill, J. C., Young, J., Mason, E., et al. (2007). Acupuncture as an adjunct to exercise based physiotherapy for osteoarthritis of the knee: Randomised controlled trial. *British Medical Journal, 335*, 436.
Furlan, R., Colombo, S., Perego, F., Atzeni, F., Diana, A., Barbic, F., et al. (2005). Abnormalities of cardiovascular neural control and reduced orthostatic tolerance in patients with primary fibromyalgia. *The Journal of Rheumatology, 32*, 1787–1793.
Gockel, M., Lindholm, H., Niemist, L., & Hurri, H. (2008). Perceived disability but not pain is connected with autonomic nervous function among patients with chronic low back pain. *Journal of Rehabilitation Medicine, 40*, 355–358.

Haake, M., Müller, H. H., Schade-Brittinger, C., Basler, H. D., Schäfer, H., Maier, C., et al. (2007). German Acupuncture Trials (GERAC) for chronic low back pain: Randomized, multicenter, blinded, parallel-group trial with 3 groups. *Archives of Internal Medicine*, *167*, 1892–1898.

Hadjipavlou, G., Dunckley, P., Behrens, T. E., & Tracey, I. (2006). Determining anatomical connectivities between cortical and brainstem pain processing regions in humans: A diffusion tensor imaging study in healthy controls. *Pain*, *123*, 169–178.

Han, J. S., & Ho, Y. S. (2011). Global trends and performances of acupuncture research. *Neuroscience and Biobehavioral Reviews*, *35*, 680–687.

Henderson, L. A., Bandler, R., Gandevia, S. C., & Macefield, V. G. (2006). Distinct forebrain activity patterns during deep versus superficial pain. *Pain*, *120*, 286–296.

Hsieh, J. C., Backdahl, M. S., Hagermark, O., Elander, S. S., Rosenquist, G., & Ingvar, M. (1995). Traumatic nociceptive pain activates the hypothalamus and the periaqueductal gray: A positron emission tomography study. *Pain*, *64*, 303–314.

Huang, S. T., Chen, G. Y., Lo, H. M., Lin, J. G., Lee, Y. S., & Kuo, C. D. (2005). Increase in the vagal modulation by acupuncture at neiguan point in the healthy subjects. *The American Journal of Chinese Medicine*, *33*, 157–164.

Huang, W., Pach, D., Napadow, V., Park, K., Long, X., Neumann, J., et al. (2012). Characterizing acupuncture stimuli using brain imaging with FMRI—A systematic review and meta-analysis of the literature. *PLoS One*, 7, e32960.

Hui, K. K., Liu, J., Marina, O., Napadow, V., Haselgrove, C., Kwong, K. K., et al. (2005). The integrated response of the human cerebro-cerebellar and limbic systems to acupuncture stimulation at ST 36 as evidenced by fMRI. *NeuroImage*, *27*, 479–496.

Hui, K. K., Marina, O., Claunch, J. D., Nixon, E. E., Fang, J., Liu, J., et al. (2009). Acupuncture mobilizes the brain's default mode and its anti-correlated network in healthy subjects. *Brain Research*, *1287*, 84–103.

Hui, K. K., Nixon, E. E., Vangel, M. G., Liu, J., Marina, O., Napadow, V., et al. (2007). Characterization of the "deqi" response in acupuncture. *BMC Complementary and Alternative Medicine*, *2007*(7), 33.

Hui, K. K., Sporko, T. N., Vangel, M. G., Li, M., Fang, J., & Lao, L. (2011). Perception of Deqi by Chinese and American acupuncturists: A pilot survey. *Chinese Medicine*, *6*, 2.

Itoh, K., Katsumi, Y., Hirota, S., & Kitakoji, H. (2007). Randomised trial of trigger point acupuncture compared with other acupuncture for treatment of chronic neck pain. *Complementary Therapies in Medicine*, *15*, 172–179.

Itoh, K., Katsumi, Y., & Kitakoji, H. (2004). Trigger point acupuncture treatment of chronic low back pain in elderly patients—A blinded RCT. *Acupuncture in Medicine*, *2004*(22), 170–177.

Janig, W. (1992). Pain and the sympathetic nervous system: Pathophysiological mechanism. In S. R. Bannister & C. J. Mathias (Eds.), *Autonomic failure* (pp. 231–251). Oxford: Oxford University Press.

Kaptchuk, T. J. (2002). The placebo effect in alternative medicine: Can the performance of a healing ritual have clinical significance? *Annals of Internal Medicine*, *136*, 817–825.

Kaptchuk, T. J., Stason, W. B., Davis, R. B., Legedza, A. R., Schnyer, R. N., Kerr, C. E., et al. (2006). Sham device v inert pill: Randomised controlled trial of two placebo treatments. *British Medical Journal*, *332*, 391–397.

Kawakita, K., & Gotoh, K. (1996). Role of polymodal receptors in the acupuncture-mediated endogenous pain inhibitory systems. *Progress in Brain Research*, *113*, 507–523.

Kawakita, K., Shinbara, H., Imai, K., Fukuda, F., Yano, T., & Kuriyama, K. (2006). How do acupuncture and moxibustion act?—Focusing on the progress in Japanese acupuncture research. *The Journal of Pharmacological Sciences*, *100*, 443–459.

Kong, J., Gollub, R., Huang, T., Polich, G., Napadow, V., Hui, K., et al. (2007). Acupuncture de qi, from qualitative history to quantitative measurement. *The Journal of Alternative and Complementary Medicine*, *13*, 1059–1070.

Kong, J., Kaptchuk, T. J., Webb, J. M., Kong, J. T., Sasaki, Y., Polich, G. R., et al. (2009). Functional neuroanatomical investigation of vision-related acupuncture point specificity—A multisession fMRI study. *Human Brain Mapping*, *30*, 38–46.

Kung, Y. Y., Chen, F. P., Chaung, H. L., Chou, C. T., Tsai, Y. Y., & Hwang, S. J. (2001). Evaluation of acupuncture effect to chronic myofascial pain syndrome in the cervical and upper back regions by the concept of Meridians. *Acupuncture & Electro-Therapeutics Research*, *26*, 195–202.

Kwan, C. L., Crawley, A. P., Mikulis, D. J., & Davis, K. D. (2000). An fMRI study of the anterior cingulate cortex and surrounding medial wall activations evoked by noxious cutaneous heat and cold stimuli. *Pain*, *85*, 359–374.

Lee, S., Lee, M. S., Choi, J. Y., Lee, S. W., Jeong, S. Y., & Ernst, E. (2010). Acupuncture and heart rate variability: A systematic review. *Autonomic Neuroscience*, *155*, 5–13.

Li, G., Cheung, R. T., Ma, Q. Y., & Yang, E. S. (2003). Visual cortical activations on fMRI upon stimulation of the vision-implicated acupoints. *Neuroreport*, *14*, 669–673.

Li, A. H., Zhang, J. M., & Xie, Y. K. (2004). Human acupuncture points mapped in rats are associated with excitable muscle/skin-nerve complexes with enriched nerve endings. *Brain Research*, *1012*, 154–159.

Liberzon, I., King, A. P., Britton, J. C., Phan, K. L., Abelson, J. L., & Taylor, S. F. (2007). Paralimbic and medial prefrontal cortical involvement in neuroendocrine responses to traumatic stimuli. *The American Journal of Psychiatry*, *164*, 1250–1258.

Loggia, M. L., Kim, J., Gollub, R. L., Vangel, M. G., Kirsch, I., Kong, J., et al. (2013). Default mode network connectivity encodes clinical pain: An arterial spin labeling study. *Pain*, *154*, 24–33.

Longhurst, J. C. (2010). Defining meridians: A modern basis of understanding. *Journal of Acupuncture and Meridian Studies*, *3*, 67–74.

Low, R. (2001). *Acupuncture techniques for successful point selection.* Woburn: Butterworth-Heinemann.

Lu, G. W. (1983). Characteristics of afferent fiber innervation on acupuncture points zusanli. *The American Journal of Physiology*, *245*, R606–R612.

Luppino, G., Matelli, M., Camarda, R., & Rizzolatti, G. (1993). Corticocortical connections of area F3 (SMA-proper) and area F6 (pre-SMA) in the macaque monkey. *The Journal of Comparative Neurology*, *338*, 114–140.

Maihöfner, C., Baron, R., DeCol, R., Binder, A., Birklein, F., Deuschl, G., et al. (2007). The motor system shows adaptive changes in complex regional pain syndrome. *Brain*, *130*, 2671–2687.

Melzack, R., Stillwell, D. M., & Fox, E. J. (1977). Trigger points and acupuncture points for pain: Correlations and implications. *Pain*, *3*, 3–23.

Morecraft, R. J., & Van-Hoesen, G. W. (1992). Cingulate input to the primary and supplementary motor cortices in the rhesus monkey: Evidence for somatotopy in areas 24c and 23c. *The Journal of Comparative Neurology*, *322*, 471–489.

Morecraft, R. J., & Van-Hoesen, G. W. (1993). Frontal granular cortex input to the cingulate (M3), supplementary (M2) and primary (M1) motor cortices in the rhesus monkey. *The Journal of Comparative Neurology*, *337*, 669–689.

Napadow, V., Lee, J., Kim, J., Cina, S., Maeda, Y., Barbieri, R., et al. (2012). Brain correlates of phasic autonomic response to acupuncture stimulation: An event-related fMRI study. *Human Brain Mapping*, *34*, 2592–2606.

Nishijo, K., Mori, H., Yosikawa, K., & Yazawa, K. (1997). Decreased heart rate by acupuncture stimulation in humans via facilitation of cardiac vagal activity and suppression of cardiac sympathetic nerve. *Neuroscience Letters*, *227*, 165–168.

Ongür, D., An, X., & Price, J. L. (1998). Prefrontal cortical projections to the hypothalamus in macaque monkeys. *The Journal of Comparative Neurology, 401*, 480–505.

Park, J. J., Akazawa, M., Ahn, J., Beckman-Harned, S., Lin, F. C., Lee, K., et al. (2011). Acupuncture sensation during ultrasound guided acupuncture needling. *Acupuncture in Medicine, 29*, 257–265.

Park, J. E., Ryu, Y. H., Liu, Y., Jung, H. J., Kim, A. R., Jung, S. Y., et al. (2013). A literature review of de qi in clinical studies. *Acupuncture in Medicine, 31*, 132–142.

Passatore, M., Filippi, G. M., & Grassi, C. (1985). Cervical sympathetic nerve stimulation can induce an intrafusal muscle fiber contraction in the rabbit. In I. A. Boyd & M. H. Gladden (Eds.), *The muscle spindle* (pp. 221–226). London: Macmillan.

Passatore, M., & Roatta, S. (2006). Influence of sympathetic nervous system on sensorimotor function: Whiplash associated disorders (WAD) as a model. *European Journal of Applied Physiology, 98*, 423–449.

Peyron, R., Schneider, F., Faillenot, I., Convers, P., Barral, F. G., Garcia-Larrea, L., et al. (2004). An fMRI study of cortical representation of mechanical allodynia in patients with neuropathic pain. *Neurology, 63*, 1838–1846.

Picard, N., & Strick, P. L. (2001). Imaging the premotor areas. *Current Opinion in Neurobiology, 11*, 663–672.

Pilkington, K., Kirkwood, G., Rampes, H., Cummings, M., & Richardson, J. (2007). Acupuncture for anxiety and anxiety disorders—A systematic literature review. *Acupuncture in Medicine, 25*, 1–10.

Puri, B. K., Agour, M., Gunatilake, K. D., Fernando, K. A., Gurusinghe, A. I., & Treasaden, I. H. (2010). Reduction in left supplementary motor area grey matter in adult female fibromyalgia sufferers with marked fatigue and without affective disorder: A pilot controlled 3-T magnetic resonance imaging voxel-based morphometry study. *The Journal of International Medical Research, 38*, 1468–1472.

Ragozzino, M. E. (2000). The contribution of cholinergic and dopaminergic afferents in the rat prefrontal cortex to learning, memory and attention. *Psychobiology, 28*, 238–247.

Raichle, M. E., MacLeod, A. M., Snyder, A. Z., Powers, W. J., Gusnard, D. A., & Shulman, G. L. (2001). A default mode of brain function. *Proceedings of National Academy of Sciences of the United States of America, 98*, 676–682.

Rathbone, J., & Xia, J. (2005). Acupuncture for schizophrenia. *Cochrane Database of Systematic Reviews, 4,* CD005475.

Rempel-Clower, N. L., & Barbas, H. (1998). Topographic organization of connections between the hypothalamus and prefrontal cortex in the rhesus monkey. *The Journal of Comparative Neurology, 398*, 393–419.

Ridderinkhof, K. R., Ullsperger, M., Crone, E. A., & Nieuwenhuis, S. (2004). The role of the medial prefrontal cortex in cognitive control. *Science, 306*, 443–447.

Rodriguez-Raecke, R., Niemeier, A., Ihle, K., Ruether, W., & May, A. (2013). Structural brain changes in chronic pain reflect probably neither damage nor atrophy. *PLoS One, 8*, e54475.

Roth, L. U., Maret-Maric, A., Adler, R. H., & Neuenschwander, B. E. (1997). Acupuncture points have subjective (needing sensation) and objective (serum cortisol increase) specificity. *Acupuncture in Medicine, 15*, 2–5.

Sakai, S., Hori, E., Umeno, K., Kitabayashi, N., Ono, T., & Nishijo, H. (2007). Specific acupuncture sensation correlates with EEGs and autonomic changes in human subjects. *Autonomic Neuroscience, 133*, 158–169.

Sakai, Y., Kumano, H., Nishikawa, M., Sakano, Y., Kaiya, H., Imabayashi, E., et al. (2006). Changes in cerebral glucose utilization in patients with panic disorder treated with cognitive-behavioral therapy. *NeuroImage, 33*, 218–226.

Scharf, H. P., Mansmann, U., Streitberger, K., Witte, S., Krämer, J., Maier, C., et al. (2006). Acupuncture and knee osteoarthritis: A three-armed randomized trial. *Annals of Internal Medicine, 145*, 12–20.
Schott, G. D. (1999). Pain and the sympathetic nervous system. In C. J. Mathias & S. R. Bannister (Eds.), *Autonomic failure* (pp. 520–526). Oxford: Oxford University Press.
Schweinhardt, P., & Bushnell, M. C. (2010). Pain imaging in health and disease—How far have we come? *Journal of Clinical Investigation, 120*, 3788–3797.
Seiyama, A., Hazeki, O., & Tamura, M. (1988). Noninvasive quantitative analysis of blood oxygenation in rat skeletal muscle. *The Journal of Biochemistry, 103*, 419–424.
Selkowitz, D. M. (1992). The sympathetic nervous system in neuromotor function and dysfunction and pain. A brief review and discussion. *Functional Neurology*, 7, 89–95.
Shi, G. X., Yang, X. M., Liu, C. Z., & Wang, L. P. (2012). Factors contributing to therapeutic effects evaluated in acupuncture clinical trials. *Trials, 13*, 42.
Shimada, S., Hiraki, K., Matsuda, G., & Oda, I. (2004). Decrease in prefrontal hemoglobin oxygenation during reaching tasks with delayed visual feedback: A near-infrared spectroscopy study. *Brain Research Cognitive Brain Research, 20*, 480–490.
Siedentopf, C. M., Golaszewski, S. M., Mottaghy, F. M., Ruff, C. C., Felber, S., & Schlager, A. (2002). Functional magnetic resonance imaging detects activation of the visual association cortex during laser acupuncture of the foot in humans. *Neuroscience Letters, 327*, 53–56.
Sonuga-Barke, E. J. S., & Castellanos, F. X. (2007). Spontaneous attentional fluctuations in impaired states and pathological conditions: A neurobiological hypothesis. *Neuroscience and Biobehavioral Reviews, 31*, 977–986.
Strangman, G., Culver, J. P., Thompson, J. H., & Boas, D. A. (2002). A quantitative comparison of simultaneous BOLD fMRI and NIRS recordings during functional brain activation. *NeuroImage, 17*, 719–731.
Streitberger, K., Ezzo, J., & Schneider, A. (2006). Acupuncture for nausea and vomiting: An update of clinical and experimental studies. *Autonomic Neuroscience, 129*, 107–117.
Sun, J., Zhu, Y., Yang, Y., Jin, L., von Deneen, K. M., Qin, W., et al. (2013). What is the de-qi-related pattern of BOLD responses? A review of acupuncture studies in fMRI. *Evidence-Based Complementary and Alternative Medicine, 2013*, 297839.
Takamoto, K., Hori, E., Urakawa, S., Sakai, S., Ishikawa, A., Kohno, S., et al. (2010). Cerebral hemodynamic responses induced by specific acupuncture sensations during needling at trigger points: A near-infrared spectroscopic study. *Brain Topography, 23*, 279–291.
Takeda, W., & Wessel, J. (1994). Acupuncture for the treatment of pain of osteoarthritic knees. *Arthritis Care and Research*, 7, 118–122.
Takeuchi, M., Hori, E., Takamoto, K., Tran, A. H., Satoru, K., Ishikawa, A., et al. (2009). Brain cortical mapping by simultaneous recording of functional near infrared spectroscopy and electroencephalograms from the whole brain during right median nerve stimulation. *Brain Topography, 22*, 197–214.
Taylor, S. F., Welsh, R. C., Chen, A. C., Velander, A. J., & Liberzon, I. (2007). Medial frontal hyperactivity in reality distortion. *Biological Psychiatry, 61*, 1171–1178.
Travell, J., & Simons, D. G. (1983). *Myofasical pain and dysfunction: The trigger point manual.* Baltimore, MD: Williams & Wilkins.
Vincent, C. A., Richardson, P. H., Black, J. J., & Pither, C. E. (1989). The significance of needle placement site in acupuncture. *Journal of Psychosomatic Research, 33*, 489–496.
Wang, Y., Shima, K., Sawamura, H., & Tanji, J. (2001). Spatial distribution of cingulate cells projecting to the primary, supplementary, and pre-supplementary motor areas: A retrograde multiple labeling study in the macaque monkey. *Neuroscience Research, 39*, 39–49.

Wang, K. M., Yao, S. M., Xian, Y. L., & Hou, Z. L. (1985). A study on the receptive field of acupoints and the relationship between characteristics of needling sensation and groups of afferent fibres. *Scientia Sinica Series B*, *28*, 963–971.

White, P., Bishop, F., Hardy, H., Abdollahian, S., White, A., Park, J., et al. (2008). Southampton needle sensation questionnaire: Development and validation of a measure to gauge acupuncture needle sensation. *The Journal of Alternative and Complementary Medicine*, *14*, 373–379.

White, P., Prescott, P., & Lewith, G. (2010). Does needling sensation (de qi) affect treatment outcome in pain? Analysis of data from a larger single-blind, randomised controlled trial. *Acupuncture in Medicine*, *28*, 120–125.

Willoch, F., Rosen, G., Tölle, T. R., Oye, I., Wester, H. J., Berner, N., et al. (2000). Phantom limb pain in the human brain: Unraveling neural circuitries of phantom limb sensations using positron emission tomography. *Annals of Neurology*, *48*, 842–849.

Wray, S., Cope, M., Delpy, D. T., Wyatt, J. S., & Reynolds, E. O. (1988). Characterization of the near infrared absorption spectra of cytochrome aa3 and haemoglobin for the non-invasive monitoring of cerebral oxygenation. *Biochimica et Biophysica Acta*, *933*, 184–192.

Wu, M. T., Sheen, J. M., Chuang, K. H., Yang, P., Chin, S. L., Tsai, C. Y., et al. (2002). Neuronal specificity of acupuncture response: A fMRI study with electroacupuncture. *NeuroImage*, *16*, 1028–1037.

Yamamoto, T., & Kato, T. (2002). Paradoxical correlation between signal in functional magnetic resonance imaging and deoxygenated hemoglobin content in capillaries: A new theoretical explanation. *Physics in Medicine and Biology*, *47*, 1121–1141.

Yasui, H., Takamoto, K., Hori, E., Urakawa, S., Nagashima, Y., Yada, Y., et al. (2010). Significant correlation between autonomic nervous activity and cerebral hemodynamics during thermotherapy on the neck. *Autonomic Neuroscience*, *156*, 96–103.

Zhang, H., Bian, Z., & Lin, Z. (2010). Are acupoints specific for diseases? A systematic review of the randomized controlled trials with sham acupuncture controls. *Chinese Medicine*, *5*, 1.

Zhou, Y., Liang, M., Tian, L., Wang, K., Hao, Y., Liu, H., et al. (2007). Functional disintegration in paranoid schizophrenia using resting-state fMRI. *Schizophrenia Research*, *97*, 194–205.

CHAPTER THREE

Acupuncture Point Specificity

Jing-jing Xing*, Bai-Yun Zeng†, Juan Li*, Yi Zhuang*, Fan-rong Liang*,[1]

*Acupuncture and Tuina School, Chengdu University of Traditional Chinese Medicine, Chengdu, China
†Neurodegenerative Disease Research Group, Institute of Pharmaceutical Science, School of Biomedical Sciences, King's College London, London, United Kingdom
[1]Corresponding author: e-mail address: acuresearch@126.com

Contents

Abstract

Acupuncture, as a modality treatment, has gained increasing popularity and acceptance between public and health-care professionals worldwide. Recently, there has been intensive debate about the efficacy of acupuncture therapy due to the conflicting outcome of clinical trials. Acupoint specificity was regarded as one of the core scientific issues with respect to acupuncture practice at the Society for Acupuncture Research international symposium held in 2007. In this chapter, we reviewed the recent development in basic science and clinical studies on the role of acupoint specificity. The evidence cumulated from brain imaging and many biological studies showed that the point specificity in acupuncture does exist, although acupoint specificity-related issues such as sham acupoint and placebo phenomenon need to be seriously considered. How to optimize the efficacy of acupoint and minimize the impact of sham acupuncture is an urgent issue faced by acupuncture community, and more studies are warranted on the subjects.

International Review of Neurobiology, Volume 111
ISSN 0074-7742
http://dx.doi.org/10.1016/B978-0-12-411545-3.00003-1

1. BACKGROUND

Acupuncture is a fundamental part of traditional Chinese medicine and has gained increasing popularity and acceptance between public and health-care professionals worldwide. In the past two decades, acupuncture basic research has been advanced rapidly and shown demonstrable physiological and biochemical effects in healthy human subjects and animal models and expanded its therapeutic outcome from initial pain-related conditions to cardiovascular, gastrointestinal, endocrine, and neurological disorders (Dhond, Kettner, & Napadow, 2007; Iwa, Tateiwa, Sakita, Fujimiya, & Takahashi, 2007; Langevin, Bouffard, Badger, Churchill, & Howe, 2006; Stener-Victorin, Jedel, & Mannerås, 2008; Zhou, Fu, Tjen-A-Looi, Li, & Longhurst, 2005). Recently, some conflicting outcomes of acupuncture clinical trials led to debate about the efficacy of acupuncture therapy. Acupoint specificity was regarded as one of the core scientific issues with respect to acupuncture practice at the Society for Acupuncture Research international symposium in 2007 and at the American Association of Acupuncture conference in 2010. Here, we reviewed the recent development in basic science and clinical studies in acupuncture and found that although sham acupuncture has effects on many condition, acupoint specificity does exist.

2. LITERATURE RESEARCH OF ACUPOINT SPECIFICITY

The ancient literatures did not clarify the difference between the acupoints on the same meridian in a direct way; however, we can find clues in Miraculous pivot (Lingshu Jing), one of the earliest medical classics in China, Chapter 4 records: "The Xing-spring and Shu-stream points are for treating the external channels and the he-sea points are for the internal hollow organs." It emphasizes that Xing-spring points and Shu-stream points are selected for peripheral disease in meridians, and he-sea points are chosen for disease in Zang-fu organs. Yellow Emperor's Canon of Internal Medicine, the most important theoretical text in the traditional Chinese medical literature, accomplished the theory about Five shu points and the clinical characteristics of Five shu points. It emphasized that the Jing-well points should be selected for fullness in the epigastrium; the Xing-spring points for febrile diseases; the Shu-stream points for heavy sensation of the body and painful joints; the Jing-river points for cough and asthma due to pathogenic cold and heat; and the He-sea points for diarrhea due to pervasive flow of qi. Although the literature studies did not specify the acupoint specificity directly, it does, in some extent, reflect the fact that verum acupoint play a most important part in meridian and collateral theory.

Recently, we carried out literature searching and screening to evaluate the acupoint specificity on certain conditions (Zhao et al., 2012). A database was set up in which it consists of over 2,600,000 eligible pieces of ancient and modern acupuncture texts on migraine, functional dyspepsia (FD), uterus-related disorders, and dysmenorrhoeal, and data processing platform was formed utilizing data-mining technology. Prescriptions relating acupoints for the four diseases were extracted for acupoints and its pertaining meridians analysis. Evaluation of relevant literature revealed the following facts: Acupoints on the Shaoyang meridian and Yangming meridian were chiefly selected for migraine; Fengchi (GB20) and Waiguan (TE5) on the Shaoyang meridian were most frequently used in migraine, though Touwei (ST8) and Zusanli (ST36) on the Yangming meridian were also used by acupuncturists (Chen et al., 2009). The acupoints on the meridian of the foot Yangming were optimal choice for FD, among which Zusanli (ST36) and Liangqiu (ST34) were especially employed. In addition, Zhongwan (CV12) and Weishu (BL21), as the alarm and transport (Fu and Mu) points of the stomach, were frequently applied points (Xie, Zhu, Wu, Chen, & Liang, 2008; Ren, Zhao, Liu, & Liang, 2009). The acupoints employed for treatment of uterus-related disorders and dysmenorrhoeal were mainly on spleen meridian, ren meridian, and kidney meridian, among which Sanyinjiao (SP6), Guanyuan (CV4), and Taixi (KI3) were most frequently used from each meridian (Chen, Xie, Zhu, Yang, & Liu, 2008; Chen, Zhu, & Xie, 2008). The results indicate that acupoints on different meridians possess specific effects on targeted disease and certain acupoints possess more specific effects than the other acupoints on the same meridian. This suggests that acupoint specificity is closely related to meridians and degree of convergence of the meridian qi. Thus, meridians are the prerequisite for achieving acupoint specificity.

3. BASIC RESEARCH

3.1. Brain imaging studies

Although acupuncture literature indicates the existing of acupoint specificity, the underlying mechanisms of point specificity are not well understood. Noninvasive brain imaging techniques such as positron emission tomography (PET) and functional magnetic resonance imaging (fMRI) have enabled direct study of the relation between acupoint stimulation and neural activity in human brain. Using fMRI, specificity of acupoint TE5 was investigated in 36 healthy subjects using fMRI (Huang et al., 2010). The data demonstrated that, after true needling at TE5, sham needling, and true needling at a sham point, TE5 needling activated the right superior frontal gyrus (BA8) and the left cerebellum when compared with sham needling, and activated the right parietal lobe, the left cerebellum, and the right inferior semilunar lobule when compared with true needling on sham point (Huang et al., 2010). It is well known that the insula regulates impulsive and aggressive behavior and that the temporal lobe regulates auditory functions. The parietal lobe receives nervous sensation from the opposite side of the body, and the superior frontal gyrus is involved in such activities as writing and movement of the upper limbs. The cerebellum regulates activities of the trunk muscles and plays an important role in maintaining balance and posture. Acupoint TE5, according to TCM theory, is indicated to treat difficult movement of the elbow and arm, pain in the upper limbs, shaking of the hand, ear disorders, irritation and susceptibility to anger, etc. These results provide validation of underlying mechanism of point specificity in acupuncture. In another fMRI study of 12 healthy subjects, EA stimulation at the left GB34 specifically activated the right putamen, caudate, thalamus, cerebellum, as well as the left caudate, ventral lateral thalamus, and cerebellum, all associated with motor function; whereas EA stimulation at the left control point (2 cm lateral in the nonmeridian to GA34) specifically activated the right BA6, BA8, BA40, BA44, thalamus, as well as the left thalamus and cerebellum (Na et al., 2009), the results of study provided further evidence for the existence of point specificity in acupuncture. Using high temporal resolution magnetoencephalography technology and applying time-frequency analysis to detect differential oscillatory brain dynamics induced by acupuncture stimulation at either ST36 or sham acupoint of on the right leg of healthy subjects, You et al. (2011) reported that ST36 stimulation specifically increased delta band power in contralateral temporal regions and decreased power in the counterparts of ipsilateral hemisphere compared to the sham point, although both verum and sham acupoints induced frequency power changes in delta band, indicating the specific modulatory effects of acupoint.

Feng et al. (2011) investigated the acupoint specificity by exploring the whole brain functional connectivity analysis on the poststimulus resting brain modulated by acupuncture at acupoint PC6 in comparison with acupoint PC7 and GB37. The brain was divided into 90 regions and analyzed the functional connectivity for each condition. The fMRI data demonstrated statistically significant differences in functional correlations throughout the entire brain following acupuncture at PC6 compared with PC7. For direct comparisons, increased correlations for PC6 compared to PC7 were primarily between the prefrontal regions and the limbic/paralimbic and subcortical regions, whereas decreased correlations were mainly between the parietal regions and the limbic/paralimbic and subcortical regions. On the other hand, increased correlations for PC6 compared to GB37 were primarily between the prefrontal regions and somatosensory regions, whereas decreased correlations were mainly related with the occipital regions (Feng et al., 2011). This suggests that acupuncture at different acupoints, either from the same meridian or from different meridians, exerts heterogeneous neural modulatory effects on the poststimulus resting brain, providing evidences for the relatively function-oriented specificity of acupuncture. It has been known that acupoints S36 and SP6 are used to treat visceral conditions, and GB34 and B57 are indicated for muscle and tendon disorders (Li, Tougas, Chiverton, & Hunt, 1992; Tsui & Leung, 2002). An fMRI study, concerning the point specificity, showed that stimulation of S36 and SP6 activated orbital frontal cortex and deactivated hippocampus and parietal BA7, regions linked to visceral disorders; on the other hand, stimulation of GB34 and B57 activated the dorsal thalamus and inhibited primary motor and premotor cortices (Cervero & Laird, 1999; Price, 2006; Zhang et al., 2004). Although all four acupoints enter into the same spinal segment, they exerted distinct though overlapped cerebral response patterns, suggesting that even acupoints, located in the same spinal segment, elicit disease-specific responses upon stimulation.

3.2. Biological studies

Many studies debating acupoint specificity assessed the role of acupuncture in pain relief. Recently, several large-scale randomized controlled trials (RCTs) in chronic pain patients failed to display specificity in acupuncture treatment (Brinkhaus et al., 2006; Harris et al., 2005; Linde et al., 2005; Melchart et al., 2005). Those studies include chronic low back pain, fibromyalgia, migraine, and tension-type headache, and the authors found that there was no significant difference of analgesia between verum and sham acupuncture-treated groups. This has led to suggestions that acupoint stimulation is no more effective than a placebo intervention; as opioid receptor, specifically, the μ-opioid receptors (MORs) are also activated by sham acupuncture treatment (Benedetti & Amanzio, 1997; Pomeranz & Chiu, 1976; Zubieta et al., 2005). More recently, Harris et al. (2009) compared the short- (first treatment) and long-term (1 month later) effects of traditional Chinese acupuncture (TA, acupoints stimulation induces de-qi sensation) with sham acupuncture (SA, nonskin penetrating pricking at acupoints induces sensation) on *in vivo* MOR binding availability in patients with fibromyalgia and reported that TA and SA have different effect on MORs. Although both TA and SA produced similar analgesic effects on patients, data of PET with 11C-carfentanil showed that TA therapy exerted an increase in MOR availability over both short and term periods, whereas there was no change or decrease in MOR availability in SA-treated patient group. The brain areas, showing increased MOR availability by TA treatments, includes those areas classically associated with the regulation of pain and stress in human (Zubieta et al., 2003, 2001), such as the amygdale, the dorsal and perigenual anterior cingulate, and insular cortex. This clearly indicated that TA treatment had a distinct neurobiological effect in different regions of brain in human. Another study, investigating neuronal specificity of acupoint at the same meridian in healthy subjects, by real EA (deep needling at acupoints) and shallow EA (subcutaneous needling at acupoints) stimulation, fMRI data, showed the real EA at GB34 and GB39 induced stronger needling sensation and modulated pain-related neuromatrix more effectively than shallow EA (Zhang et al., 2007).

Although both verum and sham acupoint stimulation showed analgesia, the studies of effect of acupuncture on blood pressure also demonstrated the acupoint specificity. A series of experiments showed that EA at acupoints P5–P6, LI4–L7, LI10–LI11, and ST36–ST37 effectively reduced blood pressure in rats or cats, which was closely linked to the reduction in the premotor sympathetic cardiovascular neuronal activity in the rostral ventral lateral medulla (rVLM), whereas EA at LI6–LI7 and G37–G39 had no effect on cardiovascular or rVLM neuronal activity (Li, Rowshan, Crisostomo, Tjen-A-Looi, & Longhurst, 2002; Tjen-A-Looi, Li, & Longhurst, 2004; Zhou, Tjen-A-Looi, & Longhurst, 2005). Further, effect of acupuncture on blood pressure or cardiovascular reflex responses to gastric distension was elicited from stimulation of deeper nerves such as the median nerve below P5–P6 acupoints rather than from needle insertion without stimulation at P5–P6 acupoints or cutaneous superficial stimulation at LI6–LI7 acupoints (Tjen-A-Looi et al., 2004; Zhou, Fu, et al., 2005). These findings indicated that the effect of standard acupoint simulation on cardiovascular responses is specific and superficial or minimal acupuncture may serve as a valid control as it does not alter the cardiovascular responses.

The central influence of acupuncture on gastric motor function was investigated in rats (Iwa et al., 2007). Stimulation at ST25 and ST36 elicited different responses in the brain stem, as shown by the transcription factor, c-Fos immunohistochemistry. Acupuncture at ST36 significantly increased the number of c-Fos immunoreactive cells at the dorsal motor nucleus of vagus and the mediocaudal and caudal nucleus tractus solitarius (NTS). Stimulation at ST25 significantly increased the number of c-Fos immunoreactive cells at the rostral ventrolateral medulla and the number of c-Fos immunopositive cells at the mediocaudal NTS (Iwa et al., 2007). This study provided anatomical evidence of possible neural pathway in mediating acupuncture-induced gastric motor responses.

Point specificity in acupuncture has been mainly supported by the studies, showing that specific acupoint stimulations elicit many central effects, such as increased MOR binding availability, increased opioid expression, activated neuronal function in specific regions in the brain, and increased cardiovascular neuronal activity in rVLM (Feng et al., 2011; Harris et al., 2009; Huang, Wang, Han, & Wan, 2002; Tjen-A-Looi et al., 2004). Very recently, studies from both rodent and human subjects, using microdialysis to sample interstitial fluids, demonstrated that analgesic effects of acupoint stimulation at ST36 were mediated by a local release of adenosine and activation of adenosine A1 receptors (Goldman et al., 2010; Takano et al., 2012). Further, sham acupuncture including nonacupoint stimulation or insertion of acupuncture needle without rotation did not induce significant increase in local adenosine concentration. The reports suggest that acupoint-induced specific local responses play an important role in acupuncture analgesia and corroborate point specificity in acupuncture. In acupuncture, ashi points are also called "reflexing points," "unfixed points," or "tender spots" which lie on painful points of body without relating to meridians and acupoints. Ashi points are mostly used to treat pain syndromes, which have been one of the issues leading to opponents of acupuncture therapy to suggest the acupoint nonspecificity. The studies of Goldman et al. (2010) and Takano et al. (2012) indicate that the regional effects through the local release of neuromodulators such as adenosine might be one of the mechanisms of ashi point analgensia.

As acupoints are always located on regions that arc abundant in nerves and blood and lymph vessels; this is where nerve endings, nerve receptors, blood vessels, andmast cells, etc., are densely distributed and neural and neuroactive components are highly concentrated (Zhang, Wang, & McAlonan, 2012). The studies investigating differences in structure between acupoints and nonacupoints in human subjects, using X-ray emission and synchrotron X-ray fluorescence analysis, showed that there were high concentrations of mast cells displaying increased degranulation as well as somewhat greater accumulation of microvessels at acupoint area. The contents of Ca, Fe, Cu, and Zn were significantly higher at three of four acupoints examined than in the nonacupoint tissue (Yan et al., 2009; Zhang et al., 2008).

Recently, a linear programming base feature selection method was used by utilizing ^{1}H nuclear magnetic resource to investigate the effect of acupuncture at several meridian points on plasma metabolite biomarkers, to understand molecular mechanism of acupuncture effect (Wang et al., 2012). The data from healthy human subjects showed that stimulation at acupoints ST3, ST21, and ST36 of Yangming meridian influences mainly plasma micromolecular metabolites, closely associated with energy metabolism pathway. Stimulation at GB34 manipulates mainly plasma macromolecular metabolites and is closely linked to lipid metabolism and transport. Acupoint BL40 had no effect on plasma metabolites (Wang et al., 2012). The metabolite biomarkers elicited from ST3, ST21, and ST36 of Yangming are different from those of GB34 and BL40, providing the metabolite evidence of acupoint specificity and can be useful to investigate point specificity in future study.

Overall, studies, involved brain imaging, electrophysiology, receptor binding and HPLC, etc, techniques, showed that point specificity does exist among acupoints not only from the same meridians but also from different meridians, and the acupoints entering the same spinal segments.

4. CLINICAL TRIAL STUDIES

Recently, several of clinical trials of acupuncture treatments have been carried out within many conditions following extensive acupuncture research during past two decades. Although acupuncture treatments have shown super effect than normal cure, there were no significant differences between verum acupuncture and sham acupuncture treatments. Point specificity in acupuncture has been one of the core issues being debated the effectiveness of acupuncture therapy. Here, we summarized recent development in acupuncture clinical trials in different conditions.

4.1. Chronic pain-related conditions

Chronic pain is a heterogeneous group of medical conditions that significantly impact on an individual's physical, emotional, and psychological distress. Acupuncture is one of the most common nonpharmacological pain-relieving alternative treatments, and many clinical trials especially randomized controlled trials in chronic patients were performed. A systematic review of individual patients data meta-analysis from 29 of 31 eligible RCTs with a total 17,992 patients, to determine the effect size of acupuncture for back and neck pain, osteoarthritis, chronic headache, and shoulder pain, reported that there is significant difference in pain relief between true and sham acupuncture, suggesting that acupuncture is more than a placebo effect (Vickers et al., 2012).

A multicenter RCTs recruited 140 patients with migraine, designed by comparing verum acupuncture plus placebo to sham acupuncture plus flunarizine and measured by proportion of responders (defined as the proportion of patients with a reduction of migraine days by at least 50%), visual analogue scale (VAS, 0–10 cm) for pain, as well as the physical and mental component summary scores of the 36-item short-form health survey (SF-36). Although the patients in the acupuncture group had significant better responder rates and fewer migraine days compared with the control group, there were no significant differences between the two groups in VAS scores and SF-36 physical and mental component summary scores (Wang, Jing, et al., 2011; Wang, Zhang, et al., 2011). In another study of 218 patients with migraine, the efficacy of acupuncture in treating acute migraine attacks was assessed (Li et al., 2009) and the data showed that verum acupuncture elicited a significant reduction in VAS scores, displayed more effective in relieving pain and preventing migraine relapse or aggravation compared to sham acupuncture treatment (Li et al., 2009). A further multicenter study of 480 patients with migraine was participated, investigating long-term analgesic effect between acupoints and nonacupoints and exploring the differences between different acupoints (Li et al., 2008). The patients were randomized into treatment group 1 (Shaoyang-specific acupuncture), treatment group 2 (Shaoyang-nonspecific acupuncture), treatment group 3 (Yangming-specific acupuncture), and the control group (nonacupoint group). After 4 weeks' treatments, the primary outcome, which was the number of days when the subjects experienced a migraine during weeks 4–8 and 13–16, was assessed. Fewer days with migraine were witnessed in three acupuncture groups compared with nonacupoint group during weeks 4–8, and with no difference between treatment groups ($P > 0.05$) and significant fewer days during weeks 13–16 (Shaoyang-specific acupuncture vs. control ($P < 0.01$); Shaoyang-nonspecific acupuncture vs. control ($P < 0.01$); Yangming-specific acupuncture vs. control ($P < 0.05$)). This study shows that acupoints are more effective than nonacupoints and acupoints from shaoyang meridian are more effective than Yangming meridian in long-term analgesia for migraine which further presents differences between acupoints and nonacupoints, as well as among acupoints (Li et al., 2008). Difference in the selection of acupoints, in needling techniques, among other factors, might be attributed to the different outcome from those studies.

Several large-scale RCTs in 568 patients with chronic pain conditions including low back pain, and tension-type headache reported that although acupuncture is more effective than no treatment, there was no significant difference between verum acupuncture and minimal (superficial) acupuncture (Brinkhaus et al., 2006; Melchart et al., 2005). Experimental and clinical studies have shown that minimal acupuncture, used as a placebo control, is not necessarily inert from physiological perspective. The minimal acupuncture was shown to evoke activity in cutaneous afferent nerve, leading to pronounced effects on the functional connectivity in the brain, resulting in deactivation of limbic structures, associated with increase in activity in hypothalamus leading to the activation of pain and sympathetic inhibition mechanisms (Deng, Hou, Holodny, & Cassileth, 2008; Fang et al., 2009; Wu et al., 1999, 2002). A comparative analysis of the effects of minimal acupuncture on chronic pain, data collected from several large-scale RCTs, revealed that both verum and minimal acupuncture are effective in migraine, whereas verum acupuncture is more effective in low back pain and knee osteoarthritis than minimal acupuncture (Lund, Naslund, & Lundeberg, 2009). Although the minimal acupuncture was originally designed to reduce bias, it in fact may just introduce a bias against the treatment being tested (Birch, 2006). As under certain conditions minimal acupuncture may have both specific and nonspecific effects, the relevance of minimal acupuncture as an inert placebo control should be cautious.

4.2. Cardiovascular system

Point specificity in acupuncture on blood pressure reduction has been evaluated in many clinical trials. The Boston SHARP clinical trial found no significant difference between verum and sham treatments in 188 patients, and blood pressure declined significantly in both groups (Macklin et al., 2006). Since then, more clinical trials are reported. An RCT of 27 subjects with normal or hypertension showed that EA treatment significantly decreased systolic blood pressure, from 117.8 mm Hg before treatment to 110.1 mm Hg after treatment ($P < 0.05$), but diastolic blood pressure was down only from 78.1 mm Hg before treatment to 74.8 mm Hg ($P > 0.05$) after treatment, whereas sham acupuncture had no effect on either systolic or diastolic blood pressure (Zhang, Ngb, & Sau, 2009). Another large single-blind RCT of 160 patients with uncomplicated arterial hypertension demonstrated that in verum acupuncture-treated group, mean 24-h ambulatory systolic and diastolic blood pressures decreased significantly after 6-week treatment by 5.4 mm Hg (95% CI, 3.2–7.6) and 3.0 mm Hg (95% CI, 1.5–4.6), respectively; in contrast, the sham acupuncture control group showed no significant response, and at the end of the course, it was 6.4 mm and 3.7 mm higher than the verum acupuncture group for systolic and diastolic readings (Brinkhaus, 2008). At 3 and 6 months after treatment, mean systolic and diastolic blood pressures returned to pretreatment levels in the verum acupuncture group (Brinkhaus, 2008). These suggested that verum acupuncture, but not sham acupuncture, significantly lowered 24-h ambulatory blood pressure and the effect disappears after cessation of acupuncture treatment. Blood pressure-lowering effect elicited by acupuncture may be due to its improvement of endothelial function. A pilot, randomized, double-blind, placebo-controlled crossover trial (Park et al., 2010) showed that flow-mediated dilation was significantly improved with the acupuncture treatment at ST36 and ST36 plus PC6, whereas floor-mediated dilation was unchanged after the acupuncture treatment at PC6 or the sham treatment. Indeed, basic experiments, investigating the mechanisms of acupuncture on regulating endothelial function in rats with hypoxia-induced pulmonary hypertension or with stress-induced hypertention, confirmed that verum acupuncture significantly improved circulation eNOS concentration and markedly increased calcitonin gene-related peptide and nitrogen monoxidum in blood plasma, paralleling to its blood pressure reduction (Pan et al., 2010; Wang, Jing, et al., 2011; Wang, Zhang, et al., 2011).

Clinical trials have been used to assess the acupoint specificity in other cardiovascular conditions. A small trial, assessing the relationship between acupuncture and glucose metabolism in cerebral functional regions in 43 poststroke patients using PET-CT techniques, showed that verum acupuncture activated specific brain regions compared to sham acupuncture and this may related to its effect on the functional recovery of patients with stroke (Huang et al., 2012). Acupuncture has been shown, by a study of 57 patients with persistent atrial fibrillation, that, compared to sham acupuncture, verum acupuncture treatment for 10 weeks had significantly anti-arrhythmic effect, persisting throughout whole treatment period, and had a markedly reduction in arrhythmic recurrence during a 12-month follow-up period (Lomuscio, Belletti, Battezzatim, & Lombardi, 2011).

4.3. Other conditions

Acupuncture has been used to treat FD, and a study of 712 FD patients and control subjects showed that the Symptom Index of Dyspepsia was reduced in verum acupoint group compared to placebo groups (Ma et al., 2012). PET-CT studies of 72 patients with FD demonstrated that verum acupoint stimulation induced an extensive deactivation in cerebral activities including brainstem, anterior cingulate cortex, insula, thalamus, and hypothalamus, parallel to the symptoms improvement, in verum acupoint group compared to sham acupoint group (Zeng et al., 2012). This suggests that the effects of acupoint stimulation were modulated through the brain homeostatic afferent network.

Series clinical trials were conducted to examine the specific effect of acupuncture on primary dysmenorrhea (PD). The 66 patients with PD were randomized into two groups (Yu et al., 2010). The treatment group received manual acupuncture bilaterally at SP6 for 5 min after needling sensation (de-qi) was elicited during the period of menstrual pain, while in control group, the needle was bilaterally at GB39 for 5 min during the period of menstrual pain. Compared with the control group, patients in the treatment group showed significant reductions in menstrual pain, values of pulsatility, resistance index, and ratio of the systolic and diastolic peaks (A/B) in the uterine arteries 5 min after treatment. These suggest that needling at SP6 can immediately improve uterine arterial blood flow in patients with PD, whereas GB39 did not achieve these effects (Yu et al., 2010). The specific pain-relieving effect of SP6 was confirmed in another trial with 50 PD patients (Ma et al., 2010). However, a large size trial with 200 eligible participants with PD showed that there was no significant difference in pain relieving of dysmenorrhea between SP6- and GB39-treated groups (Liu et al., 2011). Although all studies used the same acupoints, the different outcome measurements were employed, which makes data difficult to be interpreted.

5. FUTURE RESEARCH IN ACUPOINT SPECIFICITY

Recent developments in basic and clinical studies in acupuncture therapy have greatly enhanced our understanding of the effects and mechanisms of acupuncture. Cumulative evidence shows that stimulation of different acupoints elicits distinct local and central neural physiological responses. Meanwhile, many questions have been raised including acupuncture needling (type, depth, direction, needle number, and intensity of stimulation), acupoint location, more importantly acupoint specificity-related issues, such as sham acupuncture, for example, needle insertion at nonacupuncture points, shallow needle insertion that does not penetrate below the skin (minimal or superficial needling), and blunt needles that touch, but do not penetrate the skin (placebo needling) and placebo phenomena. How to optimize the effect of verum acupuncture or how to minimize impact of sham acupuncture is an urgent issue faced by acupuncture community. A recent study, showing that needle motion and force during acupuncture performance, can be measured and subsequently analyzed, provided preliminary evidence that it is possible to optimize acupuncture treatment parameters, an objective way for characterizing needling in acupuncture research (Davis, Churchill, Badger, Dunn, & Langevin, 2012). This is very encouraging and may lead a standardized approach or protocols for acupuncture research and clinical trials in the future. Further, an article published by Langevin et al. (2010) may have some clues for above questions. Nonetheless, further studies are urgently needed for better understanding of the mechanisms underlying the acupoint specificity and the sham acupuncture.

REFERENCES

Benedetti, F., & Amanzio, M. (1997). The neurobiology of placebo analgesia: From endogenous opioids to cholecystokinin. *Progress in Neurobiology*, *52*, 109–125.

Birch, S. (2006). A review and analysis of placebo treatments, placebo effects, and placebo controls in trials of medical procedures when sham is not inert. *Journal of Alternative and Completementary Medicine*, *12*, 303–310.

Brinkhaus, B. (2008). Acupuntura en pacientes con hipertensión arterial randomized trial of acupuncture to lower blood pressure. *Revista Internacional de Acupuntura*, *2*, 39–40.

Brinkhaus, B., Witt, C. M., Jena, S., Linde, K., Streng, A., Wagenpfeil, S., et al. (2006). Acupuncture in patients with chronic low back pain: a randomized controlled trial. *Archives of Internal Medicine*, *166*(4), 450.

Cervero, F., & Laird, J. M. (1999). Visceral pain. *The Lancet*, *353*, 2145–2148.

Chen, Q., Wu, X., Lu, S. F., Ren, Y. L., Li, S. J., Zhu, H., et al. (2009). Characteristics and related factors analysis on fourteen meridian acupoints in migraine treatment. *Liaoning Journal of Traditional Chinese Medicine*, *36*, 1477–1480.

Chen, X., Xie, J. P., Zhu, J., Yang, H., & Liu, C. Z. (2008). Acupoint selection rules in dysmenorrhea treatment. *Shanghai Journal of Acupuncture and Moxibustion*, *27*, 45–46.

Chen, X., Zhu, J., & Xie, J. (2008). Rules of selection points on treating diseases related to uterus by acupuncture and moxibustion. *Chinese Archives of Traditional Chinese Medicine*, *26*, 1905–1906.

Davis, R. T., Churchill, D. L., Badger, G. J., Dunn, J., & Langevin, H. M. (2012). A new method for quantifying the needling component of acupuncture treatments. *Acupuncture in Medicine*, *30*(2), 113–119.

Deng, G., Hou, B. L., Holodny, A. I., & Cassileth, B. R. (2008). Functional magnetic resonance imaging (fMRI) changes and saliva production associated with acupuncture at LI-2 acupuncture point: A randomized controlled study. *BMC Complementary & Alternative Medicine*, *8*, 37.

Dhond, R. P., Kettner, N., & Napadow, V. (2007). Neuroimaging acupuncture effects in the human brain. *Journal of Alternative and Complementary Medicine*, *13*(6), 603–616.

Fang, J., Jin, Z., Wang, Y., Li, K., Kong, J., Nixon, E. E., et al. (2009). The salient characteristics of the central effects of acupuncture needling: Limbic-paralimbic-neocortical network modulation. *Human Brain Mapping*, *30*(4), 1196–1206.

Feng, Y. Y., Bai, L. J., Zhang, W. S., Ren, Y. S., Xue, T., Wang, H., et al. (2011). Investigation of acupoint specificity by whole brain functional connectivity analysis from fMRI data. *Conference Proceedings—IEEE Engineering in Medicine and Biology Society*, *2011*, 2784–2787.

Goldman, N., Chen, M., Fujita, T., Xu, Q., Peng, W., Liu, W., et al. (2010). Adenosine A1 receptors mediate local anti-nociceptive effects of acupuncture. *Nature Neuroscience*, *13*, 883–888.

Harris, R. E., Tian, X., Williams, D. A., Tian, T. X., Cupps, T. R., & Petzke, F. (2005). Treatment of fibromyalgia with formula acupuncture: Investigation of needle placement, needle stimulation, and treatment frequency. *Journal of Alternative and Complementary Medicine*, *11*, 663–671.

Harris, R. E., Zubieta, J. K., Scott, D. J., Napadow, V., Gracely, R. H., & Clauw, D. J. (2009). Traditional Chinese acupuncture and placebo (sham) acupuncture are differentiated by their effects on mu-opioid receptors (MORs). *NeuroImage*, *47*(3), 1077–1085.

Huang, Y., Lai, X. S., Shan, B. C., Tang, C. Z., Yang, J. J., Zeng, T., et al. (2010). Specific cerebral activation following true and sham Waiguan (SJ 5) needling functional magnetic resonance imaging evidence. *Neural Regeneration Research*, *5*, 1712–1716.

Huang, Y., Tang, C. Z., Wang, S. X., Lu, Y. J., Shen, W., Yang, J., et al. (2012). Acupuncture regulates the glucose metabolism in cerebral functional regions in chronic stage ischemic stroke patients—A PET-CT cerebral functional imaging study. *BMC Neuroscience*, *13*, 75.

Huang, C., Wang, Y., Han, J. S., & Wan, Y. (2002). Characteristics of electroacupuncture-induced analgesia in mice: Variation with strain, frequency, intensity and opioid involvement. *Brain Research*, *945*, 20–25.

Iwa, M., Tateiwa, M., Sakita, M., Fujimiya, M., & Takahashi, T. (2007). Anatomical evidence of regional specific effects of acupuncture on gastric motor function in rats. *Autonomic Neuroscience: Basic and Clinical*, *137*, 67–76.

Langevin, H. M., Bouffard, N. A., Badger, G. J., Churchill, D. L., & Howe, A. K. (2006). Subcutaneous tissue fibroblast cytoskeletal remodeling induced by acupuncture: Evidence for a mechanotransduction-based mechanism. *Journal of Cell Physiology*, *207*(3), 767–774.

Langevin, H. M., Wayne, P. M., Macpherson, H., Schnyer, R., Milley, R. M., et al. (2010). Paradoxes in acupuncture research: Strategies for moving forward. *Evidence-Based Complementary and Alternative Medicine*, 180805.

Li, Y., Liang, F. R., Yang, X. G., Tian, X. P., Yan, J., & Sun, G. J. (2009). Acupuncture for treating acute attacks of migraine: A randomized controlled trial. *Headache*, *49*, 805–816.

Li, Y., Liang, F. R., Yu, S. G., Liu, X. G., Tang, Y., Yang, X., et al. (2008). Randomized controlled trial to treat migraine with acupuncture: Design and protocol. *Trials*, *9*, 57.

Li, P., Rowshan, K., Crisostomo, M., Tjen-A-Looi, S. C., & Longhurst, J. C. (2002). Effect of electroacupuncture on pressor reflex during gastric distension. *American Journal of Physiology: Regulatory, Integrative and Comparative Physiology*, *283*(6), R1335–R1345.

Li, Y., Tougas, G., Chiverton, S. G., & Hunt, R. H. (1992). The effect of acupuncture on gastrointestinal function and disorders. *American Journal of Gastroenterology*, *87*, 1372–1381.

Linde, K., Streng, A., Jurgens, S., Hoppe, A., Brinkhaus, B., Witt, C., et al. (2005). Acupuncture for patients with migraine: A randomized controlled trial. *Journal of the American Medical Association*, *293*, 2118–2125.

Liu, C. Z., Xie, J. P., Wang, L. P., Zheng, Y. Y., Ma, Z. B., Yang, H., et al. (2011). Immediate analgesia effect of single point acupuncture in primary dysmenorrhea: A randomized controlled trial. *Pain Medicine*, *12*, 300–307.

Lomuscio, A., Belletti, S., Battezzatim, P. M., & Lombardi, F. (2011). Efficacy of acupuncture in preventing atrial fibrillation recurrences after electrical cardioversion. *Journal of Cardiovascular Electrophysiology*, *22*(3), 241–247.

Lund, I., Naslund, J., & Lundeberg, T. (2009). Minimal acupuncture is not a valid placebo control in randomized controlled trials of acupuncture: A physiologist's perspective. *Chinese Medicine*, *4*, 1–9.

Ma, Y. X., Ma, L. X., Liu, X. L., Ma, Y. X., Lv, K., Wang, D., et al. (2010). A comparative study on the immediate effects of electroacupuncture at Sanyinjiao (SP6), xuanzhong (GB39) and a non-meridian point, on menstrual pain and uterine arterial blood flow, in primary dysmenorrhea patients. *Pain Medicine*, *11*(10), 1564–1575.

Ma, T. T., Yu, S. Y., Li, Y., Liang, F. R., Tian, X. P., Zheng, H., et al. (2012). Randomised clinical trial: An assessment of acupuncture on specific meridian or specific acupoint vs. sham acupuncture for treating functional dyspepsia. *Alimentary Pharmacology & Therapeutics*, *35*, 552–561.

Macklin, E. A., Wayne, P. M., Kalish, L. A., Valaskatgis, P., Thompson, J., Pian-Smith, M. C., et al. (2006). Stop hypertension with the acupuncture research program (SHARP). *Hypertension*, *48*(5), 838–845.

Melchart, D., Streng, A., Hoppe, A., Brinkhaus, B., Witt, C., Wagenpfeil, S., et al. (2005). Acupuncture in patients with tension-type headache: Randomised controlled trial. *BMJ*, *331*, 376–382.

Na, B. J., Jahng, G. H., Park, S. U., Jung, W. S., Moon, S. K., Park, J. M., et al. (2009). An fMRI study of neuronal specificity of an acupoint: Electroacupuncture stimulation of Yanglingquan (GB34) and its sham point. *Neuroscience Letters*, *464*(1), 1–5. http://dx.doi.org/10.1016/j.neulet.2009.08.009.

Pan, P., Zhang, X., Qian, H., Shi, W., Wang, J., Bo, Y., et al. (2010). Effects of electro-acupuncture on endothelium-derived endothelin-1 and endothelial nitric oxide synthase of rats with hypoxia-induced pulmonary hypertension. *Experimental Biology and Medicine*, *235*(5), 642–648.

Park, J. M., Shin, A. S., Park, S. U., Sohn, I. S., Jung, W. S., & Moon, S. K. (2010). The acute effect of acupuncture on endothelial dysfunction in patients with hypertension: A pilot, randomized, double-blind, placebo-controlled crossover trial. *Journal of Alternative and Complement Medicine*, *16*(8), 883–888.

Pomeranz, B., & Chiu, D. (1976). Naloxone blockade of acupuncture analgesia: Endorphin implicated. *Life Science*, *19*, 1757–1762.

Price, J. L. (2006). Prefrontal cortical networks related to visceral function and mood. *Annals of the New York Academy of Sciences*, *877*, 383–396.

Ren, Y. L., Zhao, L., Liu, M. L., & Liang, F. R. (2009). Data mining-based study on characteristics of acupoints selection on ancient acupuncture treatment of functional dyspepsia. *Liaoning Journal of Traditional Chinese Medicine*, *36*(2), 259–262.

Stener-Victorin, E., Jedel, E., & Manneràs, L. (2008). Acupuncture in polycystic ovary syndrome: Current experimental and clinical evidence. *Journal of Neuroendocrinology*, *20*, 290–298.

Takano, T., Chen, X., Luo, F., Fujita, T., Ren, Z., Goldman, N., et al. (2012). Traditional acupuncture triggers a local increase in adenosine in human subjects. *Journal of Pain*, *13*(12), 1215–1223.

Tjen-A-Looi, S. C., Li, P., & Longhurst, J. C. (2004). Medullary substrate and differential cardiovascular responses during stimulation of specific acupoints. *American Journal of Physiology: Regulatory, Integrative and Comparative Physiology*, *287*, R852–R862.

Tsui, P., & Leung, M. C. (2002). Comparison of the effectiveness between manual acupuncture and electro-acupuncture on patients with tennis elbow. *Acupuncture Electrother Research*, *27*, 107–117.

Vickers, A., Cronin, A., Maschino, A., Lewith, G., MacPherson, H., Foster, N., et al. (2012). Acupuncture for chronic pain: An individual patient data meta-analysis of randomized trials. *BMC Complementary and Alternative Medicine*, *172*(19), 1444–1453.

Wang, L., Jing, M. X., Zhi, J. M., Lu, J., Wang, C. Y., & Liu, Q. G. (2011). Effects of reinforcing and reducing methods by twirling and rotating the needle on contents of CGRP and NO in rats with stress-induced hypertension. *Zhongguo Zhen Jiu*, *31*(4), 337–341.

Wang, Y., Wu, Q. F., Chen, C., Wu, L. Y., Yan, X. Z., Yu, S. G., et al. (2012). Revealing metabolite biomarkers for acupuncture treatment by linear programming based feature selection. *BMC Systems Biology*, *6*(Suppl 1), S15.

Wang, L. P., Zhang, X. Z., Guo, J., Liu, H. L., Zhang, Y., Liu, C. Z., et al. (2011). Efficacy of acupuncture for migraine prophylaxis: A single-blinded, double-dummy, randomized controlled trial. *Pain*, *152*, 1864–1871.

Wu, M. T., Hsieh, J. C., Xiong, J., Yang, C. F., Pan, H. B., Chen, Y. C., et al. (1999). Central nervous pathway for acupuncture stimulation: Localization of processing with functional MR imaging of the brain-preliminary experience. *Radiology*, *212*, 133–141.

Wu, M. T., Sheen, J. M., Chuang, K. H., Yang, P., Chin, S. L., et al. (2002). Neuronal specificity of acupuncture response: A fMRI study with electroacupuncture. *NeuroImage*, *16*, 1028–1037.

Xie, X., Zhu, H., Wu, X., Chen, Q., & Liang, F. R. (2008). Regularity of acupoints' application in acupuncture and moxibustion clinical trials for functional dyspepsia. *Journal of Chengdu University of Traditional Chinese Medicine*, *31*, 1–4.

Yan, X., Zhang, X., Liu, C., Dang, R., Huang, Y., He, W., et al. (2009). Do acupuncture points exist? *Physics in Medicine and Biology*, *54*(9), N143–N150.

You, Y., Bai, L., Dai, R., Xue, T., Zhong, C., Feng, Y., et al. (2011). Differential neural responses to acupuncture revealed by MEG using wavelet-based time-frequency analysis: A pilot study. *Conference Proceedings: IEEE Engineering in Medicine and Biology Society*, *2011*, 7099–7102.

Yu, Y. P., Ma, L. X., Ma, Y. X., Ma, Y. X., Liu, Y. Q., Liu, C. Z., et al. (2010). Immediate effect of acupuncture at Sanyinjiao (SP6) and Xuanzhong (GB39) on uterine arterial blood flow in primary dysmenorrhea. *Journal of Alternative and Complementary Medicine*, *16*, 1073–1078.

Zeng, F., Qin, W., Ma, T. T., Sun, J. B., Tang, Y., Yuan, K., et al. (2012). Influence of acupuncture treatment on cerebral activity in functional dyspepsia patients and its relationship with efficacy. *American Journal of Gastroenterology*, *107*, 1236–1247.

Zhang, J. H., Cao, X. D., Lie, J., Tang, W. J., Liu, H. Q., & Fenga, X. Y. (2007). Neuronal specificity of needling acupoints at same meridian: A control functional magnetic resonance imaging study with electroacupuncture. *Acupuncture Electrother Research, 32*(3–4), 179–193.

Zhang, D., Ding, G., Shen, X., Yao, W., Zhang, Z., Zhang, Y., et al. (2008). Role of mast cells in acupuncture effect: A pilot study. *EXPLORE: The Journal of Science and Healing, 4*(3), 170–177.

Zhang, W. T., Jin, Z., Luo, F., Zhang, L., Zeng, Y. W., & Han, J. S. (2004). Evidence from brain imaging with fMRI supporting functional specificity of acupoints in humans. *Neuroscience Letters, 354*(1), 50–53.

Zhang, J., Ngb, D., & Sau, A. (2009). Effects of electrical stimulation of acupuncture points on blood pressure. *Journal of Chiropractic Medicine, 8*, 9–14.

Zhang, Z. J., Wang, X. M., & McAlonan, G. M. (2012). Neural acupuncture unit: A new concept for interpreting effects and mechanisms of acupuncture. *Evidence-Based Complementary and Alternative Medicine, 2012*, 429412.

Zhao, L., Chen, J., Liu, C. Z., Li, Y., Cai, D. J., Tang, Y., et al. (2012). A review of acupoint specificity research in china: Status Quo and prospects. *Evidence-Based Complementary and Alternative Medicine, 2012*, 1–16.

Zhou, W., Fu, L., Tjen-A-Looi, S. C., Li, P., & Longhurst, J. C. (2005). Afferent mechanisms underlying stimulation modality-related modulation of acupuncture-related cardiovascular responses. *Journal of Applied Physiology, 98*, 872–880.

Zhou, W. Y., Tjen-A-Looi, S. C., & Longhurst, J. C. (2005). Brain stem mechanisms underlying acupuncture modality-related modulation of cardiovascular responses in rats. *Journal of Applied Physiology, 99*, 851–860.

Zubieta, J. K., Bueller, J. A., Jackson, L. R., Scott, D. J., Xu, Y., Koeppe, R. A., et al. (2005). Placebo effects mediated by endogenous opioid activity on μ-opioid receptors. *Journal of Neuroscience, 25*, 7754–7762.

Zubieta, J. K., Ketter, T. A., Bueller, J. A., Xu, Y., Kilbourn, M. R., Young, E. A., et al. (2003). Regulation of human affective responses by anterior cingulate and limbic μ-opioid neurotransmission. *Archives of General Psychiatry, 60*, 1145–1153.

Zubieta, J. K., Smith, Y. R., Bueller, J. A., Xu, Y., Kilbourn, M. R., Jewett, D. M., et al. (2001). Regional μ-opioid receptor regulation of sensory and affective dimensions of pain. *Science, 293*, 311–315.

CHAPTER FOUR

Acupuncture Stimulation Induces Neurogenesis in Adult Brain

Min-Ho Nam*, Kwang Seok Ahn*, Seung-Hoon Choi*,†,1
*Department of Pathology, College of Korean Medicine, Kyung Hee University, Seoul, Republic of Korea
†Korea Institute of Oriental Medicine, Daejeon, Republic of Korea
[1]Corresponding author: e-mail address: choish@khu.ac.kr

Contents

Abstract

The discovery of adult neurogenesis was a turning point in the field of neuroscience. Adult neurogenesis offers an enormous possibility to open a new therapeutic paradigm of neurodegenerative diseases and stroke. Recently, several studies suggested that acupuncture may enhance adult neurogenesis. Acupuncture has long been an important treatment for brain diseases in the East Asia. The scientific mechanisms of acupuncture treatment for the diseases, such as Alzheimer's disease, Parkinson's disease, and stroke, have not been clarified yet; however, the neurogenic effect of acupuncture can be a possible reason. Here, we have reviewed the studies on the effect of stimulation at various acupoints for neurogenesis, such as ST36 and GV20. The suggested mechanisms are also discussed including upregulation of brain-derived neurotrophic factor, glial cell line-derived neurotrophic factor, basic fibroblast growth factor and neuropeptide Y, and activation of the function of primo vascular system.

International Review of Neurobiology, Volume 111
ISSN 0074-7742
http://dx.doi.org/10.1016/B978-0-12-411545-3.00004-3

1. INTRODUCTION

It was a broadly believed dogma of neuroscience in the last century that central nervous system (CNS) regeneration is not possible in adulthood (Gonzalez-Castaneda, Galvez-Contreras, Luquin, & Gonzalez-Perez, 2011). However, since the first study reporting neurogenesis in the adult brain was published in 1965 (Altman & Das, 1965), it is now accepted that new neural stem cells (NSCs) are generated continuously and they mature into functional neurons and glial cells throughout life (Eriksson et al., 1998; Gould, Reeves, Graziano, & Gross, 1999; van Praag et al., 2002). Adult neurogenesis is increased in the hippocampus after stroke (Li, Yu, Ogle, Ding, & Wei, 2008) and Alzheimer's disease (AD) (Enciu et al., 2011), meaning that it can be a compensatory action for replacing the out-of-date neurons (Imitola et al., 2004). The discovery of adult neurogenesis raises the expectancy that the brain has an intrinsic reparative power and manipulations enhancing the neurogenesis can be a new approach for the treatment of neurodegenerative diseases and stroke. Some interventions including an enriched environment (Kempermann, Wiskott, & Gage, 2004) and exercise (van Praag et al., 2002) are reported to promote the adult neurogenesis, and in addition to those, acupuncture is considered as an effective therapy for boosting neurogenesis (Nam, Yin, Soh, & Choi, 2011).

Acupuncture treatment has long been a basic therapy for neurological disorders such as stroke and AD in the East Asia, and many researchers have tried to figure out the effectiveness with modern research tools (Chou, Chu, & Lin, 2009; Wang et al., 2012; Zhou & Jin, 2008). In spite of these efforts, the effect is not conclusive due to insufficient number and poor quality of the studies (Lee, Shin, & Ernst, 2009; Wu, Mills, Moher, & Seely, 2010). Recently, a series of papers have reported that acupuncture stimulation is effective for enhancing adult neurogenesis, and a review concluded that acupuncture at ST36 is beneficial for adult neurogenesis (Nam et al., 2011). They imply that upregulating neurogenesis can be a mechanism of acupuncture efficacy for some neurological disorders. This chapter discusses all reported studies about acupuncture and adult neurogenesis and analyzes its suggested mechanisms.

2. ADULT NEUROGENESIS

Adult neurogenesis generally occurs predominantly in two regions of the brain; the subgranular zone (SGZ) in the dentate gyrus (DG) of the hippocampus and the subventricular zone (SVZ) of the lateral ventricles (Abrous, Koehl, & Le Moal, 2005; Gould et al., 1999; Kempermann et al., 2004). In addition to these well-known neurogenic areas, the evidence for adult neurogenesis in neocortex, striatum, amygdala, and substantia nigra has been reported (Gould, 2007). NSCs can be found in these areas, and they are also called as neural progenitors because their stem-cell properties have not demonstrated convincingly *in vivo* until recently (Zhao, Deng, & Gage, 2008). These NSCs have potential to be differentiated into neurons, astrocytes, and oligodendrocytes and also can undergo cell division for self-renewal with maintaining the stem-cell potential (Gonzalez-Castaneda et al., 2011; Taupin & Gage, 2002). The newly generated neurons in the adult mammalian brain have normal neuronal morphology and function such as showing passive membrane properties, firing action potentials, and synaptic actions (van Praag et al., 2002). These new neurons are expected to have a function in brain repair (van Praag et al., 2002).

Cell proliferation and differentiation of the DG is responsible for overall hippocampal neurogenesis (Kempermann & Gage, 2002), because only DG continues to develop through adulthood among the hippocampal formations. NSCs in the SGZ of DG migrate into the granular cell layer of the hippocampus where they differentiate into dentate granule cells (Gonzalez-Castaneda et al., 2011; Lledo, Alonso, & Grubb, 2006; Ming & Song, 2005). NSCs in the SVZ predominantly migrate into the olfactory bulb via the rostral migratory stream and differentiate into local interneurons (Coskun & Luskin, 2002; Gonzalez-Castaneda et al., 2011; Gritti et al., 2002), while hippocampal neurogenesis occurs only in the hippocampus.

Hippocampal neurogenesis has been observed in adult animals from birds to humans (Altman & Das, 1965; Barnea & Nottebohm, 1994; Eriksson et al., 1998; Kaplan & Hinds, 1977; Kornack & Rakic, 1999; Kuhn, Dickinson-Anson, & Gage, 1996). The first study on adult neurogenesis was reported by Altman and Das in 1965 with [^{3}H]-thymidine labeling studies (Altman & Das, 1965). After that, several new methods for labeling dividing cells were discovered; the most broadly used is incorporation of bromodeoxyuridine (BrdU), a marker of the S-phase of the cell cycle (Corotto, Henegar, & Maruniak, 1993; Taupin & Gage, 2002). BrdU incorporation method can be performed with some immunohistochemical staining by using antibodies of neuronal nuclei (NeuN), neuron-specific enolase (NSE), doublecortin (DCX), and glial fibrillary acidic protein (GFAP), depending on which cell type NSCs differentiate into or how mature the new cells are (Fuchs & Gould, 2000). For example, BrdU with

NeuN is used for demonstrating not only cell proliferation but also neuronal differentiation—BrdU with GFAP for cell proliferation and glial differentiation.

3. ACUPUNCTURE FOR ADULT NEUROGENESIS

In the East Asia, acupuncture has been used for treating a wide range of disorders including brain diseases for thousands of years. Based on the clinical effects of acupuncture treatment for neurodegenerative diseases such as AD (Wang et al., 2012; Zhou & Jin, 2008) and Parkinson's disease (PD) (Joh, Park, Kim, & Lee, 2010; Lam et al., 2008), some researchers have reported that acupuncture therapy may induce adult neurogenesis. In particular, stimulating the following acupoints by acupuncture or electroacupuncture (EA) appears to enhance brain cell proliferation and neuronal proliferation in some particular areas: ST36 (Kim et al., 2002, 2001), GV20 (Hwang et al., 2010a; Liu et al., 2007), PC6 (Lee, Shim, Lee, Yang, & Hahm, 2009), HT7 (Park et al., 2002), CV17, CV12, CV6, SP10 (Cheng et al., 2008), LI11, TE5, GB30 (Gao, Wang, Wang, & Zhu, 2011), GV16, GV8 (Yang, Shen, Guo, & Sun, 2005), CV4, CV6, CV24 (Yang, Yu, Rao, Liu, & Pi, 2008), and stomach point of auricular acupoints (Kim, Chung, & Kim, 2001). Out of 15 studies on acupuncture and adult neurogenesis, 10 studies dealt with ST36, 4 studies with GV20, and 3 studies with the other acupoints.

The methods used for confirming cell proliferation and neuronal differentiation were similar. In most of the studies, BrdU incorporation method was basically applied (Cheng et al., 2009, 2008; Chung et al., 2007; Kim, Chung, et al., 2001; Kim et al., 2002; Liu et al., 2007); in some studies, additional immunohistochemistry stainings were combined with BrdU staining, such as NeuN (Gao et al., 2011; Kim, Kim, et al., 2001), GFAP (Tao et al., 2010), Nestin, and NSE (Yang et al., 2005). In two studies, Ki67 staining was used instead of BrdU incorporation in order to confirm cell proliferation (Hwang et al., 2010a, 2010b), and one of them used DCX staining with it (Hwang et al., 2010a).

3.1. Acupoints-related neurogenesis

3.1.1 ST36

ST36, an acupoint located on the anterior tibia muscle (Fig. 4.1A), is one of the most important acupoints in clinical acupuncture. Acupuncture or EA at

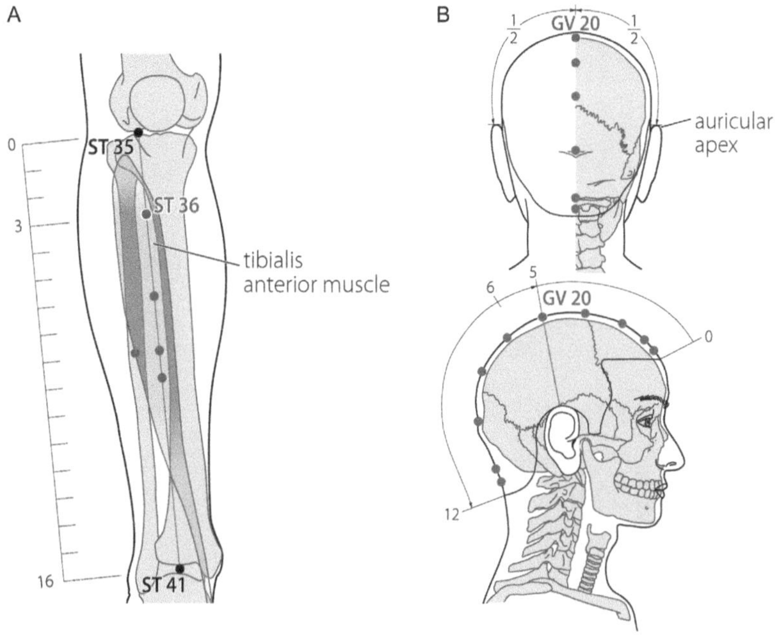

Figure 4.1 The anatomical location of acupoint ST36 (A) and GV20 (B). (A) ST36 is located on the tibialis anterior muscle of the lower limb. (B) GV20 is located on the midline of the scalp. *The figures are taken from WHO Standard Acupuncture Point Locations in the Western Pacific Region.* (For color version of this figure, the reader is referred to the online version of this chapter.)

ST36 is used for a number of pathological conditions of the digestive system (Xu et al., 2006), cardiovascular system (Oh, Yang, Choi, & Kang, 2012), and also nervous system (Zhou & Jin, 2008). In terms of the nervous system, ST36 is one of the seven acupoints used for stroke treatment (Lee, Shin, & Kim, 2004) and stimulated for treating AD and PD (Wang et al., 2011; Zhou & Jin, 2008).

Several papers have published that acupuncture stimulation at ST36 may induce adult neurogenesis in the SVZ and DG in the brain of animals with stroke, diabetes, stress, and high tendency to AD (Table 4.1). The first paper on this issue reported that acupuncture stimulation at ST36 increased cell proliferation in the DG of gerbils along with stroke. This study provided the first evidence of increased birth of dentate progenitor cells after acupuncture (Kim, Kim, et al., 2001). A study confirmed this research, reporting the effect of acupuncture at ST36, exclusive of LI4 (Chung et al., 2007). In

Table 4.1 Summary of papers on adult neurogenesis by acupuncture at ST36

	Animal			Acupuncture stimulation					
Study	Strain, age/weight, sex	Model	Acupoints	Type	Stimulation method	Time	Frequency, period	Results	Remark
Kim, Kim, et al. (2001)	Gerbils, 11–13 weeks	Transient global ischemia (CCA occlusion)	ST36	MA	–	20 min	Two times/day for 9 days	Acupuncture at ST36 increased cell proliferation in the DG	
Kim et al. (2002)	SD rats, 6 weeks	STZ-induced diabetic rats	ST36	MA	–	20 min	Two times/day for 7 days	Acupuncture at ST36 enhanced cell proliferation in the DG and promoted proliferation of neuronal precursor cells	No significant effect on cell proliferation and NPY expression was observed under normal conditions
Yun et al. (2002)	SD rats, 6 weeks male	Immobilization stress	ST36	EA	2 Hz, 1–2 mA, 0.3 ms pulse width			EA at ST36 significantly restored cell proliferation decline by immobilization stress	

Park et al. (2002)	SD rats 14 days	Maternally separated rats	ST36, HT7 (both side) (for each)	MA	3 mm depth twisting the needle at the speed of twice a second for 30 s and removed immediately		One time/day for 1 week	Acupuncture at HT7 enhanced cell proliferation in the DG significantly, whereas acupuncture at ST36 did not	HT7 is usually used for the emotional deficit such as maternal separation in traditional East Asian medicine
Chung et al. (2007)	Gerbils 11–13 weeks	Transient global ischemia (CCA occlusion)	ST36, LI4 (both side) (for each)	MA	0.3 mm diameter, 2–4 mm depth	20 min	Two times/day for 8 days	Acupuncture at ST36 enhanced cell proliferation in the DG significantly, whereas acupuncture at LI4 did not	
Cheng et al. (2008)	SAMP8 mice, 4 months, male	–	ST36, CV17, CV12, CV6, SP10	MA	–	–	One time/day for 15 days (rest on the 8th day)	Acupuncture treatment improved cell proliferation in the DG	
Tao et al. (2010)	SD rats 250 ± 30 g, male	Transient MCA occlusion	ST36, LI11	EA	Dense disperse wave of 1 and 20 Hz	30 min	One time/day for 21 days	EA may promote cells proliferation and neural/glial differentiation	

Continued

Table 4.1 Summary of papers on adult neurogenesis by acupuncture at ST36—cont'd

	Animal			Acupuncture stimulation					
Study	**Strain, age/weight, sex**	**Model**	**Acupoints**	**Type**	**Stimulation method**	**Time**	**Frequency, period**	**Results**	**Remark**
Hwang et al. (2010a)	Wistar rats, 13 weeks, male	–	ST36, GV20	MA	MA: 5 mm depth	20 min	One time/day for 21 days	MA and EA enhanced cell proliferation and neural differentiation in the DG with greater effect than that in only MA	
				EA	EA: 5 mm depth, disperse-dense waves of 5 and 20 Hz, 2–4 mA				
Hwang et al. (2010b)	Wistar rats, 13 weeks, male	–	ST36, GV20	EA	5 mm depth, disperse-dense waves of 5 and 20 Hz, 2–4 mA	20 min	One time/day for 21 days	EA enhanced cell proliferation and neural differentiation in the DG	
Gao et al. (2011)	SD rats 14 days	–	ST36, LI11, TE5, GB30	EA	2 Hz, 0.5 ms, 0.7 mV	30 min	One time/day for 1 week	EA had a long effect (e.g., minimum 4 weeks) on progenitor cell proliferation and neural differentiation	

MA, manual acupuncture; EA, electroacupuncture; MCA, middle cerebral artery; CCA, common carotid artery

2010, a study also suggested the beneficial effect of EA at ST36 and LI11 for adult neurogenesis in ischemic rats (Tao et al., 2010). EA was shown to be effective for differentiating the neural progenitors into not only mature neurons but also astrocytes by using BrdU/NeuN and BrdU/GFAP methods. In addition to stroke model, diabetic model also showed the enhancement of adult neurogenesis by acupuncture treatment at ST36 (Kim et al., 2002). Immunohistochemistry of NPY was performed instead of NeuN, because NPY was recently suggested to promote the proliferation of neuronal precursor cells (Hansel, Eipper, & Ronnett, 2001). EA at ST36 also restored a decrease in cell proliferation in stressed rats (Yun et al., 2002). In old SAMP8 mice with high tendency to AD, acupuncture treatment at ST36, CV17, CV12, CV6, and SP10 was shown to be effective on cell proliferation in the DG (Cheng et al., 2008).

Acupuncture has been found to be beneficial to enhance cell proliferation and neural differentiation in a healthy animal (Table 4.1). ST36 and GV20 of EA and acupuncture stimulation was reported to promote adult neurogenesis and the effect of EA was greater than that of manual acupuncture (Hwang et al., 2010a, 2010b). A paper analyzed the potential mechanism of this effect and suggested that EA significantly increased neuroblast plasticity via pCREB and brain-derived neurotrophic factor (BDNF) activation in the dentate gyrus (Hwang et al., 2010b). An interesting result was reported by Gao et al. that the effect of EA at ST36, LI11, TE5, and GB30 on neurogenesis lasted until 4 weeks after the last stimulation, meaning that EA treatment for enhancing adult neurogenesis is beyond a transient remedy (Gao et al., 2011). However, one study reported a beneficial effect on neurogenesis with acupuncture at HT7, whereas no effect at ST36 (Park et al., 2002). In spite of this study specifically performed on maternally separated rats, 9 of 10 reports demonstrated that both acupuncture stimulation at ST36 (Chung et al., 2007; Kim et al., 2002; Kim, Kim, et al., 2001; Yun et al., 2002) and simultaneous stimulation at several acupoints including ST36 (Cheng et al., 2008; Gao et al., 2011; Hwang et al., 2010a, 2010b; Tao et al., 2010) can enhance cell proliferation in the DG of the hippocampus.

3.1.2 GV20

GV20, an acupoint located on the scalp (Fig. 4.1B), is one of the most widely used acupoints for brain diseases, such as AD (Zhang, Guan, & Jiang, 2010), PD (Xia, Wang, Ding, Kang, & Liu, 2012), and stroke (Kim et al., 2012). GV20 is also known as one of the seven acupoints used for stroke treatment

(Lee et al., 2004). Recently, in addition to ST36, acupuncture stimulation at GV20 has been reported to induce adult neurogenesis in the DG and striatum by four papers including two that dealt with ST36 and GV20 together (Table 4.2). In 2009, a study reported that combined therapy of intranasal administration of nerve growth factor and EA at GV20 and GV26 has a synergistic effect in enhancing cell proliferation in cerebral ischemic rats, whose effect is beyond the beneficial effect of EA sole therapy (Cheng et al., 2009). This study also suggested that cell proliferation induced by EA is observed not only in the DG but also in the striatum and that the effect of short-term EA treatment lasts 26 days at least. On the other hand, a paper reported that EA at GV20 and Anmian, an extraordinary acupoints, does not show any significant enhancement of adult neurogenesis in a healthy rat, while it blocks the decrease of hippocampal progenitor cell proliferation induced by chronic stress (Liu et al., 2007). This result conflicts with studies insisting EA stimulation at both ST36 and GV20 is effective for adult neurogenesis in healthy animal (Hwang et al., 2010a, 2010b). Even though it appeared to enhance neurogenesis, there is still lack of studies on acupuncture treatment at only GV20 for adult neurogenesis to make a definite conclusion.

3.1.3 The other points

In addition to ST36 and GV20, some more acupoints were studied for adult neurogenesis (Table 4.3). Most of the acupoints were selected based on the theory of traditional East Asian medicine. A paper about efficacy of auricular acupuncture on adult neurogenesis was published by Kim, whose group has reported several papers on this issue, particularly on ST36 (Kim, Chung, et al., 2001). Given that ST36, a major acupoint for stomach disease, showed significant probability to enhance adult neurogenesis, stomach point was selected out of many auricular acupoints and showed beneficial effect for adult neurogenesis. EA stimulation at GV8 and GV16 was effective for improving neuronal regeneration and their maturation after stroke (Yang et al., 2005). GV8 and GV16, located on the Governor Vessel, which is regarded to nourish the brain and spinal cord, are often used for many kinds of brain diseases in clinics (Pang, Itano, Sumitani, Negi, & Miyamoto, 2003; Ying & Cheng, 1994). Acupuncture treatment at CV4, CV6, and CV24 was also reported to enhance cell proliferation and neural/glial differentiation in the DG (Yang et al., 2008). These acupoints from Conception Vessel have been used for treatment of stroke (Ji & Zhang, 2009), because Conception Vessel and Governor Vessel,

Table 4.2 Summary of papers on adult neurogenesis by acupuncture at GV20

	Animal			Acupuncture stimulation					
Study	Strain, age/weight, sex	Model	Acupoints	Type	Stimulation method	Time	Frequency, period	Results	Remark
Liu et al. (2007)	SD rats, 4 months	Chronic unpredictable stress	GV20, Anmian (extraordinary acupoint)	EA	3 mm depth (GV20), 2 mm depth (Anmian), alternating strings of dense–sparse frequencies (60 Hz for 5 s and 4 Hz for 2.5 s alternately)	30 min	Two times/day for 21 days	EA blocked the decrease of hippocampal progenitor cell proliferation induced by chronic unpredictable stress	Under normal condition, no enhancement of cell proliferation was observed
Cheng et al. (2009)	SD rats, 250–280 g, male	Transient MCA occlusion	GV20, GV26	EA	4 Hz, 2 mA	30 min	3 days	The intranasal administration of nerve growth factor and EA may have a synergistic effect in enhancing cell proliferation. Only EA also shows the effect	The cell proliferation effect was observed in not only DG but also striatum

Continued

Table 4.2 Summary of papers on adult neurogenesis by acupuncture at GV20—cont'd

Study	Animal		Acupoints	Acupuncture stimulation				Results	Remark
	Strain, age/weight, sex	Model		Type	Stimulation method	Time	Frequency, period		
Hwang et al. (2010a)	Wistar rats, 13 weeks, male	–	ST36, GV20	MA	MA: 5 mm depth	20 min	One time/day for 21 days	MA and EA enhanced cell proliferation and neural differentiation in the DG with greater effect than that in only MA	Same paper with ninth paper of ST36
				EA	EA: 5 mm depth, disperse-dense waves of 5 and 20 Hz, 2–4 mA				
Hwang et al. (2010b)	Wistar rats, 13 weeks, male	–	ST36, GV20	EA	5 mm depth, disperse-dense waves of 5 and 20 Hz, 2–4 mA	20 min	One time/day for 21 days	EA enhanced cell proliferation and neural differentiation in the DG	Same paper with 10th paper of ST36

Table 4.3 Summary of papers on adult neurogenesis by acupuncture at other acupoints

Study	Animal		Acupoints	Acupuncture stimulation				Results	Remarks
	Strain, age/weight, sex	Model		Type	Stimulation method	Time	Frequency, period		
Kim, Chung,and& Kim (2001)	Gerbils, 150 ± 10 g	–	Stomach point (auricular acupuncture)	MA	0.12 mm diameter 2.0 mm depth	20 min	Two times/day for 2 days	Auricular acupuncture enhances cell proliferation in the DG	
Yang et al. (2005)	SD rats, 220–250 g, male	Transient MCA occlusion	GV16, GV8	EA	60 Hz (1 s) and 2 Hz (3 s) alternately. 10 mA	20 min		EA can improve neuronal regeneration, newborn neuron migration and their maturation in the striatum after stroke	
Yang et al. (2008)	Wistar rats, 250 ± 20 g, male	–	CV4, CV6, CV24	EA	Sparse-dense wave of 1 and 20 Hz	10 min	One time/day for 28 days	EA may promote differentiation and proliferation of the nervous stem cells in the cerebral ischemia rats	

which are running along the midline of the ventral part and dorsal part of the body, respectively, are connected to each other.

3.2. Mechanism of the neurogenic effect of acupuncture

3.2.1 Mechanism under a view of traditional medical theories

The meridian is a line connecting acupoints and a mediator that makes it possible that acupuncture stimulation treats a distant part. Throughout the whole body, there are 12 main meridians and 8 extra meridians. Two of eight extra meridians, Governor Vessel (GV) and Conception Vessel (CV), have their own pathway for Qi flow, just same as 12 main meridians do. Some of them, such as GV, bladder meridian, are thought to be directly connected to the brain and others also connected to the brain indirectly due to the continuity of all meridians. Therefore, in addition to GV20 on the head, stimulation at ST36 on the lower limb and some CV acupoints on the stomach can be effective for brain diseases.

Besides some acupoints located near the brain, most of the acupoints used for enhancing adult neurogenesis have function of tonifying Qi in the whole body. Tonified Qi affects physiological functions of the whole body; it may also activate brain physiological function including neurogenesis. Therefore, acupuncture stimulation at ST36, GV20, etc., can promote adult neurogenesis.

3.2.2 Scientific mechanism

The molecular mechanisms of acupuncture on adult neurogenesis have not yet been clarified. Several studies are suggesting some possible mechanisms: first, upregulation of BDNF, glial cell line-derived neurotrophic factor (GDNF), and basic fibroblast growth factor (bFGF) (Hwang et al., 2010b; Liu et al., 2007; Tao et al., 2010); second, upregulation of neuropeptide Y (NPY); and third, activation of the function of primo vascular system (PVS).

It has been reported by several researchers that acupuncture stimulation including EA increases the expression of growth factors such as BDNF (Hwang et al., 2010b; Yun et al., 2002), GDNF (Dong, Ma, Xie, Wang, & Wu, 2005; Liang et al., 2003), and bFGF (Ou, Han, Da, Huang, & Cheng, 2001) under various conditions, such as stress, stroke, and normal condition (Manni, Albanesi, Guaragna, Barbaro Paparo, & Aloe, 2010; Tao et al., 2010). Neurotrophic factors are responsible for the growth and survival of developing neurons, and it is reported that they are necessary for adult neurogenesis (Rossi et al., 2006; Yuan et al., 2013). bFGF is also known to be beneficial for neurogenesis (Wang et al., 2008). The transcript factor, cyclic AMP response element-binding protein (CREB) that regulates the transcription of BDNF (Zhu, Lau, Liu, Wei, & Lu, 2004), can be upregulated by EA at ST36 and GV20 (Hwang et al., 2010b).

NPY has been studied in detail to understand the mechanism of adult neurogenesis induced by acupuncture. Acupuncture upregulated NPY expression in the CNS (Kim et al., 2002), which promotes the proliferation of neuronal precursor cells (Hansel et al., 2001).

At last, a hypothesis has been suggested that acupuncture effect for adult neurogenesis may be mediated by PVS (Nam et al., 2011), which is proposed as an anatomical structure of meridians (Soh, 2009). The PVS is thought to be a systemic circulatory system that forms whole-body network in which so-called primo fluid flows. The primo fluid contains primo microcells (Baik, Ogay, Jeoung, & Soh, 2009) that function like the very small embryonic-like stem cells (Ratajczak, Zuba-Surma, Shin, Ratajczak, & Kucia, 2008) and are thought to be engaged in tissue and cell regeneration, similar to the role of pluripotent stem cells (Kim, 1965). Until now, there is lack of substantial evidence supporting this hypothesis yet. However, PVS has been observed in the brain (Lee, Kim, & Soh, 2008) and the skin (Lee & Soh, 2010) and is thought to be linked in both the organs (Lee, Eom, & Soh, 2010). If it turned out to be true, it should be considered that the neurogenic effect may be mediated by the primo microcells.

4. NEUROGENESIS IN NEURODEGENERATIVE DISEASES

4.1. Alzheimer's disease

Acupuncture treatment for treating AD has been researched by some scientists. In order to improve AD, several clinical and animal researches have used ST36 and GV20, main acupoints for adult neurogenesis, and acupuncture at these acupoints shows a significant effect (Zhang et al., 2010; Zhou & Jin, 2008). In spite of these results, a review study concluded that the existing evidence does not demonstrate the effectiveness of acupuncture for AD (Lee, Shin, et al., 2009) because of lack of clinical studies of good quality.

Although the pathological changes of neurogenesis during AD are still controversial, it is accepted that neurogenesis is engaged in the pathology of AD. The deficit of learning and memory, the main symptom of AD, could partly be attributed to the impaired endogenous neurogenesis in the SGZ

(Zhao et al., 2008). Some studies reported increased neurogenesis in the brains of patients with AD (Jin et al., 2004), suggesting that neurogenesis acts as a compensation for loss of function (Imitola et al., 2004). On the other hand, some animal studies demonstrated that hippocampal neurogenesis is both increased and decreased (Gonzalez-Castaneda et al., 2011) according to the AD severity; a compensatory increase of NSC proliferation in the early stages and decreased proliferation in the advanced stages (Biscaro, Lindvall, Hock, Ekdahl, & Nitsch, 2009; Gan et al., 2008).

In this context, enhancing neurogenesis can be a therapeutic strategy. Even though the numbers of newly generated DG neurons in an aging brain are much lower than the vast number of degenerating neurons in many brain regions, however, stimulating DG adult neurogenesis might have some therapeutic potential with understanding that the hippocampus has a crucial role in learning and memory (Winner, Kohl, & Gage, 2011). Several researches have suggested that acupuncture may promote adult neurogenesis in the SVZ and DG (Cheng et al., 2008; Hwang et al., 2010a; Kim, Kim, et al., 2001). Therefore, neurogenesis can be underlying mechanism of acupuncture treatment for AD.

4.2. Parkinson's disease

For treatment of PD, the second most common neurodegenerative disorder, acupuncture treatment is also used in East Asian countries. Some reviews already reported the potential effectiveness of acupuncture for treating PD (Joh et al., 2010; Lam et al., 2008). ST36 and GV20 were selected in order to improve the symptoms of PD by some clinical and animal studies (Sun et al., 2012; Wang et al., 2011; Xia et al., 2012).

There are several studies reporting the relation between adult neurogenesis and PD. Some nonmotor symptoms of PD, such as impaired olfaction, depression, apathy, and anxiety, may partly rely on proper olfactory processing and hippocampal function, implying that they are related with adult neurogenesis (Marxreiter, Regensburger, & Winkler, 2013). A negative effect on adult neurogenesis was observed in both the DG (Crews et al., 2008; Winner et al., 2004) and SVZ/OB systems (Winner et al., 2004, 2008) of the alpha-synuclein-based PD animal model (Winner et al., 2011). In human study, a significant decrease in neural progenitor cells was also reported in the SVZ (Hoglinger et al., 2004).

The most broadly used drug for PD, L-dopa, has a positive effect on increasing the number of proliferating NSCs in the SVZ, meaning that it may have a role in neurogenesis (O'Sullivan et al., 2011). It suggests that enhancing neurogenesis can be effective for treatment of PD. For these reasons, acupuncture that enhances neurogenesis can be a suggestive therapy for PD.

4.3. Stroke

Stroke is the most frequent disease of Korean medical hospitals. Acupuncture treatment for poststroke rehabilitation in East Asia is based on a large body of preclinical and clinical research (Wu et al., 2010). Its effectiveness is also suggested by the American Heart Association in 2011 (Wu et al., 2010). ST36 and GV20 are the main acupoints for poststroke rehabilitation (Lee et al., 2004).

Neurogenesis is significantly upregulated after stroke (Arvidsson, Collin, Kirik, Kokaia, & Lindvall, 2002; Jin et al., 2001; Munnik, 2011). Cell proliferation surges after stroke (Jin et al., 2001; Zhang, Zhang, Zhang, & Chopp, 2001), and migration of new neurons to the ischemic area needs more time up to four months (Thored et al., 2006). In humans also, newborn neurons were observed in the ischemic areas due to stroke (Jin et al., 2006). During stroke, many neurogenesis-related factors start to be released in the ischemic area simultaneously with an inflammatory response. A number of epigenetic controls, growth factors, cytokines, transmembrane receptors, and signaling molecules have been reported to influence stroke-induced neurogenesis, such as BDNF (Li et al., 2010), VEGF (Sun et al., 2003), FGF2 (Leker et al., 2007), EGF (Teramoto, Qiu, Plumier, & Moskowitz, 2003), and Notch (Wang, Mao, Xie, Greenberg, & Jin, 2009) (Munnik, 2011).

Many researchers expected that adult neurogenesis may be beneficial for the poststroke recovery by replacing dead neurons (Arvidsson et al., 2002; Parent, Vexler, Gong, Derugin, & Ferriero, 2002) and have tried to investigate its potential for therapeutic use in cellular (Chen et al., 2001) and pharmacological approaches (Leker, Lasri, & Chernoguz, 2009). Above all, it is focused to enhance the natural neurogenic capacities of the brain as a therapeutic approach, maximizing the intrinsic potential of proliferation, migration, and survival of new neurons after stroke. Some therapies are being developed through animal experiments such as intraventricular infusion of bFGF, EGF, BDNF, and VEGF, which aim to amplify endogenous neurogenesis and to improve the ischemic microenvironment to be receptive to integration of migrated new neurons (Zhang & Chopp, 2011). Acupuncture can be a preferential therapy in this respect, because it is already known to be effective for promoting adult neurogenesis in the stroke models without any material addition (Cheng et al., 2009; Kim, Kim, et al., 2001; Tao et al., 2010).

5. CONCLUSION

This chapter reviewed the studies on acupuncture and adult neurogenesis, and most of the published articles demonstrated that acupuncture stimulation enhances cell proliferation and neuronal differentiation in the brain. Although the neurogenic effect of stimulation at ST36 and GV20 is most verified among a number of acupoints, additional studies of good quality should be done to guarantee the effect, especially on GV20. Acupuncture has appeared to be beneficial for neurogenesis in not only normal animals but also stroke, diabetes, and stress model. Upregulation of neurotrophic factors (i.e., BDNF and GDNF), growth factors (i.e., bFGF), and NPY and activation of the function of PVS were suggested as its plausible mechanisms of action. In this chapter, we suggest that the effect of acupuncture on neurodegenerative diseases and stroke may also be due to its neurogenic action.

ACKNOWLEDGMENT

This work was supported by the project of National Junior Research Fellowship, which National Research Foundation of Korea conducts from 2012.

REFERENCES

Abrous, D. N., Koehl, M., & Le Moal, M. (2005). Adult neurogenesis: From precursors to network and physiology. *Physiological Reviews*, *85*, 523–569.

Altman, J., & Das, G. D. (1965). Autoradiographic and histological evidence of postnatal hippocampal neurogenesis in rats. *Journal of Comparative Neurology*, *124*, 319–335.

Arvidsson, A., Collin, T., Kirik, D., Kokaia, Z., & Lindvall, O. (2002). Neuronal replacement from endogenous precursors in the adult brain after stroke. *Nature Medicine*, *8*, 963–970.

Baik, K. Y., Ogay, V., Jeoung, S. C., & Soh, K. S. (2009). Visualization of Bonghan microcells by electron and atomic force microscopy. *Journal of Acupuncture and Meridian Studies*, *2*, 124–129.

Barnea, A., & Nottebohm, F. (1994). Seasonal recruitment of hippocampal neurons in adult free-ranging black-capped chickadees. *Proceedings of the National Academy of Sciences of the United States of America*, *91*, 11217–11221.

Biscaro, B., Lindvall, O., Hock, C., Ekdahl, C. T., & Nitsch, R. M. (2009). Abeta immunotherapy protects morphology and survival of adult-born neurons in doubly transgenic APP/PS1 mice. *The Journal of Neuroscience*, *29*, 14108–14119.

Chen, J., Li, Y., Wang, L., Zhang, Z., Lu, D., Lu, M., et al. (2001). Therapeutic benefit of intravenous administration of bone marrow stromal cells after cerebral ischemia in rats. *Stroke*, *32*, 1005–1011.

Cheng, S., Ma, M., Ma, Y., Wang, Z., Xu, G., & Liu, X. (2009). Combination therapy with intranasal NGF and electroacupuncture enhanced cell proliferation and survival in rats after stroke. *Neurological Research*, *31*, 753–758.

Cheng, H., Yu, J., Jiang, Z., Zhang, X., Liu, C., Peng, Y., et al. (2008). Acupuncture improves cognitive deficits and regulates the brain cell proliferation of SAMP8 mice. *Neuroscience Letters*, *432*, 111–116.

Chou, P., Chu, H., & Lin, J. G. (2009). Effects of electroacupuncture treatment on impaired cognition and quality of life in Taiwanese stroke patients. *Journal of Alternative and Complementary Medicine*, *15*, 1067–1073.

Chung, J. H., Lee, E. Y., Jang, M. H., Kim, C. J., Kim, J., Ha, E., et al. (2007). Acupuncture decreases ischemia-induced apoptosis and cell proliferation in dentate gyrus of gerbils. *Neurological Research*, *29*(Suppl. 1), S23–S27.

Corotto, F. S., Henegar, J. A., & Maruniak, J. A. (1993). Neurogenesis persists in the subependymal layer of the adult mouse brain. *Neuroscience Letters*, *149*, 111–114.

Coskun, V., & Luskin, M. B. (2002). Intrinsic and extrinsic regulation of the proliferation and differentiation of cells in the rodent rostral migratory stream. *Journal of Neuroscience Research*, *69*, 795–802.

Crews, L., Mizuno, H., Desplats, P., Rockenstein, E., Adame, A., Patrick, C., et al. (2008). Alpha-synuclein alters Notch-1 expression and neurogenesis in mouse embryonic stem cells and in the hippocampus of transgenic mice. *The Journal of Neuroscience*, *28*, 4250–4260.

Dong, Z. Q., Ma, F., Xie, H., Wang, Y. Q., & Wu, G. C. (2005). Changes of expression of glial cell line-derived neurotrophic factor and its receptor in dorsal root ganglions and spinal dorsal horn during electroacupuncture treatment in neuropathic pain rats. *Neuroscience Letters*, *376*, 143–148.

Enciu, A. M., Nicolescu, M. I., Manole, C. G., Muresanu, D. F., Popescu, L. M., & Popescu, B. O. (2011). Neuroregeneration in neurodegenerative disorders. *BMC Neurology*, *11*, 75.

Eriksson, P. S., Perfilieva, E., Bjork-Eriksson, T., Alborn, A. M., Nordborg, C., Peterson, D. A., et al. (1998). Neurogenesis in the adult human hippocampus. *Nature Medicine*, *4*, 1313–1317.

Fuchs, E., & Gould, E. (2000). Mini-review: In vivo neurogenesis in the adult brain: Regulation and functional implications. *European Journal of Neuroscience*, *12*, 2211–2214.

Gan, L., Qiao, S., Lan, X., Chi, L., Luo, C., Lien, L., et al. (2008). Neurogenic responses to amyloid-beta plaques in the brain of Alzheimer's disease-like transgenic (pPDGF-APPSw, Ind) mice. *Neurobiology of Disease*, *29*, 71–80.

Gao, J., Wang, S., Wang, X., & Zhu, C. (2011). Electroacupuncture enhances cell proliferation and neuronal differentiation in young rat brains. *Neurological Sciences*, *32*, 369–374.

Gonzalez-Castaneda, R. E., Galvez-Contreras, A. Y., Luquin, S., & Gonzalez-Perez, O. (2011). Neurogenesis in Alzheimer s disease: A realistic alternative to neuronal degeneration? *Current Signal Transduction Therapy*, *6*, 314–319.

Gould, E. (2007). How widespread is adult neurogenesis in mammals? *Nature Reviews Neuroscience*, *8*, 481–488.

Gould, E., Reeves, A. J., Graziano, M. S., & Gross, C. G. (1999). Neurogenesis in the neocortex of adult primates. *Science*, *286*, 548–552.

Gritti, A., Bonfanti, L., Doetsch, F., Caille, I., Alvarez-Buylla, A., Lim, D. A., et al. (2002). Multipotent neural stem cells reside into the rostral extension and olfactory bulb of adult rodents. *The Journal of Neuroscience*, *22*, 437–445.

Hansel, D. E., Eipper, B. A., & Ronnett, G. V. (2001). Neuropeptide Y functions as a neuroproliferative factor. *Nature, 410*, 940–944.

Hoglinger, G. U., Rizk, P., Muriel, M. P., Duyckaerts, C., Oertel, W. H., Caille, I., et al. (2004). Dopamine depletion impairs precursor cell proliferation in Parkinson disease. *Nature Neuroscience*, 7, 726–735.

Hwang, I. K., Chung, J. Y., Yoo, D. Y., Yi, S. S., Youn, H. Y., Seong, J. K., et al. (2010a). Comparing the effects of acupuncture and electroacupuncture at Zusanli and Baihui on cell proliferation and neuroblast differentiation in the rat hippocampus. *Journal of Veterinary Medical Science, 72*, 279–284.

Hwang, I. K., Chung, J. Y., Yoo, D. Y., Yi, S. S., Youn, H. Y., Seong, J. K., et al. (2010b). Effects of electroacupuncture at Zusanli and Baihui on brain-derived neurotrophic factor and cyclic AMP response element-binding protein in the hippocampal dentate gyrus. *Journal of Veterinary Medical Science, 72*, 1431–1436.

Imitola, J., Raddassi, K., Park, K. I., Mueller, F. J., Nieto, M., Teng, Y. D., et al. (2004). Directed migration of neural stem cells to sites of CNS injury by the stromal cell-derived factor 1alpha/CXC chemokine receptor 4 pathway. *Proceedings of the National Academy of Sciences of the United States of America, 101*, 18117–18122.

Ji, X. Q., & Zhang, Z. L. (2009). Observation on therapeutic effect of nuchal acupuncture and abdominal acupuncture for treatment of stroke patients with spastic hemiplegia. *Zhongguo Zhen Jiu, 29*, 961–965.

Jin, K., Minami, M., Lan, J. Q., Mao, X. O., Batteur, S., Simon, R. P., et al. (2001). Neurogenesis in dentate subgranular zone and rostral subventricular zone after focal cerebral ischemia in the rat. *Proceedings of the National Academy of Sciences of the United States of America, 98*, 4710–4715.

Jin, K., Peel, A. L., Mao, X. O., Xie, L., Cottrell, B. A., Henshall, D. C., et al. (2004). Increased hippocampal neurogenesis in Alzheimer's disease. *Proceedings of the National Academy of Sciences of the United States of America, 101*, 343–347.

Jin, K., Wang, X., Xie, L., Mao, X. O., Zhu, W., Wang, Y., et al. (2006). Evidence for stroke-induced neurogenesis in the human brain. *Proceedings of the National Academy of Sciences of the United States of America, 103*, 13198–13202.

Joh, T. H., Park, H. J., Kim, S. N., & Lee, H. (2010). Recent development of acupuncture on Parkinson's disease. *Neurological Research, 32*(Suppl 1), 5–9.

Kaplan, M. S., & Hinds, J. W. (1977). Neurogenesis in the adult rat: Electron microscopic analysis of light radioautographs. *Science, 197*, 1092–1094.

Kempermann, G., & Gage, F. H. (2002). Genetic determinants of adult hippocampal neurogenesis correlate with acquisition, but not probe trial performance, in the water maze task. *European Journal of Neuroscience, 16*, 129–136.

Kempermann, G., Wiskott, L., & Gage, F. H. (2004). Functional significance of adult neurogenesis. *Current Opinion in Neurobiology, 14*, 186–191.

Kim, B.-H. (1965). The Kyungrak system. *Journal of Jo Sun Medicine, 108*.

Kim, M. W., Chung, Y. C., Jung, H. C., Park, M. S., Han, Y. M., Chung, Y. A., et al. (2012). Electroacupuncture enhances motor recovery performance with brain-derived neurotrophic factor expression in rats with cerebral infarction. *Acupuncture in Medicine, 30*, 222–226.

Kim, E. H., Chung, J. H., & Kim, C. J. (2001). Auricular acupuncture increases cell proliferation in the dentate gyrus of Sprague-Dawley rats. *Acupuncture and Electro-Therapeutics Research, 26*, 187–194.

Kim, E. H., Jang, M. H., Shin, M. C., Lim, B. V., Kim, H. B., Kim, Y. J., et al. (2002). Acupuncture increases cell proliferation and neuropeptide Y expression in dentate gyrus of streptozotocin-induced diabetic rats. *Neuroscience Letters, 327*, 33–36.

Kim, E. H., Kim, Y. J., Lee, H. J., Huh, Y., Chung, J. H., Seo, J. C., et al. (2001). Acupuncture increases cell proliferation in dentate gyrus after transient global ischemia in gerbils. *Neuroscience Letters, 297*, 21–24.

Kornack, D. R., & Rakic, P. (1999). Continuation of neurogenesis in the hippocampus of the adult macaque monkey. *Proceedings of the National Academy of Sciences of the United States of America, 96*, 5768–5773.

Kuhn, H. G., Dickinson-Anson, H., & Gage, F. H. (1996). Neurogenesis in the dentate gyrus of the adult rat: Age-related decrease of neuronal progenitor proliferation. *The Journal of Neuroscience, 16*, 2027–2033.

Lam, Y. C., Kum, W. F., Durairajan, S. S., Lu, J. H., Man, S. C., Xu, M., et al. (2008). Efficacy and safety of acupuncture for idiopathic Parkinson's disease: A systematic review. *Journal of Alternative and Complementary Medicine, 14*, 663–671.

Lee, B. C., Eom, K. H., & Soh, K. S. (2010). Primo-vessels and primo-nodes in rat brain, spine and sciatic nerve. *Journal of Acupuncture and Meridian Studies, 3*, 111–115.

Lee, B. C., Kim, S., & Soh, K. S. (2008). Novel anatomic structures in the brain and spinal cord of rabbit that may belong to the Bonghan system of potential acupuncture meridians. *Journal of Acupuncture and Meridian Studies, 1*, 29–35.

Lee, B., Shim, I., Lee, H. J., Yang, Y., & Hahm, D. H. (2009). Effects of acupuncture on chronic corticosterone-induced depression-like behavior and expression of neuropeptide Y in the rats. *Neuroscience Letters, 453*, 151–156.

Lee, M. S., Shin, B. C., & Ernst, E. (2009). Acupuncture for Alzheimer's disease: A systematic review. *International Journal of Clinical Practice, 63*, 874–879.

Lee, S., Shin, K., & Kim, J. (2004). Effect of Seven Points of CVA Acupuncture on cerebral blood flow. *Journal of Korean Acupuncture and Moxibustion Society, 21*, 83–98.

Lee, B. C., & Soh, K. S. (2010). Visualization of acupuncture meridians in the hypodermis of rat using Trypan blue. *Journal of Acupuncture and Meridian Studies, 3*, 49–52.

Leker, R. R., Lasri, V., & Chernoguz, D. (2009). Growth factors improve neurogenesis and outcome after focal cerebral ischemia. *Journal of Neural Transmission, 116*, 1397–1402.

Leker, R. R., Soldner, F., Velasco, I., Gavin, D. K., Androutsellis-Theotokis, A., & McKay, R. D. (2007). Long-lasting regeneration after ischemia in the cerebral cortex. *Stroke, 38*, 153–161.

Li, B., Piao, C. S., Liu, X. Y., Guo, W. P., Xue, Y. Q., Duan, W. M., et al. (2010). Brain self-protection: The role of endogenous neural progenitor cells in adult brain after cerebral cortical ischemia. *Brain Research, 1327*, 91–102.

Li, W. L., Yu, S. P., Ogle, M. E., Ding, X. S., & Wei, L. (2008). Enhanced neurogenesis and cell migration following focal ischemia and peripheral stimulation in mice. *Developmental Neurobiology, 68*, 1474–1486.

Liang, X. B., Luo, Y., Liu, X. Y., Lu, J., Li, F. Q., Wang, Q., et al. (2003). Electro-acupuncture improves behavior and upregulates GDNF mRNA in MFB transected rats. *NeuroReport, 14*, 1177–1181.

Liu, Q., Yu, J., Mi, W. L., Mao-Ying, Q. L., Yang, R., Wang, Y. Q., et al. (2007). Electroacupuncture attenuates the decrease of hippocampal progenitor cell proliferation in the adult rats exposed to chronic unpredictable stress. *Life Sciences, 81*, 1489–1495.

Lledo, P. M., Alonso, M., & Grubb, M. S. (2006). Adult neurogenesis and functional plasticity in neuronal circuits. *Nature Reviews Neuroscience, 7*, 179–193.

Manni, L., Albanesi, M., Guaragna, M., Barbaro Paparo, S., & Aloe, L. (2010). Neurotrophins and acupuncture. *Autonomic Neuroscience: Basic and Clinical, 157*, 9–17.

Marxreiter, F., Regensburger, M., & Winkler, J. (2013). Adult neurogenesis in Parkinson's disease. *Cellular and Molecular Life Sciences, 70*, 459–473.

Ming, G. L., & Song, H. (2005). Adult neurogenesis in the mammalian central nervous system. *Annual Review of Neuroscience, 28*, 223–250.

Munnik, W. A. d. (2011). *The role of neurogenesis in stroke recovery, Master's thesis. Neuroscience & cognition.* Utrecht, The Netherlands: Utrecht University.

Nam, M. H., Yin, C. S., Soh, K. S., & Choi, S. H. (2011). Adult neurogenesis and acupuncture stimulation at ST36. *Journal of Acupuncture and Meridian Studies, 4*, 153–158.

Oh, Y. I., Yang, E. J., Choi, S. M., & Kang, C. W. (2012). The effect of electroacupuncture on insulin-like growth factor-I and oxidative stress in an animal model of renal failure-induced hypertension. *Kidney and Blood Pressure Research, 35*, 634–643.

O'Sullivan, S. S., Johnson, M., Williams, D. R., Revesz, T., Holton, J. L., Lees, A. J., et al. (2011). The effect of drug treatment on neurogenesis in Parkinson's disease. *Movement Disorders, 26*, 45–50.

Ou, Y. W., Han, L., Da, C. D., Huang, Y. L., & Cheng, J. S. (2001). Influence of acupuncture upon expressing levels of basic fibroblast growth factor in rat brain following focal cerebral ischemia—Evaluated by time-resolved fluorescence immunoassay. *Neurological Research, 23*, 47–50.

Pang, J., Itano, T., Sumitani, K., Negi, T., & Miyamoto, O. (2003). Electroacupuncture attenuates both glutamate release and hyperemia after transient ischemia in gerbils. *The American Journal of Chinese Medicine, 31*, 295–303.

Parent, J. M., Vexler, Z. S., Gong, C., Derugin, N., & Ferriero, D. M. (2002). Rat forebrain neurogenesis and striatal neuron replacement after focal stroke. *Annals of Neurology, 52*, 802–813.

Park, H. J., Lim, S., Lee, H. S., Lee, H. J., Yoo, Y. M., Lee, H. J., et al. (2002). Acupuncture enhances cell proliferation in dentate gyrus of maternally-separated rats. *Neuroscience Letters, 319*, 153–156.

Ratajczak, M. Z., Zuba-Surma, E. K., Shin, D. M., Ratajczak, J., & Kucia, M. (2008). Very small embryonic-like (VSEL) stem cells in adult organs and their potential role in rejuvenation of tissues and longevity. *Experimental Gerontology, 43*, 1009–1017.

Rossi, C., Angelucci, A., Costantin, L., Braschi, C., Mazzantini, M., Babbini, F., et al. (2006). Brain-derived neurotrophic factor (BDNF) is required for the enhancement of hippocampal neurogenesis following environmental enrichment. *European Journal of Neuroscience, 24*, 1850–1856.

Soh, K. S. (2009). Bonghan circulatory system as an extension of acupuncture meridians. *Journal of Acupuncture and Meridian Studies, 2*, 93–106.

Sun, Z., Jia, J., Gong, X., Jia, Y., Deng, J., Wang, X., et al. (2012). Inhibition of glutamate and acetylcholine release in behavioral improvement induced by electroacupuncture in parkinsonian rats. *Neuroscience Letters, 520*, 32–37.

Sun, Y., Jin, K., Xie, L., Childs, J., Mao, X. O., Logvinova, A., et al. (2003). VEGF-induced neuroprotection, neurogenesis, and angiogenesis after focal cerebral ischemia. *The Journal of Clinical Investigation, 111*, 1843–1851.

Tao, J., Xue, X. H., Chen, L. D., Yang, S. L., Jiang, M., Gao, Y. L., et al. (2010). Electroacupuncture improves neurological deficits and enhances proliferation and differentiation of endogenous nerve stem cells in rats with focal cerebral ischemia. *Neurological Research, 32*, 198–204.

Taupin, P., & Gage, F. H. (2002). Adult neurogenesis and neural stem cells of the central nervous system in mammals. *Journal of Neuroscience Research, 69*, 745–749.

Teramoto, T., Qiu, J., Plumier, J. C., & Moskowitz, M. A. (2003). EGF amplifies the replacement of parvalbumin-expressing striatal interneurons after ischemia. *The Journal of Clinical Investigation, 111*, 1125–1132.

Thored, P., Arvidsson, A., Cacci, E., Ahlenius, H., Kallur, T., Darsalia, V., et al. (2006). Persistent production of neurons from adult brain stem cells during recovery after stroke. *Stem Cells, 24*, 739–747.

van Praag, H., Schinder, A. F., Christie, B. R., Toni, N., Palmer, T. D., & Gage, F. H. (2002). Functional neurogenesis in the adult hippocampus. *Nature, 415*, 1030–1034.

Wang, Z. L., Cheng, S. M., Ma, M. M., Ma, Y. P., Yang, J. P., Xu, G. L., et al. (2008). Intranasally delivered bFGF enhances neurogenesis in adult rats following cerebral ischemia. *Neuroscience Letters, 446*, 30–35.

Wang, X., Mao, X., Xie, L., Greenberg, D. A., & Jin, K. (2009). Involvement of Notch1 signaling in neurogenesis in the subventricular zone of normal and ischemic rat brain in vivo. *Journal of Cerebral Blood Flow & Metabolism, 29*, 1644–1654.

Wang, Z., Nie, B., Li, D., Zhao, Z., Han, Y., Song, H., et al. (2012). Effect of acupuncture in mild cognitive impairment and Alzheimer disease: A functional MRI study. *PLoS One, 7*, e42730.

Wang, H., Pan, Y., Xue, B., Wang, X., Zhao, F., Jia, J., et al. (2011). The antioxidative effect of electro-acupuncture in a mouse model of Parkinson's disease. *PLoS One, 6*, e19790.

Winner, B., Kohl, Z., & Gage, F. H. (2011). Neurodegenerative disease and adult neurogenesis. *European Journal of Neuroscience, 33*, 1139–1151.

Winner, B., Lie, D. C., Rockenstein, E., Aigner, R., Aigner, L., Masliah, E., et al. (2004). Human wild-type alpha-synuclein impairs neurogenesis. *Journal of Neuropathology and Experimental Neurology, 63*, 1155–1166.

Winner, B., Rockenstein, E., Lie, D. C., Aigner, R., Mante, M., Bogdahn, U., et al. (2008). Mutant alpha-synuclein exacerbates age-related decrease of neurogenesis. *Neurobiology of Aging, 29*, 913–925.

Wu, P., Mills, E., Moher, D., & Seely, D. (2010). Acupuncture in poststroke rehabilitation: A systematic review and meta-analysis of randomized trials. *Stroke, 41*, e171–e179.

Xia, Y., Wang, H. D., Ding, Y., Kang, B., & Liu, W. G. (2012). Parkinson's disease combined with depression treated with electroacupuncture and medication and its effect on serum BDNF. *Zhongguo Zhen Jiu, 32*, 1071–1074.

Xu, S., Hou, X., Zha, H., Gao, Z., Zhang, Y., & Chen, J. D. (2006). Electroacupuncture accelerates solid gastric emptying and improves dyspeptic symptoms in patients with functional dyspepsia. *Digestive Diseases and Sciences, 51*, 2154–2159.

Yang, Z. J., Shen, D. H., Guo, X., & Sun, F. Y. (2005). Electroacupuncture enhances striatal neurogenesis in adult rat brains after a transient cerebral middle artery occlusion. *Acupuncture and Electro-Therapeutics Research, 30*, 185–199.

Yang, Z., Yu, H., Rao, X., Liu, Y., & Pi, M. (2008). Effects of electroacupuncture at the conception vessel on proliferation and differentiation of nerve stem cells in the inferior zone of the lateral ventricle in cerebral ischemia rats. *Journal of Traditional Chinese Medicine, 28*, 58–63.

Ying, S., & Cheng, J. (1994). Effects of electro-acupuncture on EEG during transient global ischemia and reperfusion in gerbils. *Zhen Ci Yan Jiu, 19*, 29–32.

Yuan, M., Wen, S. J., Yang, C. X., Pang, Y. G., Gao, X. Q., Liu, X. Q., et al. (2013). Transplantation of neural stem cells overexpressing glial cell line-derived neurotrophic factor enhances Akt and Erk1/2 signaling and neurogenesis in rats after stroke. *Chinese Medical Journal, 126*, 1302–1309.

Yun, S. J., Park, H. J., Yeom, M. J., Hahm, D. H., Lee, H. J., & Lee, E. H. (2002). Effect of electroacupuncture on the stress-induced changes in brain-derived neurotrophic factor expression in rat hippocampus. *Neuroscience Letters, 318*, 85–88.

Zhang, R. L., & Chopp, M. (2011). Neurogenesis in stroke treatment. *European Neurological Review, 6*, 246–248.

Zhang, P., Guan, S. S., & Jiang, G. H. (2010). Effects of electroacupuncture on expression of Abeta positive cells of the hippocampus and SOD activity in rats with streptozocin-Alzheimer's disease. *Zhongguo Zhen Jiu, 30*, 1007–1010.

Zhang, R. L., Zhang, Z. G., Zhang, L., & Chopp, M. (2001). Proliferation and differentiation of progenitor cells in the cortex and the subventricular zone in the adult rat after focal cerebral ischemia. *Neuroscience, 105*, 33–41.

Zhao, C., Deng, W., & Gage, F. H. (2008). Mechanisms and functional implications of adult neurogenesis. *Cell, 132*, 645–660.

Zhou, Y., & Jin, J. (2008). Effect of acupuncture given at the HT 7, ST 36, ST 40 and KI 3 acupoints on various parts of the brains of Alzheimer's disease patients. *Acupuncture and Electro-Therapeutics Research*, *33*, 9–17.

Zhu, D. Y., Lau, L., Liu, S. H., Wei, J. S., & Lu, Y. M. (2004). Activation of cAMP-response-element-binding protein (CREB) after focal cerebral ischemia stimulates neurogenesis in the adult dentate gyrus. *Proceedings of the National Academy of Sciences of the United States of America*, *101*, 9453–9457.

CHAPTER FIVE

Acupuncture and Neurotrophin Modulation

Marzia Soligo*, Stefania Lucia Nori†, Virginia Protto*, Fulvio Florenzano‡, Luigi Manni*,1

*Institute of Translational Pharmacology, National Research Council (CNR), Rome, Italy
†Department of Pharmaceutical and Biomedical Sciences (FARMABIOMED) Nanomates, University of Salerno, Fisciano, Italy
‡Confocal Microscopy Unit, European Brain Research Institute (EBRI)—Institute of Cellular Biology and Neurobiology, National Research Council (CNR), Rome, Italy
[1]Corresponding author: e-mail address: luigi.manni@ift.cnr.it

Contents

Abstract

The Western explanation for acupuncture effectiveness is based upon more than half a century of basic and clinical research, which identified the activation of sensory system and the subsequent activity-dependent regulation of neurotransmitters, neurohormones, and several classes of neuromodulators as plausible mechanism for the acupuncture's therapeutic properties. The regulation of neurotrophins' expression and activity is one of the possible neurophysiological mechanisms underlying acupuncture's effects on neuropathic pain, nerve injury, neurodegeneration, and even in the regulation of gonadal functions. The present work will review the scientific literature produced by a decade of investigations on the relationship between acupuncture and neurotrophins. This scientific production and the current knowledge about the neural and neurohormonal activity-dependent regulation of neurotrophin expression/action suggest the existence of a link between the ability of acupuncture in regulating neural physiology and its effects on the neurotrophic milieu in different disease states.

International Review of Neurobiology, Volume 111
ISSN 0074-7742
http://dx.doi.org/10.1016/B978-0-12-411545-3.00005-5

1. INTRODUCTION

1.1. Neurotrophins

The neurotrophins (NTs) are a family of growth and survival factors regulating the development and the maintenance of functional phenotype of neuronal cells (Ernfors, 2001; McAllister, 2001; Poo, 2001; Rush et al., 1997). NTs are indeed capable of supporting neurogenesis, cell migration, and differentiation in developing nervous system as well as neuronal survival, synaptogenesis, and synaptic plasticity in adults (Aloe, Rocco, Bianchi, & Manni, 2012; Chao, Rajagopal, & Lee, 2006; Hempstead, 2006; McAllister, 2001; Poo, 2001). NTs have also been proposed as potential therapeutics for the regeneration of damaged axons after peripheral and central nervous system injuries (Chao et al., 2006; Connor & Dragunow, 1998; Olson, 1993; Terenghi, 1999) and the protection of neurons from sufferance and damages occurring during the development and progression of neurodegenerative diseases (Allen & Dawbarn, 2006; Aloe et al., 2012; Semkova & Krieglstein, 1999). Nerve growth factor (NGF), brain-derived neurotrophic factor (BDNF), neurotrophin 3 (NT-3), and neurotrophin 4 (NT-4) constitute the mammalian NTs, all descending from a common NT ancestor gene (Hallbook, 1999), each with distinct and/or overlapping activities within the developing and mature peripheral and central nervous systems (Barde, 1990).

The NGF is the first discovered NT (Cohen, Levi-Montalcini, & Hamburger, 1954; Levi-Montalcini, 1952). Together with BDNF, NGF is undoubtedly the most studied among NTs. The NGF exerts its trophic function on NGF-responsive neurons both during development and in adulthood. Indeed, NGF is essential for the development and maintenance of neurons in the peripheral nervous system (PNS) and for the functional integrity of specific neurons in the central nervous system (CNS) (Alleva, Aloe, & Bigi, 1993; Aloe, Bracci-Laudiero, Bonini, & Manni, 1997; McAllister, 2001; Rush et al., 1997). For over 35 years, NGF has been considered a very powerful and selective growth factor for sympathetic and sensory neurons and for cells deriving from the neuronal crest (Cowan, 2001; Rush et al., 1997), until its presence has been detected in the CNS, mostly in the circuitry linking the basal forebrain cholinergic nuclei to the cortex and hippocampus (Seiler & Schwab, 1984; Shelton & Reichardt, 1986). Today, we know that in the CNS, the NGF is mainly produced in the cortex, the hippocampus and the pituitary gland, the basal ganglia, the thalamus, the spinal cord, and the retina (McAllister, 2001). The functional role of NGF varies depending on the selected brain areas. NGF regulates the functional integrity of the cholinergic basal forebrain nuclei (BFN) with their main innervation targets in the cortex and hippocampus (Fischer et al., 1987; Kawaja, Walsh, Tovich, & Julien, 1998). It has also been demonstrated that the NT is present in the hypothalamus, where NGF synthesis is regulated by stressful psychosocial events (Aloe, Alleva, & Fiore, 2002; Aloe, Tirassa, & Alleva, 1994; Maestripieri, De Simone, Aloe, & Alleva, 1990; Taglialatela, Angelucci, Scaccianoce, Foreman, & Perez-Polo, 1991), having a modulatory role in the hypothalamus–pituitary–adrenal (HPA) axis function (Aloe et al., 2002). NGF also exerts its action on differentiation and function of several kinds of nonneuronal cells. Indeed, NGF regulates the functions of the immune-hematopoietic system (Aloe, 2001; Aloe et al., 1997) and of the neuroendocrine system (Aloe et al., 1997). The current view depicts NGF as a main regulator of the balanced interplay between the nervous, endocrine, and immune systems (Alleva et al., 1993; Aloe et al., 2002).

BDNF is found in a wide range of tissues, in the CNS, PNS, and peripheral tissues (Mufson, Kroin, Sendera, & Sobreviela, 1999). It is active in the hippocampus, cortex, and basal forebrain, where it plays a crucial role in the process of learning and memory (Mufson et al., 1999). BDNF knockout mice suffer developmental defects in the brain and sensory nervous system and usually die soon after birth, suggesting an important role of BDNF in normal neural development (Bartoletti et al., 2002). Deregulation of the brain BDNF seems to be of relevance for the development of conditions such as depression, schizophrenia, Alzheimer's disease (AD), Huntington's disease, Rett syndrome, and dementia (Allen & Dawbarn, 2006). Stress and the stress hormone corticosterone have been shown to reduce brain BDNF levels in rats (Smith, 1996), leading to hippocampus atrophy, a condition always described in humans suffering from chronic depression (Sapolsky, 2000). BDNF plays a crucial role in the regulation of activity-dependent synaptic plasticity and in the process of hippocampus-dependent learning and memory consolidation (Soule, Messaoudi, & Bramham, 2006). Indeed, long-term potentiation (LTP) is greatly impaired in the hippocampus after disruption of the BDNF gene (Korte et al., 1995). Unlike NGF, BDNF is released in activity-dependent manner in the synaptic cleft, directly modulating the behavior of the postsynaptic side (Gottmann, Mittmann, & Lessmann, 2009; Soule et al., 2006). Thus, BDNF release in response to synaptic stimulation could act as a decisive factor for the conversion of high-frequency synaptic activity into long-term synaptic memories, especially at glutamatergic and GABAergic synapses (Gottmann et al., 2009; Soule et al., 2006).

NT-3 has a critical role in the development of the PNS and CNS and is involved in the maintenance of the adult nervous system. NT-3 is the first NT to be expressed in the PNS through embryogenesis and can sustain either mitogenesis or the exit of neuronal progenitors from the cell cycle (Chalazonitis, 1996). NT-3 supports the survival and the differentiation of sensory neurons, which become dependent on NGF during later stages of development (Lewin & Barde, 1996). NT-3 also shows the highest expression levels in the CNS during the perinatal development and is mostly expressed in the hippocampus, the neocortex, and the cerebellum (Zhou & Rush, 1994). In adults, expression levels of NT-3 and BDNF are comparable in most areas of the brain (Ernfors, Wetmore, Olson, & Persson, 1990; Hochhaus et al., 2001; Maisonpierre et al., 1990) and in the skin of diabetic patients (Apfel, 1999a). Preclinical and clinical trials suggest its possible pharmacological use in the treatment of Charcot–Marie–Tooth disease type 1A (Sahenk et al., 2005) and in functional constipation (Coulie et al., 2000).

NT-4 is the last NT to be discovered (Hallbook, Ibanez, & Persson, 1991; Ip et al., 1992). NT-4 is expressed in the postnatal and adult hippocampus, neocortex, and cerebellum, and thalamic nuclei (Friedman, Black, & Kaplan, 1998). Trophic effects of NT-4 have been described on cultures of noradrenergic neurons from locus coeruleus (Friedman et al., 1993), suggesting a role for NT-4 in the central regulation of autonomic activity. NT-4 is also produced in rat skeletal muscle in an activity-dependent way, suggesting its active involvement in the growth and remodeling of adult motor neuron innervation (Funakoshi et al., 1995). However, while all other NT knockouts have proven lethal during early postnatal development, mice lacking NT-4 only display minor deficits and develop normally to adulthood (Ibanez, 1996).

NTs are all synthesized as large precursor proteins and then cleaved to mature molecules (Teng, Felice, Kim, & Hempstead, 2010). Both the mature and the precursor form of NTs (ProNT) are physiologically active and have important roles during development and in adult life (Schweigreiter, 2006). ProNTs are synthesized in the endoplasmic reticulum and directed toward the extracellular space via both the constitutive and the activity-dependent secretory pathways (Hempstead, 2006). The cleavage to mature forms can occur in the trans-Golgi network, and executed by furin protease, or in the secretory vesicles, by proconvertases (Seidah, Benjannet, Pareek, Chretien, & Murphy, 1996; Seidah, Benjannet, Pareek, Savaria, et al., 1996). ProNTs can be also be processed to their mature forms after their release by extracellular proteinases like plasmin and metalloproteinases (Lee, Kermani, Teng, & Hempstead, 2001) (Fig. 5.1). The mature NTs exert

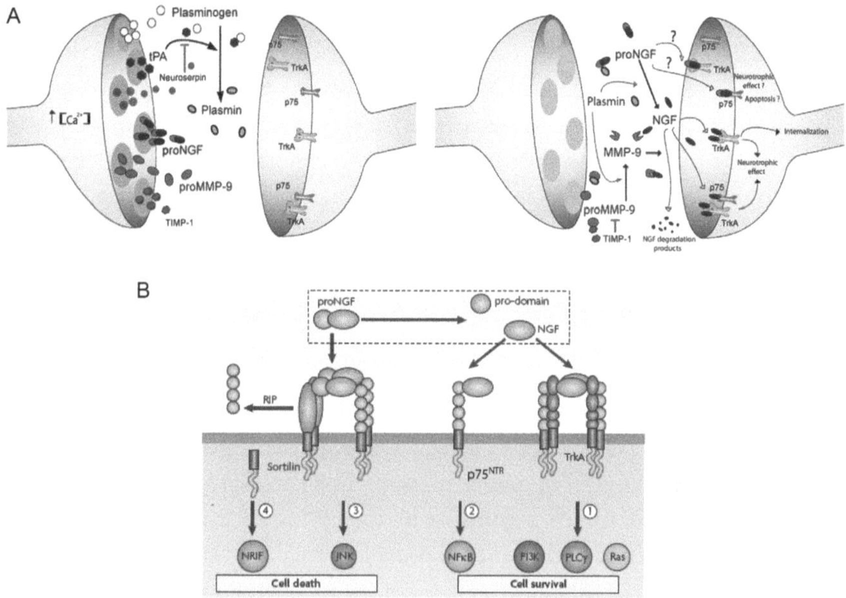

Figure 5.1 (A) Schematic representations of events leading to ProNT, in this example proNGF, conversion into mature NTs and their degradation. Neuronally stored proNGF, plasminogen, tPA, neuroserpin, proMMP-9, and TIMP-1 would be released into the extracellular space upon neuronal stimulation. Released tPA would induce the conversion of plasminogen to plasmin, where its activity is tightly regulated by secreted neuroserpin. The generated plasmin would convert proNGF into mature NGF and activate proMMP-9 into active MMP-9. Mature NGF would interact with its cognate receptors or suffer degradation by activated MMP-9. (B) The binding of mature NGF to the receptor complex TrkA/p75^{NTR} induces receptor dimerization and TrkA transphosphorylation and stimulates three major signaling pathways: the phosphatidylinositol 3-kinase (PI3K)–protein kinase B pathway, the phospholipase Cγ (PLCγ) pathway, and the Ras–mitogen-activated protein kinase pathway. p75^{NTR} also facilitates the binding of BDNF and NT4 to TrkB and that of NT3 to TrkC (not shown). p75^{NTR} also activates nuclear factor-κB (NF-κB) pathways, independent of Trk. Whereas all of the previously mentioned pathways promote cell survival, binding of proNGF to sortilin–p75^{NTR} complexes activates cell death through the stimulation of Jun N-terminal kinase (JNK) or through ligand-dependent regulated intramembrane proteolysis (RIP) of p75^{NTR}. Such stimulation of RIP results in release of the cytoplasmic tail of the receptor followed by nuclear translocation of the p75^{NTR} adaptor neurotrophin receptor-interacting factor (NRIF). *(A) Reprinted from Bruno and Cuello (2006), copyright (2006) the National Academy of Sciences, USA. (B) Reprinted from Willnow, Petersen, and Nykjaer (2008) by permission from Macmillan Publishers Ltd.:* Nature Review Neuroscience, *copyright (2008).* (For color version of this figure, the reader is referred to the online version of this chapter.)

their biological action by challenging specific tropomyosin kinase (Trk) receptors: TrkA is specific for NGF, TrkB for BDNF and NT-4, and TrkC for NT-3 (Huang & Reichardt, 2003). Stimulation of Trks is promoted by Trk forming a complex with the p75 pan-neurotrophin receptor ($p75^{NTR}$), which is a member of the tumor necrosis factor (TNF) receptor superfamily (Friedman & Greene, 1999; Reichardt, 2006; Schor, 2005). The major intracellular signaling pathways activated by the Trk/$p75^{NTR}$ complex are Ras–mitogen-activated protein kinase (MAPK), extracellular signal-regulated kinase (ERK), phosphatidylinositol 3-kinase (PI3K)-Akt, and phospholipase C (PLC)-γ (Chao et al., 2006; Klesse & Parada, 1999; Reichardt, 2006) (Fig. 5.1). The binding of NTs to $p75^{NTR}$ alone could activate nuclear factor-κB (NF-κB) pathways, independent of Trks (Mamidipudi & Wooten, 2002). All of the previously mentioned pathways promote cell survival and/or differentiation. The ProNTs, on the other hand, bind to a receptor complex composed by the sortilin that is a member of the Vps10p-domain family of transmembrane receptors and the receptor $p75^{NTR}$ (Nykjaer et al., 2004), activating cell death through the stimulation of the Jun N-terminal kinase (JNK) (Teng et al., 2010) or by the ligand-dependent regulated intramembrane proteolysis (RIP) of $p75^{NTR}$ (Skeldal, Matusica, Nykjaer, & Coulson, 2011; Willnow et al., 2008) (Fig. 5.1). Such stimulation of RIP results in release of the cytoplasmic tail of the receptor followed by nuclear translocation of the $p75^{NTR}$ adaptor neurotrophin receptor-interacting factor (NRIF). This complex picture points to the physiological activity of the ProNT/NT system as a result of several events regulating the NT production by cells that are targets of innervation for the NT-receptive neurons; the amount and characteristic of NT secretion via both the constitutive and the activity-dependent secretory pathways (Hempstead, 2006; Schweigreiter, 2006); the process of ProNT maturation and eventually the subsequent mature NT degradation by extracellular proteases (Bruno & Cuello, 2006); and the modulation of Trk, $p75^{NTR}$, and sortilin receptor expression and activity (Teng et al., 2010; Terry, Kutiyanawalla, & Pillai, 2011).

1.2. Acupuncture: Western explanation for the mechanism of action

Acupuncture is a therapeutic technique, part of the traditional Chinese medicine (TCM). Founded on philosophical and empirical basis, TCM has been used in the treatment and prevention of diseases for centuries in China and more recently also in Western countries. The scientific research on acupuncture extensively explored the possible physiological correlates of this ancient technique, and actually both the NIH (Acupuncture: National Institutes of Health Consensus Development Conference statement 2000) and the WHO (WHO, 2003) officially recognize the effectiveness of acupuncture for a wide range of pain syndrome and for pathological conditions linked to dysfunctions in the sensory and autonomic nervous system.

Since the early times of scientific research on acupuncture, it has been demonstrated that it represents a potent form of sensory stimulation (Andersson & Lundeberg, 1995). This is the main perspective point from which the Western scientific community has studied the mechanisms of acupuncture treatment, first on pain relief and then in a wide spectrum of diseases, including infections and inflammations, dysfunction of the autonomic nervous system, peripheral and central nervous system diseases, and metabolic disorders (Andersson & Lundeberg, 1995). Needle insertion at the body surface elicits regular patterns of afferent activity in peripheral nerves innervating the skin and deeper tissues (Andersson & Lundeberg, 1995; Kagitani, Uchida, Hotta, & Aikawa, 2005; Ulett, Han, & Han, 1998). The inserted needles are traditionally stimulated by manual rotation (manual acupuncture: MA) but could be also challenged by electrical stimulation, generally referred to as electroacupuncture (EA). It has been reported that MA and EA effects can be correlated to activation of a group of receptors in the skeletal muscles, which have both low- and high-threshold for mechanical stimulation, and are innervated by A-delta fibers and possibly C-fibers (Andersson & Lundeberg, 1995; Kagitani et al., 2005). They sense and convey information to the CNS about strong muscle contractions, and their stimulation by acupuncture results in the activation of physiological processes similar to those resulting from physical exercise (Andersson & Lundeberg, 1995). Polymodal receptors have also been proposed as possible mediators for the effects of acupuncture as well as moxibustion (Kawakita, 1991; Kawakita & Gotoh, 1996). They are free nerve endings receptors characterized by responsiveness to both mechanical and thermal stimulation not necessarily within a range of noxious stimulus level (Kawakita & Gotoh, 1996). The activation of this kind of receptor could explain the effects of different stimulation techniques applied at the same anatomical point over body surface and based on mechanical (acupuncture) or thermal (moxibustion) stimuli. Besides the type of receptors that could be activated by needle insertion and stimulation, it has been proposed that all kinds of type A (alpha, beta, gamma, and delta) and C sensory nerve fibers could participate in the production of regular impulses that, once directed to the spinal cord and the higher centers in the CNS, triggers the neurophysiological events that generate the therapeutic outcome (Fig. 5.2) (Andersson & Lundeberg, 1995; Kagitani et al., 2005; Kimura & Sato, 1997; Lund & Lundeberg, 2006; Lundeberg, Eriksson, Lundeberg, & Thomas, 1989).

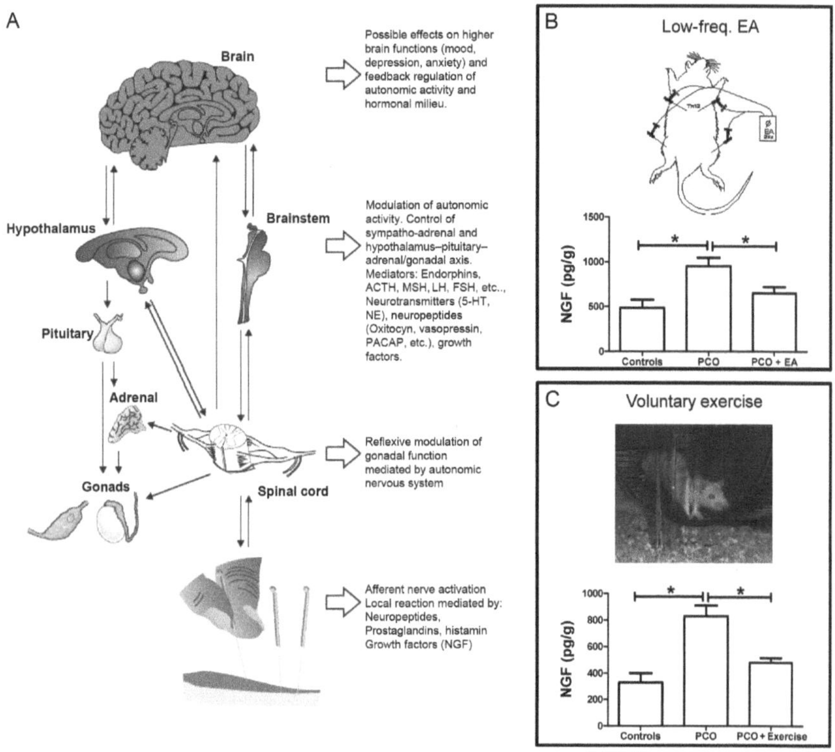

Figure 5.2 Schematic diagram (A) illustrating the neurophysiological mechanisms possibly mediating acupuncture effects. In this diagram, the end organ is the ovary, which has been the subject of investigation in a series of works that studied the effects of NGF blockade by physical therapies, namely, EA and physical exercise, in rats affected by estradiol-induced polycystic ovaries (Manni, Cajander, et al., 2005; Manni, Lundeberg, Holmang, Aloe, & Stener-Victorin, 2005; Stener-Victorin et al., 2003; Stener-Victorin, Lundeberg, et al., 2000). The effects of EA (B) and physical exercise (C) in this experimental model were quite the same. The estradiol treatment (PCO group) increased ovarian NGF with a parallel increase of sympathetic drive to the ovaries and a deregulation of ovarian sympathetic markers (tyrosine hydroxylase: TH) and adrenergic receptors (ARs). Both EA and exercise were able to counteract the effect of estradiol, decreasing ovarian NGF to baseline levels (B and C) and normalizing the ovarian expression of TH and ARs. Data presented in (B) and (C) are means $\pm$ SEM, $^{*}P<0.05$. *(A) Reprinted with modifications from Franconi et al. (2011), copyright (2011) with permission from Editrice Kurtis s.r.l.; (B) parts of this figure were originally published in Stener-Victorin et al. (2003); (C) part of this figure was originally published in L. Manni Ph.D. thesis (ISBN: 91-628-6484-X).* (For color version of this figure, the reader is referred to the online version of this chapter.)

Acupuncture therapeutic efficacy has been firstly based on the release of endogenous opioids (Hole & Berge, 1981; Kaptchuk, 2002; Ulett et al., 1998). Indeed, the identification of β-endorphin as the soluble factor mediating the pain-relieving effects of acupuncture has represented a milestone in the history of acupuncture research (Kaptchuk, 2002). During several decades of research on the neurophysiological correlates of acupuncture, other systems have been identified as mediators for the acupuncture modulation of, among others, pain relief, mood regulation and attenuation of stress-induced HPA axis activation, control of inflammation, central neurogenesis, improvement of epilepsy, amelioration of diabetes symptoms, and induction of peripheral nerve regeneration (Han, 1997; Kaptchuk, 2002). Indeed, by means of MA or EA, it is possible to affect the synthesis, release, and action of several neurotransmitters (catecholamine, glutamate, acetylcholine, GABA, serotonin, and adenosine) and neuropeptides (among others, oxytocin, NPY, CCK, VIP, SP, CGRP, and PACAP) in both the CNS and PNS (Carlsson, 2002; Hole & Berge, 1981; Kaptchuk, 2002; Ulett et al., 1998; Zhao, 2008; Zijlstra, van den Berg-de Lange, Huygen, & Klein, 2003) (Fig. 5.2). In recent times, it has emerged the evidence that the action of acupuncture can be also mediated by NTs.

2. NEUROTROPHINS AND ACUPUNCTURE

All of the components of the NT family of proteins seem to be potential candidates for mediating the biological action of acupuncture and/or EA, and the opportunity to use acupuncture-induced NT modulation in animal models of neurological and neurodegenerative disease has been investigated in the last few years. In this chapter, the literature describing the link between NTs and acupuncture will be reviewed, following an "anatomical path" that starts from peripheral organs and neurons and climbs up toward the spinal cord and the brain (Fig. 5.2). Furthermore, the possible mechanism(s) regulating NT modulation by acupuncture will be analyzed.

2.1. Peripheral field

The first data about the correlation between acupuncture and NTs come from works published in the early 2000s, exploring the effects of EA on NGF and the sympathetic nervous system in the estrogen-induced rat model of polycystic ovary (PCO) (Bai et al., 2004; Manni, Lundeberg, et al., 2005; Stener-Victorin et al., 2003; Stener-Victorin, Lundeberg, et al., 2000). This experimental model is characterized by the development of ovarian cysts, a deep deregulation of estrous cycle associated to an increase of sympathetic drive to the ovary and an augmented ovarian NGF expression (Lara et al., 2000). EA treatments in PCO rats counteract the disease-associated increase of ovarian NGF (Fig. 5.2) (Stener-Victorin, Lundeberg, et al., 2000). The studies also indirectly demonstrated that EA inhibits hyperactivity in the sympathetic nervous system (Cao, Xu, & Lu, 1983), which has been indicated as a major pathogenetic cause in PCO (Lara et al., 2002, 1993). Moreover, the efficacy of EA in reducing sympathetic activity was also demonstrated by its inhibitory effects on ovarian and hypothalamic sympathetic activation markers such as endothelin-1 (ET-1) (Stener-Victorin et al., 2003). Further studies on the effect of EA on brain and ovarian NGF content in estradiol valerate (EV)-induced PCO (Bai et al., 2004) indicated that the EA effects were restricted to the peripheral field, since they reduced ovarian but not hypothalamic NGF overexpression induced by EV. As for the mechanism(s) generating the effects of EA on ovarian NGF expression and activity in experimental PCO, it was shown that an interplay among NGF/NGF receptor system and adrenergic responsiveness characterized the development of ovarian pathology in this rat model (Manni, Cajander, et al., 2005, Manni, Holmang, Lundeberg, Aloe, & Stener-Victorin, 2005; Manni et al., 2006; Manni, Lundeberg, et al., 2005). Indeed, EV injection altered the ovarian content of TrkA and $p75^{NTR}$ and deregulated ovarian α_1- and α_2-adrenergic receptors (ARs). Moreover, it increased the ovarian content of tyrosine hydroxylase (TH) that is the rate-limiting enzyme in noradrenaline biosynthesis and an indirect marker of sympathetic activation. Both EA and the neutralization of endogenous NGF reverted most of the abnormalities in ovarian ARs and NGF receptors (Manni, Cajander, et al., 2005; Manni et al., 2006), demonstrating the sympathetic tone lowering capacity of the technique and that this is most probably achieved through EA action on the NGF system. It is worth noting that very consistent results were also obtained when EV-injected rats where treated by a protocol of five weeks voluntary physical exercise (Fig. 5.2) (Manni, Cajander, et al., 2005). These results constitute a further confirmation that the effects of acupuncture are based on physiological and anatomical substrates that are common with physical exercise and probably deep massage (Andersson & Lundeberg, 1995; Bucinskaite et al., 1994, 1996; Lundeberg, 1984; Lundeberg et al., 1989). More recently, EA was also tested in a model of PCO that more strictly resemble the development and progression of the human PCO syndrome (PCOS). The group of Stener-Victorin at Gothenburg University (Sweden) developed a dihydrotestosterone (DHT)-induced PCO model and further demonstrated that both EA and physical training are effective in improving PCO-related metabolic disturbances (Manneras, Jonsdottir, Holmang, Lonn, & Stener-Victorin, 2008) and alterations in sympathetic markers and ovarian morphology (Manneras, Cajander, Lonn, & Stener-Victorin, 2009), normalizing the DHT-induced increase of NGF mRNA (Manneras et al., 2009). Overall, the previous data, obtained using different models of PCO(S), consistently confirmed that physical therapies, like acupuncture or nonstressful aerobic physical exercise, based on the stimulation of repetitive afferents activity, caused a long-lasting depression of the sympathetic nervous system (Cao et al., 1983; Stener-Victorin, Jedel, Janson, & Sverrisdottir, 2009; Stener-Victorin, Waldenstrom, et al., 2000) and demonstrated that this effect is connected to a downregulation of the NGF organ content.

The link between EA and the NGF in the peripheral field has been also recently investigated in a model of diabetic neuropathy (Manni, Florenzano, & Aloe, 2011; Nori et al., 2013) (Fig. 5.3). These works demonstrate that EA is effective in regulating the activity of and possibly in neuroprotecting peripheral sensory neurons. Using the streptozotocin (STZ) model of diabetes, the authors confirmed that the early development of diabetic sensory neuropathy is associated with thermal hyperalgesia (Manni et al., 2011) and demonstrated that this is associated with increased NGF protein content in both the peripheral and central fields of innervation of the DRG neurons, the paw skin and spinal cord, respectively; increased expression of $p75^{NTR}$ in diabetic skin (Fig. 5.3); and increased TrkA in the spinal cord (Manni et al., 2011; Nori et al., 2013). Moreover, the STZ also increased NGF and NGF receptors expression, activated c-Jun N-terminal kinase (JNK) (Fig. 5.3) and p38 kinase, and increased the transient receptor-potential vanilloid receptor-1 (TRPV1) in DRGs (Nori et al., 2013). When low-frequency EA was applied in diabetic rats for three weeks starting one week after diabetes induction, it normalized neuropathic symptoms, decreased both NGF and NGF receptors, normalized JNK and p38 activation, decreased TRPV1, and activated the transcription factor NF-κB in the DRGs (Manni et al., 2011; Nori et al., 2013) (Fig. 5.3). Overall, these data indicate that the NGF could be primarily involved in the establishment of diabetic sensory neuropathy and that EA effects, being mediated through the modulation of NGF signaling, could have not only an effect on sensory functions but also a neuroprotective value for sensory neurons undergoing the first stage of diabetes-induced degeneration (Nori et al., 2013).

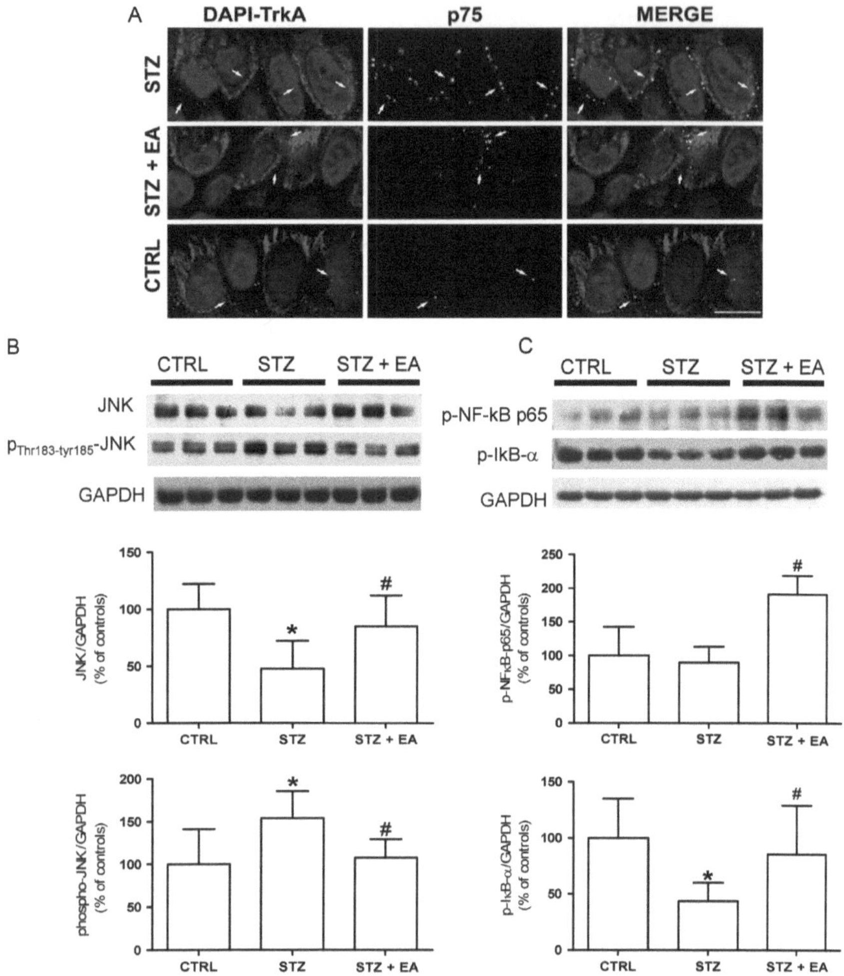

Figure 5.3 Electroacupuncture modulates NTs receptor p75^{NTR} in the skin and p75^{NTR} signaling in dorsal root ganglia (DRG) of diabetic rats. (A) Confocal microscopy of TrkA and p75^{NTR} double immunofluorescence in the in the basal epidermal layer. First column, DAPI/TrKA (blue and red); second column, p75^{NTR} (green); and third column, DAPI/TrKA/p75^{NTR} (MERGE). The amount of p75^{NTR} cytoplasmatic immunoreactive vesicles increases in the basal epidermal layer of the STZ group (first row, arrows) on very low fluorescence background. After EA treatment p75^{NTR} cytoplasmatic immunoreactive vesicles were decreased (second row, arrows), while in the in the basal epidermal layer of the control group p75^{NTR}, only few immunoreactive vesicles were detected (third row, arrows). TrkA did not show variations in the three experimental groups. Scale bar: 8 μm. (B, C) Variations of downstream p75^{NTR} signaling after STZ and EA in DRGs. Representative Western blot of three samples for each experimental group is presented

Finally, it has also been reported that the anti-inflammatory properties of acupuncture are associated with the acupuncture-induced variation of NT peripheral and spinal content. By means of a carrageenan-induced model of acute paw edema, it was indeed demonstrated that acupuncture inhibition of local inflammatory response was parallel to a decrease of tissue NGF (Chae et al., 2007). More recently, it was also demonstrated that EA's analgesic effects in a model of peripheral inflammatory pain, induced by complete Freund's adjuvant, were mediated by upregulation of spinal NT-3, which was in turn responsible for the EA-induced suppression of glial activation and control over inflammatory cytokine overexpression (Mi et al., 2011).

2.2. Spinal cord

The NT modulation exerted by EA in the peripheral field seems to take place not only in target organs or in PNS neurons but also in CNS structures directly or closely wired with PNS neurons, such as in the spinal cord. By means of an experimental model of spinal neuroplasticity, it has been demonstrated that the NGF mRNA and a significantly higher number of NGF, BDNF, and NT-3-positive neurons were observed in the spinal cord after partial dorsal rhizotomy and high-frequency EA (Wang, Wang, Li, Chen, & Wu, 2007), suggesting that NTs are involved in the EA-induced spinal plasticity after injury. The observation that EA is also able to increase both NGF, BDNF, NT-3 (Chen et al., 2007), and NT-4 (Liu et al., 2009) contents in

in each panel, together with data from densitometry analysis of three separate gel/blot runs ($n=9$). The JNK, which is known to be part of the $p75^{NTR}$ downstream signaling machinery (B), was decreased 4 weeks after STZ, while $phospho_{Thr183\text{-}Tyr185}$-JNK (p-JNK) was increased. EA normalized both variations. The presence of the $p75^{NTR}$ downstream signaling molecule, phosphorylated NF-κB-p65 complex—representing an index of the nuclear translocation activity of the transcription factor NF-κB—was unaffected by STZ (C), while EA greatly enhanced phospho-NF-κB-p65 presence in DRGs of diabetic rats, suggesting an augmented activity of the factor. Coherently, STZ lowered the phosphorylation of the IκB-α below controls level, suggesting an increase of its repressive activity upon NF-κB (C); EA in diabetic rats counteracted such effect, significantly improving IκB-α phosphorylation versus STZ group, further indicating a decreased repression of NF-κB activity induced by EA. Data presented in (B) and (C) are percent variations from the mean of Control group, obtained after normalization with GAPDH band integrated optical density. Data are expressed as % of the mean of controls ($n=9$) ± SD; $^{*}P<0.05$ versus control group. $^{\#}P<0.05$ versus STZ group. *(A–C) Reprinted with modifications from Nori et al. (2013), copyright (2013) to the authors under the terms of the Creative Commons Attribution License.* (For interpretation of the references to color in this figure legend, the reader is referred to the online version of this chapter.)

spared L6 DRG neurons after removal of adjacent DRGs extends to primary sensory neurons the hypothesis of NT-mediated effect of EA on damaged neuron plasticity and suggests an involvement of the entire NT family of molecules in mediating EA-induced spinal cord plasticity after peripheral nerve lesion.

The effects of EA on the survival and migration of stem cells transplanted in injured spinal cord in rats have also been studied and correlated with the spinal and serum NT content, in several works published in the last few years. A mixed-frequency, with alternating strings of dense-sparse 2 and 60 Hz frequencies, in EA treatment was applied in a work correlating spinal NT-3 content and the survival and migration of neural stem cells (NSC) transplanted in injured spinal cord injury in rats after EA (Chen et al., 2008). An increased number of survived NSC and an improvement in their migration length toward caudal tissue were found and correlated to an EA-induced increase of spinal NT-3 content in spinally transected rats (Chen et al., 2008). The same experimental model and EA stimulation protocol were further applied and mesenchymal stem cells (MSCs) transplanted in spinal cord transected animals (Ding et al., 2009; Yan et al., 2011) (Fig. 5.4). The results indicated that EA treatment may promote grafted MSC survival and differentiation and promote axonal regeneration and partial locomotor functional recovery as well as increase of spinal NT-3 in the transected spinal cord in rats (Ding et al., 2009; Yan et al., 2011).

Of relevance is another study performed on human healthy volunteer, demonstrating that after manual acupuncture, applied following a protocol used in spinal cord-injured patients, the presence of a class of endogenous stem cells, namely, CD133+CD34−, representing endothelial progenitors with neurogenic potential, in the serum was doubled (Moldenhauer et al., 2010). This effect on mobilization of stem cells correlated with decreased serum levels of BDNF and no changes in basal serum NGF. Expecting an increase rather than a decrease in NTs after acupuncture, the authors speculated over the need to perform quantitative NT assays over cerebrospinal fluid instead of serum, but did not give any further explanation about this apparently controversial finding (Moldenhauer et al., 2010).

EA was also applied in a model of spinal demyelination induced by ethidium bromide (Huang et al., 2011) (Fig. 5.4). By this model, the authors aimed at studying the proliferation and differentiation of oligodendrocyte precursor cells (OPCs) that could represent a tool in the treatment of multiple sclerosis. The results demonstrated that, especially with long duration EA treatment (30 days), there was an EA-induced increase of spinal NT-3

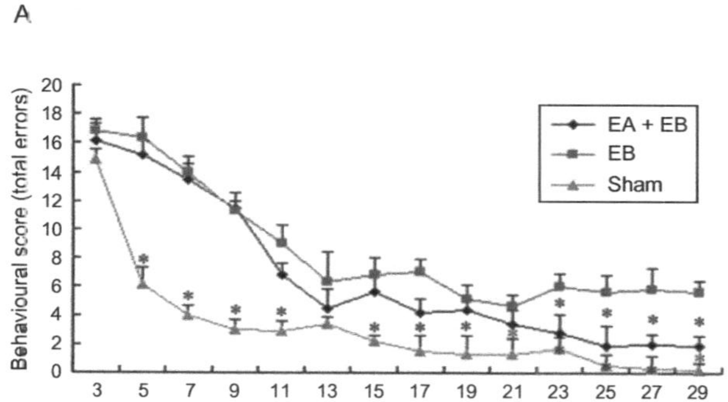

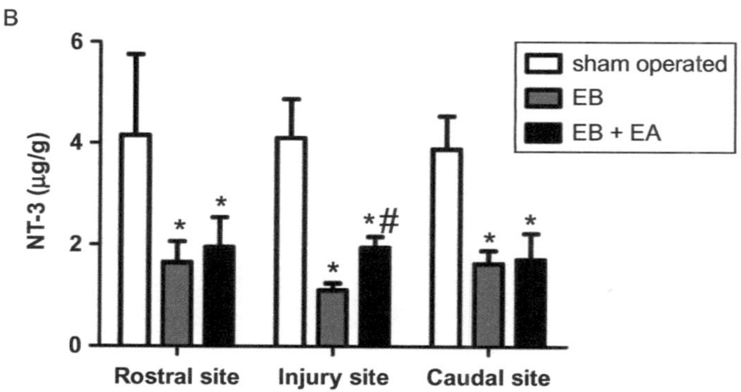

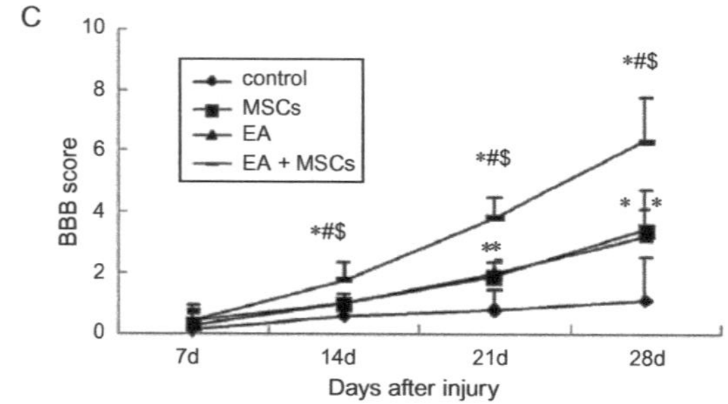

D

Comparisons of NT-3 content (ng/g, mean ± SD) in and around the injured site of spinal cord (T6-L1 segments).

Groups	Number	Content of NT-3
Control	4	1.52 ± 0.38
EA	4	2.93 ± 0.54
MSCs	4	1.72 ± 0.59*
EA + MSCs	4	3.83 ± 1.27*

One-way ANOVA: *$P < 0.05$ versus the control and EA groups.

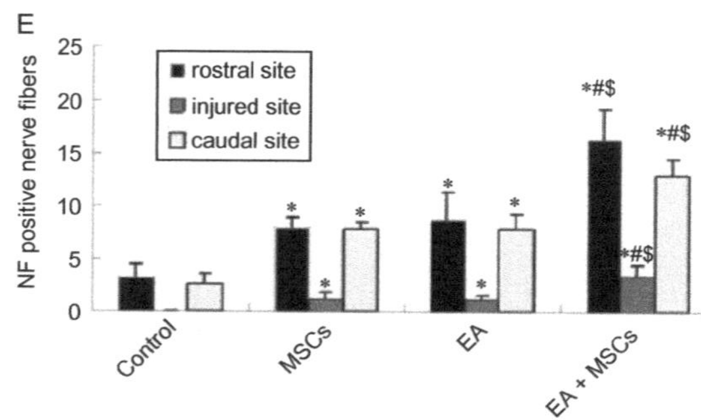

Figure 5.4 For lengend see next page.

and an increased OPC number and differentiation that correlated with remyelination and functional recovery of demyelinated spinal cord (Huang et al., 2011) (Fig. 5.4).

2.3. Brain

The effects of acupuncture on cells belonging to the brain have been studied using a wide range of techniques, encompassing classical histology and histochemistry, proteomics, genomics, and neuroimaging (Campbell, 2006; Guo et al., 1996; Hole & Berge, 1981; Hui et al., 2000; Kwon et al., 2000; Wu et al., 1999; Zhang et al., 2004). The link between acupuncture

Figure 5.4 The effects of EA treatment were analyzed on functional and structural recovery as well as on NT-3 tissue content in two models of spinal cord injury. In the first one (Huang et al., 2011) (A, B), 1 ml of 0.1 mg/ml ethidium bromide (EB) was stereotaxically injected into the T10 spinal dorsal funiculus to induce death of oligodendrocytes and demyelination. (A) Comparison of behavioral scores among the EA+EB group, EB group, and sham group. Rats receiving EB injection showed many error footsteps and higher error behavioral scores. After EA treatment, the animal behavior gradually recovered with increasing time so that normal function was attained by the end of the experiment at 30 days. The behavioral scores between the EA+EB and EB groups were significantly different beginning at 23 days. $^{*}P<0.05$ as compared with the EB group. Data = means ± SD. (B) Comparison of the NT-3 contents (ng/g) in the rostral segment, demyelination segment, and caudal segment among three groups (means ± SD). As compared to the sham group in the one-way ANOVA, the NT-3 level in three segments at day 15 after EB was significantly decreased in the EB and EA+EB ($^{*}P<0.05$ as compared with the Sham group). In the demyelination segment, the NT-3 level in the EA+EB was significantly higher than that in the EB ($^{\#}P<0.05$ as compared with the EB group). In a second model (Yan et al., 2011), a complete spinal transection was performed in rats at T10 level followed by local transplantation of mesenchymal stem cells (MSCs) and EA therapy. (C) Basso, Beattie, and Bresnahan (BBB) scores were obtained during the course of recovery. The scores increased with time in each group. Compared with the control group, three treatment groups had higher scores. Especially, the EA+MSC group had the highest score. $^{*}P<0.05$ versus the control group; $^{\#}P<0.05$ versus the MSC group; $^{\$}P<0.05$ versus the EA group. (D) The table summarizes the results obtained by ELISA quantification of NT-3 in the experimental groups. (E) The number of neurofilament (NF) positive nerve fibers, as an index of nerve fibers regeneration at the injury site, rostral site, and caudal site near injured spinal cord, was analyzed for four groups. The data indicated that EA and MSC transplantation had positive effects over nerve regeneration with an additive effect when performed together. $^{*}P<0.05$ versus the control group; $^{\#}P<0.05$ versus the MSC group; $^{\$}P<0.05$ versus the EA group. *(A, B) Reprinted with modifications from Huang et al. (2011), copyright (2011), with permission of Elsevier. (C–E) Reprinted with modifications from Yan et al. (2011), copyright (2011), with permission of Elsevier.* (For color version of this figure, the reader is referred to the online version of this chapter.)

and NT modulation in the brain and other CNS structures has been addressed studying the effects of EA on NGF and PCO in EV-injected rats (Stener-Victorin, Lundeberg, et al., 2000). In this study, we did not find significant differences of NGF protein content in hypothalamus and hippocampus of EA-treated or untreated EV-injected rats. However, EA stimulation at acupoint ST-36 has been proven effective in restoring hippocampus BDNF mRNA, declined by immobilization stress, suggesting that EA may relieve neuropathologic effects of stress by modulating brain NTs (Yun et al., 2002). In the same period, the link between EA and brain BDNF has also been studied in a mouse model of Parkinson's disease (PD) (Liang et al., 2002), comparing the effects of long-term EA applied at different frequencies on the survival and function of dopaminergic neurons in the ventral midbrain after transection of the medial forebrain bundle (MFB). It was found that high-frequency EA reduced the degeneration of dopaminergic neurons and that this could be supported by the EA-induced upregulation of BDNF in the ventral midbrain (Liang et al., 2002). Using a rat model of stroke with loss of motor function, the effects of EA on brain BDNF were further analyzed (Kim et al., 2009). The study had a negative outcome, since at least in the studied early recovery stage, any additional effect of low-frequency electroacupuncture on motor recovery as well as brain BDNF and TrkB levels was demonstrated (Kim et al., 2009). Based on the previously reported controversial results, it appears that modulating NTs by acupuncture in the CNS is not as simple as in the peripheral field, but the reason for such controversy is still not clear. Both the duration and the characteristic of the stimulation protocol (frequency and intensity) have been proposed as possible critical factors for the success or failure of experiments aimed at modulating NTs in the brain (Kim et al., 2009).

Not only NT presence but also NT activity in the brain could be affected by acupuncture in animal models of brain trauma. Receptor-activation studies demonstrated that EA treatment, applied to a rat model of cerebral ischemia–reperfusion, reversed the ischemia-induced high expression of brain NR1 subunit of the glutamate NMDA receptor and upregulated the local levels of TrkA (Sun et al., 2005). Inhibiting the specific downstream signaling pathways by protein kinase inhibitors, it was found that the neuroprotective effects of EA are mediated by the stimulation of the PI3K pathway (Sun et al., 2005). By the same stroke experimental model, it was also demonstrated that EA stimulation can reverse the ischemia-induced increase of the transient receptor-potential cation channel, subfamily M, member 7 (TRPM7), a Ca^{2+} ion channel that can act as a mediator of anoxic neuronal death once activated by oxidative stress (Aarts & Tymianski, 2005), by enhancing TrkA activation, which in turn triggers the downstream PI3K pathway (Zhao et al., 2005, 2007). More recently, the activation of the PI3K/Akt pathway has also been studied in a model of cerebral ischemia–reperfusion injury in rats after EA (Chen et al., 2012). It was found that EA improved neurological deficit and increased the neuroprotective activation of the PI3K/Akt pathway in the ischemic brain, with parallel increase of the ratio between the antiapoptotic Bcl-2 and the apoptotic Bax in the brain and increase of circulating BDNF (Chen et al., 2012). Supporting these observations, a further work on rats with occlusion of the middle cerebral artery demonstrated that after EA, the brain content of both BDNF and TrkB was elevated, correlating with the observed improvement in motor performance (Kim et al., 2012).

The neuroprotective and NT modulatory capacity of low-frequency EA was also demonstrated in a rat model of inherited retinopathy, strongly resembling human retinitis pigmentosa (Pagani, Manni, & Aloe, 2006). The authors showed an increased NGF synthesis in the retina of the Royal College of Surgeons (RCS) rats (Strauss, Stumpff, Mergler, Wienrich, & Wiederholt, 1998) after EA. This effect was parallel to a partial recovery of the impaired retinal morphological features and associated to an improved retina vascularization parallel to a local increase of the vascular endothelial growth factor (VEGF), which has been previously demonstrated to be induced by NGF (Manni, Antonelli, Costa & Aloe, 2005).

Previous studies showed that early diabetic rats suffered from decreased brain NGF, TrkA, and choline acetyltransferase (ChAT), an enzyme part of the process of acetylcholine biosynthesis and known to be under NGF control (Hefti, Dravid, & Hartikka, 1984), and from increased phosphorylation of the microtubule-associated protein tau, which is a hallmark in AD (Hanger, Anderton, & Noble, 2009). The effects of EA have also been evaluated on the progression of diabetic encephalopathy in rats and three weeks treatment with low-frequency EA was effective in reverting almost all of the described diabetes-induced deregulations in the brain cholinergic system and in tau hyperphosphorylation (Rocco et al., 2013).

Two further works recently addressed the link between acupuncture and the central regulation of memory/cognitive function in relation to NTs. It was indeed demonstrated that BDNF was increased after EA in the dentate gyrus in rats, with a parallel increase in neuroblasts proliferation that was blocked by an antagonist of Trk receptor activation (Hwang et al., 2010). Moreover, the effects of EA were investigated on the brain cholinergic system in rats with spatial cognitive impairment induced by repeated corticosterone treatments (Lee et al., 2012). It was found that MA improved the corticosterone impairing effects on spatial memory and increased immunoreactivity for ChAT in the medial septum and for acetylcholinesterase in the hippocampus and BDNF mRNA in the hippocampus (Lee et al., 2012). These results suggest that acupuncture could be valuable, through the modulation of brain BDNF expression and activity, for the neuroprotection of injured neurons and in stimulating neurogenesis in selected brain areas.

2.4. How could neurotrophins be regulated by acupuncture?

In all of the previously reviewed studies, it was mainly investigated if MA or EA treatment generated a modulation of tissue NT content and/or activity in different models of disease, but there were not any of them investigating how the needling could elicit such a modulation. As stated in the introductory section, the modulation of NTs encompasses different possible mechanisms regulating cellular NTs and NT receptor expression, NT intracellular trafficking and maturation, NT secretion and extracellular processing, and finally the challenge of different receptor complexes with consequent activation of selected and functionally opposite intracellular signaling pathways (Hempstead, 2006). Since there is no work, to the best of our knowledge, that have addressed the topic of NT modulation by needling therapies from a mechanistic point of view, it should only be possible to draw a speculative picture about the main possible factors or neurophysiological events participating in such modulation. We would like to highlight two of them, among the several possible, that in our opinion deserve a priority of investigation: *Activity-dependent and neurohormonal regulation.*

NT production, release, and maturation are controlled in an activity-dependent manner (Blochl & Thoenen, 1995, 1996; Bruno & Cuello, 2006; Castren, Berninger, Leingartner, & Lindholm, 1998; Lauterborn, Isackson, & Gall, 1994; Lindholm, Castren, Berzaghi, Blochl, & Thoenen, 1994; Zafra, Castren, Thoenen, & Lindholm, 1991). The different NTs may differ preferentially in their release site, that is, BDNF is released at synaptic cleft acting in the structural and functional synapse modeling events (Gottmann et al., 2009; Lessmann, 1998), while NGF could be released all along the neuronal processes, in particular dendrites (Blochl & Thoenen, 1996). It is conceivable that by regular patterns of neuronal activity in spinal and supraspinal circuits, generated by controlled and continuous afferent stimulation, it could be possible to exert a reasonable control over some of the regulatory mechanisms for NT expression and activity. It has yet been demonstrated by *ex vivo* models that BDNF release could be achieved in the spinal cord by short bursts of high-frequency electrical stimulation (Lever et al., 2001). Since the NMDA antagonist d-AP-5 inhibited electrically evoked BDNF release, it was concluded that afferent stimulation and activation of glutamatergic synapses in the dorsal horns mediate BDNF release (Lever et al., 2001). Catecholamine, glutamate, GABA, or neuropeptide receptor challenge is increased by acupuncture or EA treatment (Zhao, 2008) and also effective in the modulation of NTs in neuronal and peripheral tissues (Blochl & Thoenen, 1995; Furukawa & Furukawa, 1990; Heese et al., 2000; Tirassa, Manni, Aloe, & Lundeberg, 2002; Zafra et al., 1991) and could reasonably represent the physiological link between acupuncture stimulation and the modulation of NTs. Experiments addressing this topic, by specific inhibition of neurotransmitters action in acupuncture-treated animals or by blockade of selective afferent and/or efferent activities or even by using genetically modified animals lacking or overexpressing neurotransmitters receptors, could shed light on the possible activity-dependent NT modulation by acupuncture.

Hormones and neurohormones, known to be released or inhibited after acupuncture treatment, could also play a role in regulating NTs. As a few examples for this proposed mechanism, the action of estradiol or cortisone can be mentioned, both known to be regulated by acupuncture (Chen, 1997; Lim & Wong, 2010) and both exert a modulatory role over NT expression (Aloe, Bracci-Laudiero, et al., 1994; Campeau, Liberzon, Morilak, & Ressler, 2011; Haraguchi et al., 2012). On the other hand, even the release and activity of hormones can be regulated by neurotransmitters and neuropeptides release, thus suggesting that all of the action of acupuncture could at the end descend from its ability to positively or negatively modulate the activity in selected neuronal cells and circuitries.

3. RELEVANCE OF THE LINK BETWEEN NTs AND ACUPUNCTURE FOR THE TREATMENT OF DISEASES

Since their relevant function in the regulation of survival and phenotypic features of cells in the PNS, CNS, and also nonneuronal tissue, the development of clinical therapies based on NTs has attracted great interest and resources during the last three decades (Allen & Dawbarn, 2006; Aloe et al., 2012). Preclinical and clinical trials have been conducted to investigate the potentiality of NTs in the care of central and peripheral neurodegenerative diseases, depression, peripheral neuropathies, allergic diseases, and, more recently, epithelial derangements with neurotrophic origin and in vasculopathies (Allen & Dawbarn, 2006; Aloe et al., 2012). The clinical blockade of NTs, such as NGF, could also be useful for treatment of neuropathic pain, of asthma, and in some kind of dysautonomias (Allen & Dawbarn, 2006; Hefti et al., 2006). Despite the great efforts and expectations, at least two major obstacles have hampered the development of NT-based therapies: the occurrence of undesired side effects that have limited the NTs' dosage to levels not always ensuring effective therapeutic benefits and the difficulty in setting a delivery procedure that route NTs to specific CNS targets, taking also into account that NTs are not able to cross the blood–brain barrier (Pan, Banks, & Kastin, 1998).

Clinical studies with the NGF are in some way paradigmatic (Aloe et al., 2012; Apfel, 2002). It is known that the NGF can have therapeutic value, especially for patients affected by peripheral neuropathies secondary to the onset of metabolic disturbances, such as those observed in diabetes patients (Apfel, 1999b; Apfel et al., 2000), viral infections (McArthur et al., 2000; Schifitto et al., 2001), or during cancer-addressed chemotherapies (Aloe, Manni, Properzi, De Santis, & Fiore, 2000; Apfel & Kessler, 1996). Clinical trials have been performed in the 1990s on a relevant number of patients affected by diabetic and HIV-associated peripheral neuropathies (Apfel, 2002; Apfel & Kessler, 1996; Apfel et al., 2000; Schifitto et al., 2001). Despite the promising early results, most of the trials revealed the occurrence of limiting side effects such as weight loss, pain, and hyperalgesia, especially following systemic delivery of the NT (Apfel, 2002), leading to discontinuation of the development of NGF-based therapies for peripheral neuropathies.

NGF has also been tested in patients affected by central neurodegenerative pathologies, such as AD. The first clinical trials on NGF performed in patients with AD (Eriksdotter Jonhagen et al., 1998) showed not only partial beneficial effects after chronic intracerebroventricular (ICV) administration of NGF, such as an increase in central nicotine binding, cognitive improvement, and mild neuropsychological amelioration, but also negative side effects such as peripheral muscular pain and body weight loss. Other studies using ICV infusion and/or grafting of NGF-producing cells directly into the patient brain (Tuszynski et al., 2005) evidenced that the this approach might entail serious surgical risks and high costs, suggesting that less invasive strategies for delivering NGF into the brain of patients with AD should be tested.

BDNF was also tested in human studies, for the care of amyotrophic lateral sclerosis (A controlled trial of recombinant methionyl human BDNF in ALS: The BDNF Study Group (Phase III) 1999; Ochs et al., 2000), Guillain–Barrè syndrome (Bensa, Hadden, Hahn, Hughes, & Willison, 2000), colonic constipation (Coulie et al., 2000), and in diabetic peripheral neuropathy (Wellmer, Misra, Sharief, Kopelman, & Anand, 2001). NT-3 was tested in human trials for colonic constipation (Coulie et al., 2000) and in healthy subjects to assess the tolerability rate (Chaudhry et al., 2000). Besides the positive outcomes, which were in some cases present but not so striking to justify further clinical investigations, several of the previously cited trials reported side effects such as pain at the injection site, altered bowel functions, and paresthesias.

In light of the earlier-reviewed data, the rationale exists to pursue alternative therapeutic strategies, such as physical therapies based on regular and controlled sensory stimulation, aimed either at the modulation of endogenous NTs or at counteracting the NT-associated side effects, without interfering with their therapeutic action. Many basic studies describing NTs induction by MA or EA in animal models of neurodegeneration, neurotraumas, metabolic diseases, and inflammatory conditions have been reported. Only one study so far reported NT modulation by acupuncture in humans (Moldenhauer et al., 2010), but clues about the possible efficacy of acupuncture-based therapies in the modulation of endogenous NTs with subsequent positive therapeutic effect on human pathologies come from studies on the effects of regular, aerobic physical exercise (Seifert et al., 2010). As stated before, the Western standpoint about the mechanism(s) underlying acupuncture efficacy is based on the assumption that regular, modulated sensory afferent discharge is at the basis of a sequela of events in the PNS and CNS that in turn generate the therapeutic outcome (Andersson & Lundeberg, 1995). Such sensory stimulation could be achieved by different techniques, encompassing, among others, MA, EA, and physical exercise (Andersson & Lundeberg, 1995; Lundeberg, 1984; Manneras et al., 2008; Manni, Cajander, et al., 2005).

It is known that reduced circulating BDNF is a hallmark of major depression (Autry & Monteggia, 2012) and of type 2 diabetes, a condition that has been linked to an increased risk of cognitive impairment and dementia (Zhen et al., 2013). In a recent study, it has been demonstrated that brain release of BDNF was increased in healthy humans experiencing 3 months of endurance training, indicating physical exercise promotes brain health (Seifert et al., 2010). Modulation of circulating BDNF has been also reported in studies demonstrating the positive effects of regular exercise in various condition of dementia (Baker et al., 2010a, 2010b), and BDNF modulation has been indicated as a major cause of anti-inflammatory effects of physical exercise (Archer, Fredriksson, Schutz, & Kostrzewa, 2011). It is conceivable that most of the earlier-cited results could be also achieved by acupuncture-based therapies, while specific human trials are needed to further explore this possibility. An indirect confirmation of this hypothesis has been produced comparing the effects of acupuncture and physical exercise in rats affected by experimental PCO. By these studies (Manneras et al., 2008; Manni, Cajander, et al., 2005; Manni, Lundeberg, et al., 2005), it has been indeed indicated that the two therapeutic approaches reached very similar results, in terms of modulation of peripheral NGF and consequent modulation of sympathetic activity.

As for the possible use of acupuncture together with NTs treatment, to the better of our knowledge, only two studies addressed the topic of the synergistic action of NTs and acupuncture. It was demonstrated that the action of EA on sensory neurons could allow the integration between NGF administration and EA (Aloe & Manni, 2009). Indeed, EA overcomes the development of NGF-associated hyperalgesia in rats, facilitating the pharmacological use of the NT. Moreover, combined intranasal NGF and EA treatments have been applied in a model of cerebral ischemia (Cheng et al., 2009). The combination treatment led to significant improvement in neurological function, reduction in infarct volume, and increased neural progenitor cell proliferation and survival. Overall, these data indicate that combined NT/acupuncture therapies could allow the increase of NTs dosage, since acupuncture can counteract the described NT undesired side effects, and could also have a synergistic effect improving the therapeutic outcome.

4. CONCLUSION

The possibility to generate therapeutic effects on damaged nervous system, as well as on the immune and endocrine systems, by acupuncture has been confined for a long time in the field of complementary or alternative medicine. Today, a big amount of basic and human experimental data built up a scientific rationale, based on the acupuncture-induced regular patterns of afferent activity with consequent triggering of therapeutic effects. In this context, the link between acupuncture and NT modulation has been established in animal studies since the year 2000 (Stener-Victorin,

Lundeberg, et al., 2000). Despite the lack of research about the basic mechanism(s) underlying NT modulation by MA or EA, such a functional connection represents a promising avenue for the setting of supportive therapies based on sensory stimulation, for major diseases characterized by deregulation of NTs expression and/or action.

REFERENCES

A controlled trial of recombinant methionyl human BDNF in ALS: The BDNF Study Group (Phase III). (1999). *Neurology, 52*(7), 1427–1433.

Aarts, M. M., & Tymianski, M. (2005). TRPM7 and ischemic CNS injury. *The Neuroscientist, 11*(2), 116–123.

Acupuncture: National Institutes of Health Consensus Development Conference statement (2000). *Dermatology Nursing, 12*(2), 126–133.

Allen, S. J., & Dawbarn, D. (2006). Clinical relevance of the neurotrophins and their receptors. *Clinical Science (London, England), 110*(2), 175–191.

Alleva, E., Aloe, L., & Bigi, S. (1993). An updated role for nerve growth factor in neurobehavioural regulation of adult vertebrates. *Reviews in the Neurosciences, 4*(1), 41–62.

Aloe, L. (2001). Nerve growth factor and neuroimmune responses: Basic and clinical observations. *Archives of Physiology and Biochemistry, 109*(4), 354–356.

Aloe, L., Alleva, E., & Fiore, M. (2002). Stress and nerve growth factor: Findings in animal models and humans. *Pharmacology, Biochemistry, and Behavior, 73*(1), 159–166.

Aloe, L., Bracci-Laudiero, L., Alleva, E., Lambiase, A., Micera, A., & Tirassa, P. (1994). Emotional stress induced by parachute jumping enhances blood nerve growth factor levels and the distribution of nerve growth factor receptors in lymphocytes. *Proceedings of the National Academy of Sciences of the United States of America, 91*(22), 10440–10444.

Aloe, L., Bracci-Laudiero, L., Bonini, S., & Manni, L. (1997). The expanding role of nerve growth factor: From neurotrophic activity to immunologic diseases. *Allergy, 52*(9), 883–894.

Aloe, L., & Manni, L. (2009). Low-frequency electro-acupuncture reduces the nociceptive response and the pain mediator enhancement induced by nerve growth factor. *Neuroscience Letters, 449*(3), 173–177.

Aloe, L., Manni, L., Properzi, F., De Santis, S., & Fiore, M. (2000). Evidence that nerve growth factor promotes the recovery of peripheral neuropathy induced in mice by cisplatin: Behavioral, structural and biochemical analysis. *Autonomic Neuroscience, 86*(1–2), 84–93.

Aloe, L., Rocco, M. L., Bianchi, P., & Manni, L. (2012). Nerve growth factor: From the early discoveries to the potential clinical use. *Journal of Translational Medicine, 10*(1), 239.

Aloe, L., Tirassa, P., & Alleva, E. (1994). Cold water swimming stress alters NGF and low-affinity NGF receptor distribution in developing rat brain. *Brain Research Bulletin, 33*(2), 173–178.

Andersson, S., & Lundeberg, T. (1995). Acupuncture—From empiricism to science: Functional background to acupuncture effects in pain and disease. *Medical Hypotheses, 45*(3), 271–281.

Apfel, S. C. (1999a). Neurotrophic factors and diabetic peripheral neuropathy. *European Neurology, 41*(Suppl 1), 27–34.

Apfel, S. C. (1999b). Neurotrophic factors in the therapy of diabetic neuropathy. *The American Journal of Medicine, 107*(2B), 34S–42S.

Apfel, S. C. (2002). Nerve growth factor for the treatment of diabetic neuropathy: What went wrong, what went right, and what does the future hold? *International Review of Neurobiology, 50*, 393–413.

Apfel, S. C., & Kessler, J. A. (1996). Neurotrophic factors in the treatment of peripheral neuropathy. *Ciba Foundation Symposium*, *196*, 98–108.

Apfel, S. C., Schwartz, S., Adornato, B. T., Freeman, R., Biton, V., Rendell, M., et al. (2000). Efficacy and safety of recombinant human nerve growth factor in patients with diabetic polyneuropathy: A randomized controlled trial. rhNGF Clinical Investigator Group. *JAMA*, *284*(17), 2215–2221.

Archer, T., Fredriksson, A., Schutz, E., & Kostrzewa, R. M. (2011). Influence of physical exercise on neuroimmunological functioning and health: Aging and stress. *Neurotoxicity Research*, *20*(1), 69–83.

Autry, A. E., & Monteggia, L. M. (2012). Brain-derived neurotrophic factor and neuropsychiatric disorders. *Pharmacological Reviews*, *64*(2), 238–258.

Bai, Y. H., Lim, S. C., Song, C. H., Bae, C. S., Jin, C. S., Choi, B. C., et al. (2004). Electroacupuncture reverses nerve growth factor abundance in experimental polycystic ovaries in the rat. *Gynecologic and Obstetric Investigation*, *57*(2), 80–85.

Baker, L. D., Frank, L. L., Foster-Schubert, K., Green, P. S., Wilkinson, C. W., McTiernan, A., et al. (2010a). Aerobic exercise improves cognition for older adults with glucose intolerance, a risk factor for Alzheimer's disease. *Journal of Alzheimer's Disease*, *22*(2), 569–579.

Baker, L. D., Frank, L. L., Foster-Schubert, K., Green, P. S., Wilkinson, C. W., McTiernan, A., et al. (2010b). Effects of aerobic exercise on mild cognitive impairment: A controlled trial. *Archives of Neurology*, *67*(1), 71–79.

Barde, Y. A. (1990). The nerve growth factor family. *Progress in Growth Factor Research*, *2*(4), 237–248.

Bartoletti, A., Cancedda, L., Reid, S. W., Tessarollo, L., Porciatti, V., Pizzorusso, T., et al. (2002). Heterozygous knock-out mice for brain-derived neurotrophic factor show a pathway-specific impairment of long-term potentiation but normal critical period for monocular deprivation. *The Journal of Neuroscience*, *22*(23), 10072–10077.

Bensa, S., Hadden, R. D., Hahn, A., Hughes, R. A., & Willison, H. J. (2000). Randomized controlled trial of brain-derived neurotrophic factor in Guillain-Barre syndrome: A pilot study. *European Journal of Neurology*, 7(4), 423–426.

Blochl, A., & Thoenen, H. (1995). Characterization of nerve growth factor (NGF) release from hippocampal neurons: Evidence for a constitutive and an unconventional sodium-dependent regulated pathway. *The European Journal of Neuroscience*, 7(6), 1220–1228.

Blochl, A., & Thoenen, H. (1996). Localization of cellular storage compartments and sites of constitutive and activity-dependent release of nerve growth factor (NGF) in primary cultures of hippocampal neurons. *Molecular and Cellular Neurosciences*, 7(3), 173–190.

Bruno, M. A., & Cuello, A. C. (2006). Activity-dependent release of precursor nerve growth factor, conversion to mature nerve growth factor, and its degradation by a protease cascade. *Proceedings of the National Academy of Sciences of the United States of America*, *103*(17), 6735–6740.

Bucinskaite, V., Lundeberg, T., Stenfors, C., Ekblom, A., Dahlin, L., & Theodorsson, E. (1994). Effects of electro-acupuncture and physical exercise on regional concentrations of neuropeptides in rat brain. *Brain Research*, *666*(1), 128–132.

Bucinskaite, V., Theodorsson, E., Crumpton, K., Stenfors, C., Ekblom, A., & Lundeberg, T. (1996). Effects of repeated sensory stimulation (electro-acupuncture) and physical exercise (running) on open-field behaviour and concentrations of neuropeptides in the hippocampus in WKY and SHR rats. *The European Journal of Neuroscience*, *8*(2), 382–387.

Campbell, A. (2006). Point specificity of acupuncture in the light of recent clinical and imaging studies. *Acupuncture in Medicine*, *24*(3), 118–122.

Campeau, S., Liberzon, I., Morilak, D., & Ressler, K. (2011). Stress modulation of cognitive and affective processes. *Stress*, *14*(5), 503–519.

Cao, X. D., Xu, S. F., & Lu, W. X. (1983). Inhibition of sympathetic nervous system by acupuncture. *Acupuncture & Electro-Therapeutics Research, 8*, 25–35.
Carlsson, C. (2002). Acupuncture mechanisms for clinically relevant long-term effects—Reconsideration and a hypothesis. *Acupuncture in Medicine, 20*(2–3), 82–99.
Castren, E., Berninger, B., Leingartner, A., & Lindholm, D. (1998). Regulation of brain-derived neurotrophic factor mRNA levels in hippocampus by neuronal activity. *Progress in Brain Research, 117*, 57–64.
Chae, Y., Hong, M. S., Kim, G. H., Hahm, D. H., Park, H. J., Ha, E., et al. (2007). Protein array analysis of cytokine levels on the action of acupuncture in carrageenan-induced inflammation. *Neurological Research, 29*(Suppl. 1), S55–S58.
Chalazonitis, A. (1996). Neurotrophin-3 as an essential signal for the developing nervous system. *Molecular Neurobiology, 12*(1), 39–53.
Chao, M. V., Rajagopal, R., & Lee, F. S. (2006). Neurotrophin signalling in health and disease. *Clinical Science (London, England), 110*(2), 167–173.
Chaudhry, V., Giuliani, M., Petty, B. G., Lee, D., Seyedsadr, M., Hilt, D., et al. (2000). Tolerability of recombinant-methionyl human neurotrophin-3 (r-metHuNT3) in healthy subjects. *Muscle & Nerve, 23*(2), 189–192.
Chen, B. Y. (1997). Acupuncture normalizes dysfunction of hypothalamic-pituitary-ovarian axis. *Acupuncture & Electro-Therapeutics Research, 22*(2), 97–108.
Chen, A., Lin, Z., Lan, L., Xie, G., Huang, J., Lin, J., et al. (2012). Electroacupuncture at the Quchi and Zusanli acupoints exerts neuroprotective role in cerebral ischemia-reperfusion injured rats via activation of the PI3K/Akt pathway. *International Journal of Molecular Medicine, 30*(4), 791–796.
Chen, J., Qi, J. G., Zhang, W., Zhou, X., Meng, Q. S., Zhang, W. M., et al. (2007). Electro-acupuncture induced NGF, BDNF and NT-3 expression in spared L6 dorsal root ganglion in cats subjected to removal of adjacent ganglia. *Neuroscience Research, 59*(4), 399–405.
Chen, Y. Y., Zhang, W., Chen, Y. L., Chen, S. J., Dong, H., & Zeng, Y. S. (2008). Electro-acupuncture improves survival and migration of transplanted neural stem cells in injured spinal cord in rats. *Acupuncture & Electro-Therapeutics Research, 33*(1–2), 19–31.
Cheng, S., Ma, M., Ma, Y., Wang, Z., Xu, G., & Liu, X. (2009). Combination therapy with intranasal NGF and electroacupuncture enhanced cell proliferation and survival in rats after stroke. *Neurological Research, 31*(7), 753–758.
Cohen, S., Levi-Montalcini, R., & Hamburger, V. (1954). A nerve growth-stimulating factor isolated from sarcom as 37 and 180. *Proceedings of the National Academy of Sciences of the United States of America, 40*(10), 1014–1018.
Connor, B., & Dragunow, M. (1998). The role of neuronal growth factors in neurodegenerative disorders of the human brain. *Brain Research Brain Research Reviews, 27*(1), 1–39.
Coulie, B., Szarka, L. A., Camilleri, M., Burton, D. D., McKinzie, S., Stambler, N., et al. (2000). Recombinant human neurotrophic factors accelerate colonic transit and relieve constipation in humans. *Gastroenterology, 119*(1), 41–50.
Cowan, W. M. (2001). Viktor Hamburger and Rita Levi-Montalcini: The path to the discovery of nerve growth factor. *Annual Review of Neuroscience, 24*, 551–600.
Ding, Y., Yan, Q., Ruan, J. W., Zhang, Y. Q., Li, W. J., Zhang, Y. J., et al. (2009). Electro-acupuncture promotes survival, differentiation of the bone marrow mesenchymal stem cells as well as functional recovery in the spinal cord-transected rats. *BMC Neuroscience, 10*, 35.
Eriksdotter Jonhagen, M., Nordberg, A., Amberla, K., Backman, L., Ebendal, T., Meyerson, B., et al. (1998). Intracerebroventricular infusion of nerve growth factor in three patients with Alzheimer's disease. *Dementia and Geriatric Cognitive Disorders, 9*(5), 246–257.

Ernfors, P. (2001). Local and target-derived actions of neurotrophins during peripheral nervous system development. *Cellular and Molecular Life Sciences*, *58*(8), 1036–1044.
Ernfors, P., Wetmore, C., Olson, L., & Persson, H. (1990). Identification of cells in rat brain and peripheral tissues expressing mRNA for members of the nerve growth factor family. *Neuron*, *5*(4), 511–526.
Fischer, W., Wictorin, K., Bjorklund, A., Williams, L. R., Varon, S., & Gage, F. H. (1987). Amelioration of cholinergic neuron atrophy and spatial memory impairment in aged rats by nerve growth factor. *Nature*, *329*(6134), 65–68.
Franconi, G., Manni, L., Aloe, L., Mazzilli, F., Giambalvo Dal Ben, G., Lenzi, A., et al. (2011). Acupuncture in clinical and experimental reproductive medicine: A review. *Journal of Endocrinological Investigation*, *34*(4), 307–311.
Friedman, W. J., Black, I. B., & Kaplan, D. R. (1998). Distribution of the neurotrophins brain-derived neurotrophic factor, neurotrophin-3, and neurotrophin-4/5 in the postnatal rat brain: An immunocytochemical study. *Neuroscience*, *84*(1), 101–114.
Friedman, W. J., & Greene, L. A. (1999). Neurotrophin signaling via Trks and p75. *Experimental Cell Research*, *253*(1), 131–142.
Friedman, W. J., Ibanez, C. F., Hallbook, F., Persson, H., Cain, L. D., Dreyfus, C. F., et al. (1993). Differential actions of neurotrophins in the locus coeruleus and basal forebrain. *Experimental Neurology*, *119*(1), 72–78.
Funakoshi, H., Belluardo, N., Arenas, E., Yamamoto, Y., Casabona, A., Persson, H., et al. (1995). Muscle-derived neurotrophin-4 as an activity-dependent trophic signal for adult motor neurons. *Science*, *268*(5216), 1495–1499.
Furukawa, S., & Furukawa, Y. (1990). Nerve growth factor synthesis and its regulatory mechanisms: An approach to therapeutic induction of nerve growth factor synthesis. *Cerebrovascular and Brain Metabolism Reviews*, *2*(4), 328–344.
Gottmann, K., Mittmann, T., & Lessmann, V. (2009). BDNF signaling in the formation, maturation and plasticity of glutamatergic and GABAergic synapses. *Experimental Brain Research*, *199*(3–4), 203–234.
Guo, H.-F., Tian, J., Wang, X., Fang, Y., Yueping, H., & Jisheng, H. (1996). Brain substrates activated by electroacupuncture of different frequencies (1): Comparative study on the expression of oncogene c-fos and genes coding for three opioid peptides. *Molecular Brain Research*, *43*, 157–166.
Hallbook, F. (1999). Evolution of the vertebrate neurotrophin and Trk receptor gene families. *Current Opinion in Neurobiology*, *9*(5), 616–621.
Hallbook, F., Ibanez, C. F., & Persson, H. (1991). Evolutionary studies of the nerve growth factor family reveal a novel member abundantly expressed in Xenopus ovary. *Neuron*, *6*(5), 845–858.
Han, J. S. (1997). Physiology of acupuncture: Review of thirty years of research. *Journal of Alternative and Complementary Medicine*, *3*(Suppl. 1), 101–108.
Hanger, D. P., Anderton, B. H., & Noble, W. (2009). Tau phosphorylation: The therapeutic challenge for neurodegenerative disease. *Trends in Molecular Medicine*, *15*(3), 112–119.
Haraguchi, S., Sasahara, K., Shikimi, H., Honda, S., Harada, N., & Tsutsui, K. (2012). Estradiol promotes purkinje dendritic growth, spinogenesis, and synaptogenesis during neonatal life by inducing the expression of BDNF. *Cerebellum*, *11*(2), 416–417.
Heese, K., Otten, U., Mathivet, P., Raiteri, M., Marescaux, C., & Bernasconi, R. (2000). GABA(B) receptor antagonists elevate both mRNA and protein levels of the neurotrophins nerve growth factor (NGF) and brain-derived neurotrophic factor (BDNF) but not neurotrophin-3 (NT-3) in brain and spinal cord of rats. *Neuropharmacology*, *39*(3), 449–462.
Hefti, F., Dravid, A., & Hartikka, J. (1984). Chronic intraventricular injections of nerve growth factor elevate hippocampal choline acetyltransferase activity in adult rats with partial septo-hippocampal lesions. *Brain Research*, *293*(2), 305–311.

Hefti, F. F., Rosenthal, A., Walicke, P. A., Wyatt, S., Vergara, G., Shelton, D. L., et al. (2006). Novel class of pain drugs based on antagonism of NGF. *Trends in Pharmacological Sciences, 27*(2), 85–91.

Hempstead, B. L. (2006). Dissecting the diverse actions of pro- and mature neurotrophins. *Current Alzheimer Research, 3*(1), 19–24.

Hochhaus, F., Koehne, P., Schaper, C., Butenandt, O., Felderhoff-Mueser, U., Ring-Mrozik, E., et al. (2001). Elevated nerve growth factor and neurotrophin-3 levels in cerebrospinal fluid of children with hydrocephalus. *BMC Pediatrics, 1*, 2.

Hole, K., & Berge, O. G. (1981). Regulation of pain sensitivity in the central nervous system. *Cephalalgia, 1*(1), 51–59.

Huang, S. F., Ding, Y., Ruan, J. W., Zhang, W., Wu, J. L., He, B., et al. (2011). An experimental electro-acupuncture study in treatment of the rat demyelinated spinal cord injury induced by ethidium bromide. *Neuroscience Research, 70*(3), 294–304.

Huang, E. J., & Reichardt, L. F. (2003). Trk receptors: Roles in neuronal signal transduction. *Annual Review of Biochemistry, 72*, 609–642.

Hui, K. K., Liu, J., Makris, N., Gollub, R. L., Chen, A. J., Moore, C. I., et al. (2000). Acupuncture modulates the limbic system and subcortical gray structures of the human brain: Evidence from fMRI studies in normal subjects. *Human Brain Mapping, 9*(1), 13–25.

Hwang, I. K., Chung, J. Y., Yoo, D. Y., Yi, S. S., Youn, H. Y., Seong, J. K., et al. (2010). Effects of electroacupuncture at Zusanli and Baihui on brain-derived neurotrophic factor and cyclic AMP response element-binding protein in the hippocampal dentate gyrus. *The Journal of Veterinary Medical Science, 72*(11), 1431–1436.

Ibanez, C. F. (1996). Neurotrophin-4: The odd one out in the neurotrophin family. *Neurochemical Research, 21*(7), 787–793.

Ip, N. Y., Ibanez, C. F., Nye, S. H., McClain, J., Jones, P. F., Gies, D. R., et al. (1992). Mammalian neurotrophin-4: Structure, chromosomal localization, tissue distribution, and receptor specificity. *Proceedings of the National Academy of Sciences of the United States of America, 89*(7), 3060–3064.

Kagitani, F., Uchida, S., Hotta, H., & Aikawa, Y. (2005). Manual acupuncture needle stimulation of the rat hindlimb activates groups I, II, III and IV single afferent nerve fibers in the dorsal spinal roots. *The Japanese Journal of Physiology, 55*(3), 149–155.

Kaptchuk, T. J. (2002). Acupuncture: Theory, efficacy, and practice. *Annals of Internal Medicine, 136*(5), 374–383.

Kawaja, M. D., Walsh, G. S., Tovich, P. R., & Julien, J. P. (1998). Effects of elevated levels of nerve growth factor on the septohippocampal system in transgenic mice. *The European Journal of Neuroscience, 10*(7), 2207–2216.

Kawakita, K. (1991). Role of polymodal receptors in the peripheral mechanisms of acupuncture and moxibustion stimulation. In S. K. Manchanda, W. Selvamurthy, & V. Mohan Kumar (Eds.), *Advances in physiology science* (pp. 731–739). New Delhi: Macmillan.

Kawakita, K., & Gotoh, K. (1996). Role of polymodal receptors in the acupuncture-mediated endogenous pain inhibitory systems. *Progress in Brain Research, 113*, 507–523.

Kim, M. W., Chung, Y. C., Jung, H. C., Park, M. S., Han, Y. M., Chung, Y. A., et al. (2012). Electroacupuncture enhances motor recovery performance with brain-derived neurotrophic factor expression in rats with cerebral infarction. *Acupuncture in Medicine, 30*(3), 222–226.

Kim, W. S., Kim, I. S., Kim, S. J., Wei, P., Hyung Choi, D., & Han, T. R. (2009). Effect of electroacupuncture on motor recovery in a rat stroke model during the early recovery stage. *Brain Research, 1248*, 176–183.

Kimura, A., & Sato, A. (1997). Somatic regulation of autonomic functions in anesthetized animals—Neural mechanisms of physical therapy including acupuncture. *The Japanese Journal of Veterinary Research, 45*(3), 137–145.

Klesse, L. J., & Parada, L. F. (1999). Trks: Signal transduction and intracellular pathways. *Microscopy Research and Technique*, *45*(4–5), 210–216.

Korte, M., Carroll, P., Wolf, E., Brem, G., Thoenen, H., & Bonhoeffer, T. (1995). Hippocampal long-term potentiation is impaired in mice lacking brain-derived neurotrophic factor. *Proceedings of the National Academy of Sciences of the United States of America*, *92*(19), 8856–8860.

Kwon, Y. B., Kang, M. S., Son, S. S., Kim, J. T., Lee, Y. H., Han, H. J., et al. (2000). Different frequencies of electroacupuncture modified the cellular activity of serotonergic neurons in brainstem. *The American Journal of Chinese Medicine*, *28*(3–4), 435–441.

Lara, H. E., Dissen, G. A., Leyton, V., Paredes, A., Fuenzalida, H., Fiedler, J. L., et al. (2000). An increased intraovarian synthesis of nerve growth factor and its low affinity receptor is a principal component of steroid-induced polycystic ovary in the rat. *Endocrinology*, *141*(3), 1059–1072.

Lara, H. E., Dorfman, M., Venegas, M., Luza, S. M., Luna, S. L., Mayerhofer, A., et al. (2002). Changes in sympathetic nerve activity of the mammalian ovary during a normal estrous cycle and in polycystic ovary syndrome: Studies in norepinephrine release. *Microscopy Research and Technique*, *59*, 495–502.

Lara, H. E., Ferruz, J. L., Luza, S., Bustamante, D. A., Borges, Y., & Ojeda, S. R. (1993). Activation of ovarian sympathetic nerves in polycystic ovary syndrome. *Endocrinology*, *133*(6), 2690–2695.

Lauterborn, J. C., Isackson, P. J., & Gall, C. M. (1994). Seizure-induced increases in NGF mRNA exhibit different time courses across forebrain regions and are biphasic in hippocampus. *Experimental Neurology*, *125*(1), 22–40.

Lee, R., Kermani, P., Teng, K. K., & Hempstead, B. L. (2001). Regulation of cell survival by secreted proneurotrophins. *Science*, *294*(5548), 1945–1948.

Lee, B., Sur, B. J., Kwon, S., Jung, E., Shim, I., Lee, H., et al. (2012). Acupuncture stimulation alleviates corticosterone-induced impairments of spatial memory and cholinergic neurons in rats. *Evidence-Based Complementary and Alternative Medicine*, *2012*, 670536.

Lessmann, V. (1998). Neurotrophin-dependent modulation of glutamatergic synaptic transmission in the mammalian CNS. *General Pharmacology*, *31*(5), 667–674.

Lever, I. J., Bradbury, E. J., Cunningham, J. R., Adelson, D. W., Jones, M. G., McMahon, S. B., et al. (2001). Brain-derived neurotrophic factor is released in the dorsal horn by distinctive patterns of afferent fiber stimulation. *The Journal of Neuroscience*, *21*(12), 4469–4477.

Levi-Montalcini, R. (1952). Effects of mouse tumor transplantation on the nervous system. *Annals of the New York Academy of Sciences*, *55*(2), 330–344.

Lewin, G. R., & Barde, Y. A. (1996). Physiology of the neurotrophins. *Annual Review of Neuroscience*, *19*, 289–317.

Liang, X. B., Liu, X. Y., Li, F. Q., Luo, Y., Lu, J., Zhang, W. M., et al. (2002). Long-term high-frequency electro-acupuncture stimulation prevents neuronal degeneration and up-regulates BDNF mRNA in the substantia nigra and ventral tegmental area following medial forebrain bundle axotomy. *Brain Research Molecular Brain Research*, *108*(1–2), 51–59.

Lim, C. E., & Wong, W. S. (2010). Current evidence of acupuncture on polycystic ovarian syndrome. *Gynecological Endocrinology*, *26*(6), 473–478.

Lindholm, D., Castren, E., Berzaghi, M., Blochl, A., & Thoenen, H. (1994). Activity-dependent and hormonal regulation of neurotrophin mRNA levels in the brain—Implications for neuronal plasticity. *Journal of Neurobiology*, *25*(11), 1362–1372.

Liu, F., Sun, W. W., Wang, Y., Hu, L. Q., Dai, P., Tian, C. F., et al. (2009). Effects of electro-acupuncture on NT-4 expression in spinal dorsal root ganglion and associated segments of the spinal dorsal horn in cats subjected to adjacent dorsal root ganglionectomy. *Neuroscience Letters*, *450*(2), 158–162.

Lund, I., & Lundeberg, T. (2006). Are minimal, superficial or sham acupuncture procedures acceptable as inert placebo controls? *Acupuncture in Medicine*, *24*(1), 13–15.
Lundeberg, T. (1984). A comparative study of the pain alleviating effect of vibratory stimulation, transcutaneous electrical nerve stimulation, electroacupuncture and placebo. *The American Journal of Chinese Medicine*, *12*(1–4), 72–79.
Lundeberg, T., Eriksson, S., Lundeberg, S., & Thomas, M. (1989). Acupuncture and sensory thresholds. *The American Journal of Chinese Medicine*, *17*(3–4), 99–110.
Maestripieri, D., De Simone, R., Aloe, L., & Alleva, E. (1990). Social status and nerve growth factor serum levels after agonistic encounters in mice. *Physiology & Behavior*, *47*(1), 161–164.
Maisonpierre, P. C., Belluscio, L., Squinto, S., Ip, N. Y., Furth, M. E., Lindsay, R. M., et al. (1990). Neurotrophin-3: A neurotrophic factor related to NGF and BDNF. *Science*, *247*(4949 Pt 1), 1446–1451.
Mamidipudi, V., & Wooten, M. W. (2002). Dual role for p75(NTR) signaling in survival and cell death: Can intracellular mediators provide an explanation? *Journal of Neuroscience Research*, *68*(4), 373–384.
Manneras, L., Cajander, S., Lonn, M., & Stener-Victorin, E. (2009). Acupuncture and exercise restore adipose tissue expression of sympathetic markers and improve ovarian morphology in rats with dihydrotestosterone-induced PCOS. *The American Journal of Physiology—Regulatory, Integrative and Comparative Physiology*, *296*(4), R1124–R1131.
Manneras, L., Jonsdottir, I. H., Holmang, A., Lonn, M., & Stener-Victorin, E. (2008). Low-frequency electro-acupuncture and physical exercise improve metabolic disturbances and modulate gene expression in adipose tissue in rats with dihydrotestosterone-induced polycystic ovary syndrome. *Endocrinology*, *149*(7), 3559–3568.
Manni, L., Antonelli, A., Costa, N., & Aloe, L. (2005). Stress alters vascular-endothelial growth factor expression in rat arteries: Role of nerve growth factor. *Basic Research in Cardiology*, *100*(2), 121–130.
Manni, L., Cajander, S., Lundeberg, T., Naylor, A. S., Aloe, L., Holmang, A., et al. (2005). Effect of exercise on ovarian morphology and expression of nerve growth factor and alpha(1)- and beta(2)-adrenergic receptors in rats with steroid-induced polycystic ovaries. *Journal of Neuroendocrinology*, *17*(12), 846–858.
Manni, L., Florenzano, F., & Aloe, L. (2011). Electroacupuncture counteracts the development of thermal hyperalgesia and the alteration of nerve growth factor and sensory neuromodulators induced by streptozotocin in adult rats. *Diabetologia*, *54*(7), 1900–1908.
Manni, L., Holmang, A., Cajander, S., Lundeberg, T., Aloe, L., & Stener-Victorin, E. (2006). Effect of anti-NGF on ovarian expression of alpha1- and beta2-adrenoceptors, TrkA, p75NTR, and tyrosine hydroxylase in rats with steroid-induced polycystic ovaries. *American Journal of Physiology Regulatory, Integrative and Comparative Physiology*, *290*(3), R826–R835.
Manni, L., Holmang, A., Lundeberg, T., Aloe, L., & Stener-Victorin, E. (2005). Ovarian expression of alpha (1)- and beta (2)-adrenoceptors and p75 neurotrophin receptors in rats with steroid-induced polycystic ovaries. *Autonomic Neuroscience*, *118*(1–2), 79–87.
Manni, L., Lundeberg, T., Holmang, A., Aloe, L., & Stener-Victorin, E. (2005). Effect of electro-acupuncture on ovarian expression of alpha (1)- and beta (2)-adrenoceptors, and p75 neurotrophin receptors in rats with steroid-induced polycystic ovaries. *Reproductive Biology and Endocrinology*, *3*, 21.
McAllister, A. K. (2001). Neurotrophins and neuronal differentiation in the central nervous system. *Cellular and Molecular Life Sciences*, *58*(8), 1054–1060.
McArthur, J. C., Yiannoutsos, C., Simpson, D. M., Adornato, B. T., Singer, E. J., Hollander, H., et al. (2000). A phase II trial of nerve growth factor for sensory neuropathy associated with HIV infection. AIDS Clinical Trials Group Team 291 [see comments]. *Neurology*, *54*(5), 1080–1088 (published erratum appears in *Neurology* 2000, *55*(1), 162).

Mi, W. L., Mao-Ying, Q. L., Wang, X. W., Li, X., Yang, C. J., Jiang, J. W., et al. (2011). Involvement of spinal neurotrophin-3 in electroacupuncture analgesia and inhibition of spinal glial activation in rat model of monoarthritis. *The Journal of Pain*, *12*(9), 974–984.

Moldenhauer, S., Burgauner, M., Hellweg, R., Lun, A., Hohenboken, M., Dietz, E., et al. (2010). Mobilization of CD133(+)CD34(−) cells in healthy individuals following whole-body acupuncture for spinal cord injuries. *Journal of Neuroscience Research*, *88*(8), 1645–1650.

Mufson, E. J., Kroin, J. S., Sendera, T. J., & Sobreviela, T. (1999). Distribution and retrograde transport of trophic factors in the central nervous system: Functional implications for the treatment of neurodegenerative diseases. *Progress in Neurobiology*, *57*(4), 451–484.

Nori, S. L., Rocco, M. L., Florenzano, F., Ciotti, M. T., Aloe, L., & Manni, L. (2013). Increased nerve growth factor signaling in sensory neurons of early diabetic rats is corrected by electroacupuncture. *Evidence-Based Complementary and Alternative Medicine*, *2013*, 652735.

Nykjaer, A., Lee, R., Teng, K. K., Jansen, P., Madsen, P., Nielsen, M. S., et al. (2004). Sortilin is essential for proNGF-induced neuronal cell death. *Nature*, *427*(6977), 843–848.

Ochs, G., Penn, R. D., York, M., Giess, R., Beck, M., Tonn, J., et al. (2000). A phase I/II trial of recombinant methionyl human brain derived neurotrophic factor administered by intrathecal infusion to patients with amyotrophic lateral sclerosis. *Amyotrophic Lateral Sclerosis and Other Motor Neuron Disorders*, *1*(3), 201–206.

Olson, L. (1993). Reparative strategies in the brain: Treatment strategies based on trophic factors and cell transfer techniques. *Acta Neurochirurgica Supplement (Wien)*, *58*, 3–7.

Pagani, L., Manni, L., & Aloe, L. (2006). Effects of electroacupuncture on retinal nerve growth factor and brain-derived neurotrophic factor expression in a rat model of retinitis pigmentosa. *Brain Research*, *1092*(1), 198–206.

Pan, W., Banks, W. A., & Kastin, A. J. (1998). Permeability of the blood-brain barrier to neurotrophins. *Brain Research*, *788*(1–2), 87–94.

Poo, M. M. (2001). Neurotrophins as synaptic modulators. *Nature Reviews Neuroscience*, *2*(1), 24–32.

Reichardt, L. F. (2006). Neurotrophin-regulated signalling pathways. *Philosophical Transactions of the Royal Society of London Series B, Biological Sciences*, *361*(1473), 1545–1564.

Rocco, M. L., Pristera, A., Pistillo, L., Aloe, L., Canu, N., & Manni, L. (2013). Brain cholinergic markers and tau phosphorylation are altered in experimental type 1 diabetes: Normalization by electroacupuncture. *Journal of Alzheimer's Disease*, *33*(3), 767–773.

Rush, R. A., Chie, E., Liu, D., Tafreshi, A., Zettler, C., & Zhou, X. F. (1997). Neurotrophic factors are required by mature sympathetic neurons for survival, transmission and connectivity. *Clinical and Experimental Pharmacology & Physiology*, *24*(8), 549–555.

Sahenk, Z., Nagaraja, H. N., McCracken, B. S., King, W. M., Freimer, M. L., Cedarbaum, J. M., et al. (2005). NT-3 promotes nerve regeneration and sensory improvement in CMT1A mouse models and in patients. *Neurology*, *65*(5), 681–689.

Sapolsky, R. M. (2000). Glucocorticoids and hippocampal atrophy in neuropsychiatric disorders. *Archives of General Psychiatry*, *57*(10), 925–935.

Schifitto, G., Yiannoutsos, C., Simpson, D. M., Adornato, B. T., Singer, E. J., Hollander, H., et al. (2001). Long-term treatment with recombinant nerve growth factor for HIV-associated sensory neuropathy. *Neurology*, *57*(7), 1313–1316.

Schor, N. F. (2005). The p75 neurotrophin receptor in human development and disease. *Progress in Neurobiology*, 77(3), 201–214.

Schweigreiter, R. (2006). The dual nature of neurotrophins. *Bioessays*, *28*(6), 583–594.

Seidah, N. G., Benjannet, S., Pareek, S., Chretien, M., & Murphy, R. A. (1996). Cellular processing of the neurotrophin precursors of NT3 and BDNF by the mammalian proprotein convertases. *FEBS Letters*, *379*(3), 247–250.

Seidah, N. G., Benjannet, S., Pareek, S., Savaria, D., Hamelin, J., Goulet, B., et al. (1996). Cellular processing of the nerve growth factor precursor by the mammalian pro-protein convertases. *The Biochemical Journal*, *314*(Pt 3), 951–960.

Seifert, T., Brassard, P., Wissenberg, M., Rasmussen, P., Nordby, P., Stallknecht, B., et al. (2010). Endurance training enhances BDNF release from the human brain. *American Journal of Physiology Regulatory, Integrative and Comparative Physiology*, *298*(2), R372–R377.

Seiler, M., & Schwab, M. E. (1984). Specific retrograde transport of nerve growth factor (NGF) from neocortex to nucleus basalis in the rat. *Brain Research*, *300*(1), 33–39.

Semkova, I., & Krieglstein, J. (1999). Neuroprotection mediated via neurotrophic factors and induction of neurotrophic factors. *Brain Research Brain Research Reviews*, *30*(2), 176–188.

Shelton, D. L., & Reichardt, L. F. (1986). Studies on the expression of the beta nerve growth factor (NGF) gene in the central nervous system: Level and regional distribution of NGF mRNA suggest that NGF functions as a trophic factor for several distinct populations of neurons. *Proceedings of the National Academy of Sciences of the United States of America*, *83*(8), 2714–2718.

Skeldal, S., Matusica, D., Nykjaer, A., & Coulson, E. J. (2011). Proteolytic processing of the p75 neurotrophin receptor: A prerequisite for signalling?: Neuronal life, growth and death signalling are crucially regulated by intra-membrane proteolysis and trafficking of p75(NTR). *Bioessays*, *33*(8), 614–625.

Smith, M. A. (1996). Hippocampal vulnerability to stress and aging: Possible role of neurotrophic factors. *Behavioural Brain Research*, *78*(1), 25–36.

Soule, J., Messaoudi, E., & Bramham, C. R. (2006). Brain-derived neurotrophic factor and control of synaptic consolidation in the adult brain. *Biochemical Society Transactions*, *34*(Pt 4), 600–604.

Stener-Victorin, E., Jedel, E., Janson, P. O., & Sverrisdottir, Y. B. (2009). Low-frequency electro-acupuncture and physical exercise decrease high muscle sympathetic nerve activity in polycystic ovary syndrome. *American Journal of Physiology Regulatory, Integrative and Comparative Physiology*, *297*(2), R387–R395.

Stener-Victorin, E., Lundeberg, T., Cajander, S., Aloe, L., Manni, L., Waldenstrom, U., et al. (2003). Steroid-induced polycystic ovaries in rats: Effect of electro-acupuncture on concentrations of endothelin-1 and nerve growth factor (NGF), and expression of NGF mRNA in the ovaries, the adrenal glands, and the central nervous system. *Reproductive Biology and Endocrinology*, *1*, 33.

Stener-Victorin, E., Lundeberg, T., Waldenstrom, U., Manni, L., Aloe, L., Gunnarsson, S., et al. (2000). Effects of electro-acupuncture on nerve growth factor and ovarian morphology in rats with experimentally induced polycystic ovaries. *Biology of Reproduction*, *63*(5), 1497–1503.

Stener-Victorin, E., Waldenstrom, U., Tagnfors, U., Lundeberg, T., Lindstedt, G., & Janson, P. O. (2000). Effects of electro-acupuncture on anovulation in women with polycystic ovary syndrome. *Acta Obstetricia et Gynecologica Scandinavica*, *79*(3), 180–188.

Strauss, O., Stumpff, F., Mergler, S., Wienrich, M., & Wiederholt, M. (1998). The Royal College of Surgeons rat: An animal model for inherited retinal degeneration with a still unknown genetic defect. *Acta Anatomica (Basel)*, *162*(2–3), 101–111.

Sun, N., Zou, X., Shi, J., Liu, X., Li, L., & Zhao, L. (2005). Electroacupuncture regulates NMDA receptor NR1 subunit expression via PI3-K pathway in a rat model of cerebral ischemia-reperfusion. *Brain Research*, *1064*(1–2), 98–107.

Taglialatela, G., Angelucci, L., Scaccianoce, S., Foreman, P. J., & Perez-Polo, J. R. (1991). Nerve growth factor modulates the activation of the hypothalamo-pituitary-adrenocortical axis during the stress response. *Endocrinology*, *129*(4), 2212–2218.

Teng, K. K., Felice, S., Kim, T., & Hempstead, B. L. (2010). Understanding proneurotrophin actions: Recent advances and challenges. *Developmental Neurobiology*, *70*(5), 350–359.

Terenghi, G. (1999). Peripheral nerve regeneration and neurotrophic factors. *Journal of Anatomy, 194*(Pt 1), 1–14.
Terry, A. V., Jr., Kutiyanawalla, A., & Pillai, A. (2011). Age-dependent alterations in nerve growth factor (NGF)-related proteins, sortilin, and learning and memory in rats. *Physiology & Behavior, 102*(2), 149–157.
Tirassa, P., Manni, L., Aloe, L., & Lundeberg, T. (2002). Cholecystokinin-8 and nerve growth factor: Two endogenous molecules working for the upkeep and repair of the nervous system. *Current Drug Targets CNS and Neurological Disorders, 1*(5), 495–510.
Tuszynski, M. H., Thal, L., Pay, M., Salmon, D. P., U, H. S., Bakay, R., et al. (2005). A phase 1 clinical trial of nerve growth factor gene therapy for Alzheimer disease. *Nature Medicine, 11*(5), 551–555.
Ulett, G. A., Han, S., & Han, J. S. (1998). Electroacupuncture: Mechanisms and clinical application. *Biological Psychiatry, 44*(2), 129–138.
Wang, T. H., Wang, X. Y., Li, X. L., Chen, H. M., & Wu, L. F. (2007). Effect of electroacupuncture on neurotrophin expression in cat spinal cord after partial dorsal rhizotomy. *Neurochemical Research, 32*(8), 1415–1422.
Wellmer, A., Misra, V. P., Sharief, M. K., Kopelman, P. G., & Anand, P. (2001). A double-blind placebo-controlled clinical trial of recombinant human brain-derived neurotrophic factor (rhBDNF) in diabetic polyneuropathy. *Journal of the Peripheral Nervous System, 6*(4), 204–210.
WHO (2003). *Acupuncture: Review and analysis of reports on controlled clinical trials.* From http://apps.who.int/medicinedocs/en/d/Js4926e/5.html#Js4926e.5.
Willnow, T. E., Petersen, C. M., & Nykjaer, A. (2008). VPS10P-domain receptors—Regulators of neuronal viability and function. *Nature Reviews Neuroscience, 9*(12), 899–909.
Wu, M. T., Hsieh, J. C., Xiong, J., Yang, C. F., Pan, H. B., Chen, Y. C., et al. (1999). Central nervous pathway for acupuncture stimulation: Localization of processing with functional MR imaging of the brain—Preliminary experience. *Radiology, 212*(1), 133–141.
Yan, Q., Ruan, J. W., Ding, Y., Li, W. J., Li, Y., & Zeng, Y. S. (2011). Electro-acupuncture promotes differentiation of mesenchymal stem cells, regeneration of nerve fibers and partial functional recovery after spinal cord injury. *Experimental and Toxicologic Pathology, 63*(1–2), 151–156.
Yun, S. J., Park, H. J., Yeom, M. J., Hahm, D. H., Lee, H. J., & Lee, E. H. (2002). Effect of electroacupuncture on the stress-induced changes in brain-derived neurotrophic factor expression in rat hippocampus. *Neuroscience Letters, 318*(2), 85–88.
Zafra, F., Castren, E., Thoenen, H., & Lindholm, D. (1991). Interplay between glutamate and gamma-aminobutyric acid transmitter systems in the physiological regulation of brain-derived neurotrophic factor and nerve growth factor synthesis in hippocampal neurons. *Proceedings of the National Academy of Sciences of the United States of America, 88*(22), 10037–10041.
Zhang, W. T., Jin, Z., Luo, F., Zhang, L., Zeng, Y. W., & Han, J. S. (2004). Evidence from brain imaging with fMRI supporting functional specificity of acupoints in humans. *Neuroscience Letters, 354*(1), 50–53.
Zhao, Z. Q. (2008). Neural mechanism underlying acupuncture analgesia. *Progress in Neurobiology, 85*(4), 355–375.
Zhao, L., Shi, J., Sun, N., Tian, S., Meng, X., Liu, X., et al. (2005). Effect of electroacupuncture on TRPM7 mRNA expression after cerebral ischemia/reperfusion in rats via TrkA pathway. *Journal of Huazhong University of Science and Technology Medical Sciences, 25*(3), 247–250.
Zhao, L., Wang, Y., Sun, N., Liu, X., Li, L., & Shi, J. (2007). Electroacupuncture regulates TRPM7 expression through the trkA/PI3K pathway after cerebral ischemia-reperfusion in rats. *Life Sciences, 81*(15), 1211–1222.

Zhen, Y. F., Zhang, J., Liu, X. Y., Fang, H., Tian, L. B., Zhou, D. H., et al. (2013). Low BDNF is associated with cognitive deficits in patients with type 2 diabetes. *Psychopharmacology*, *227*(1), 93–100.

Zhou, X. F., & Rush, R. A. (1994). Localization of neurotrophin-3-like immunoreactivity in the rat central nervous system. *Brain Research*, *643*(1–2), 162–172.

Zijlstra, F. J., van den Berg-de Lange, I., Huygen, F. J., & Klein, J. (2003). Anti-inflammatory actions of acupuncture. *Mediators of Inflammation*, *12*(2), 59–69.

CHAPTER SIX

Acupuncture Stimulation and Neuroendocrine Regulation

Jung-Sheng Yu[*,†], Bai-Yun Zeng[‡], Ching-Liang Hsieh[§,¶,||,1]
[*]Department of Chinese Medicine, Chi Mei Medical Center, Tainan, Taiwan, ROC
[†]Department of Cosmetic Science and Institute of Cosmetic Science, College of Pharmacy and Science, Chia Nan University of Pharmacy & Science, Tainan, Taiwan, ROC
[‡]Neurodegenerative Disease Research Group, Institute of Pharmaceutical Science, School of Biomedical Sciences, King's College London, London, United Kingdom
[§]Graduate Institute of Integrated Medicine, College of Chinese Medicine, China Medical University, Taichung, Taiwan, ROC
[¶]Acupuncture Research Center, China Medical University, Taichung, Taiwan, ROC
[||]Department of Chinese Medicine, China Medical University Hospital, Taichung, Taiwan, ROC
[1]Corresponding author: e-mail address: clhsieh@mail.cmuh.org.tw

Contents

Abstract

Acupuncture has been used to treat different conditions for at least 3000 years in China and has gained increasing acceptance worldwide. The acupuncture needle inserted into the muscle layer at the acupoint produces the so-called obtaining qi sensation that causes the excitation of A-δ and C-fibers of the muscle tissue, resulting in afferent signals. The afferent signals pass through the dorsal horn cells of the spinal cord ascending to the brain, such as the hypothalamus, enhancing the release of neuropeptides and hormones, and these afferent signals in the spinal segment may innervate the visceral organ, inducing effect on visceral function. Here, we reviewed the effect of acupuncture stimulation on neuropeptides and hormones, including β-endorphin, serotonin, oxytocin, adrenocorticotropic hormone, gonadotropin-releasing hormone, corticotrophin-releasing hormone, cholecystokinin, and acetylcholine, as well as insulin sensitivity, immunomodulation (anti-inflammation), and autonomic nerve activity.

International Review of Neurobiology, Volume 111
ISSN 0074-7742
http://dx.doi.org/10.1016/B978-0-12-411545-3.00006-7

1. INTRODUCTION

Acupuncture has been used for treating various diseases for the past 3000 years in China and has gained increasing popularity and acceptance worldwide. The benefits of acupuncture stimulation for treating postoperative and chemotherapy nausea and vomiting in adults, as well as relieving postoperative dental pain, have been confirmed through consensus (N.I.H. Consensus Conference, 1998). According to traditional Chinese medicine, acupuncture can balance physiological functions between yin and yang and between qi and blood in the human body, which is similar to the neuroendocrine regulation in human physiology. The spinal nerve segmental effect plays an important role in acupuncture according to our previous study (Jan, Li, & Hsieh, 2010). Both the autonomic nerve and the spinal segmental effect are involved in the neural regulation that elicited by acupuncture stimulation. So, acupuncture exerts its neural modulation effect through the peripheral and central pathways.

Hormones can be categorized into local hormones and general hormones. Cholecystokinin (CCK) is a hormone exist both the gastrointestinal system and central nervous system and an antiopioid peptide. Electroacupuncture (EA) mediates the increase in CCK expression and vasoactive intestine peptide-containing cell distribution in the duodenum and the sphincter of Oddi to regulate biliary motility (Kuo, Chiu, Lin, Hsieh, & Wu, 2005). General hormones, such as the growth hormone and the thyroid hormone, are involved in cellular metabolism. EA induces the release of the corticotrophin-releasing hormone (CRH) in the paraventricular nucleus (PVN) of the hypothalamus (Zhao, Tian, & Chen, 2003). The CRH may modulate the hypothalamus–pituitary–adrenal axis (HPAA) and the hypothalamus–pituitary–adrenal gland axis (HPGA) to affect the physiological functions of the human body (Chrousos, Torpy, & Gold, 1998). Here, we focus on acupuncture stimulation and the neuroendocrine system to assess the effects of acupuncture on the HPGA, CCK, acetylcholine, and autonomic system.

2. ACUPUNCTURE STIMULATION AND THE HYPOTHALAMUS AND PITUITARY GLAND

Neurotransmitters and neuropeptides regulate brain function. Previous studies (Han, 2003) showed that EA can induce the β-endorphin release in the arcuate nucleus of the hypothalamus, affecting the periaqueductal gray (PAG) matter of the midbrain, medulla oblongata, and spinal cord. This neural pathway is closely related to the analgesic effect of low-frequency EA (Han, 2003). Hsieh et al. (2001), using positron emission tomography, showed that acupuncture stimulation at Hegu activated endogenous, antinociceptive pathways in the hypothalamus and the midbrain because of increasing in endorphinergic neurons in the hypothalamus. These endorphinergic neurons descend to the PAG and raphe nucleus; however, this increase was not found at the sham acupoint indicating the specificity of analgesic acupoints (Hsieh et al., 2001).

Previous studies showed that either EA or noxious thermal stimulations in anesthetized rats induced c-fos protein expression in most anterior lobe cells, which colocalize with ACTH or beta-endorphin immunoreactivity, and ACTH and beta-endorphin concentrations were markedly increased in plasma in rats (Pan, Castro-Lopes, & Coimbra, 1996). In the same study, hypothalamic c-fos expression was increased only in the mediobasal nuclei, that is, the arcuate nucleus and paraventricular nucleus, but more in the former than the latter. However, the same stimulations did not induce c-fos expression in pituitary intermediate lobe (Pan et al., 1996). This suggests that electroacupuncture stimulation has specific activation of mediobasal hypothalamic nuclei of HPAA and induces opiate release from pituitary gland, which may contribute to the acupuncture analgesia effect. In a subsequent study, the effect of EA and noxious stimulations on c-fos expression was investigated in rats deprived of nociceptive primary afferent input following neonatal capsaicin treatment (Pan, Castro-Lopes, & Coimbra, 1997). There was no increase in c-fos expression in the paraventricular, arcuate, and other hypothalamic nuclei and the pituitary gland and in the plasma release of ACTH following noxious stimulation or EA treatment. However, c-fos activation provoked by immobilization stress though markedly decreased was not abolished by capsaicin, whereas plasma release of ACTH remained undiminished (Pan et al., 1997). These findings suggest that EA acts on the hypothalamic pituitary corticotropic axis through an exclusively physical effect depending on the noxious signal elicited in the somatosensory pathway.

The effects of acupuncture on CRH have been investigated in rodents to explore its mechanisms in many conditions. Previous study reported that EA at the CV3, ST30, and SP6 acupoints induced a greater increase of CRH levels, in ovariectomized (OVX) rats than in control groups, in the PVN of the hypothalamus (Zhao et al., 2003). In another study (Li et al.,

2008), EA at GB30 acupoint increased the serum ACTH compared to sham EA 5 h after complete Freund's adjuvant (CFA)-induced hind paw inflammation and hyperalgesia in rats, and EA also activated the CRH neuron in the PVN of the hypothalamus, suggesting that EA mediated CRH to increase ACTH, producing an antiedema effect (Li et al., 2008).

3. ACUPUNCTURE STIMULATION AND CHOLECYSTOKININ

CCK, a peptide hormone of gastrointestinal system responsible for stimulating digestion of fat and protein, is abundant in the brain in addition to its secretion in the duodenum. The previous studies, used the tail-flick latency (TFL) response to determine the response or nonresponse to pain, induced by radiant-heat tests in rats, to assess whether expression of hypothalamic CCK and its receptors is related to the responsiveness to EA analgesia (Ko et al., 2006). The data showed that although EA at Zusanli acupoint in rats increased CCK mRNA expression to a greater degree in the nonresponder group than in the responder group, no significant difference occurred in the thalamus between the two groups; however, CCK-A and CCK-B receptor mRNA expression increased in the nonresponder group to a greater degree than in the responder group, showing that the CCK receptor in the hypothalamus may be more important than CCK in the analgesic effect of high-frequency EA (Ko et al., 2006). In another study (Lee et al., 2003), EA at Zusanli acupoint, in CCK-A receptor gene knockout Otsuka Long-Evans Tokushima Fatty (OLETF) rats and Long-Evans Tokushima Otsuka (LETO) rats, increased the TLF time of the radiant-heat test in both groups, indicating that the expression of the CCK-A receptor is related to the EA analgesia (Lee et al., 2003). In radiant-heat-treated rat model, the same EA treatment increased CCK-A receptor expression in the hypothalamus in the nonresponder group compared to the responder group, whereas CCK-B receptor expression in the hypothalamus was similar between the responder and nonresponder groups, suggesting that CCK receptor density also plays a vital role in acupuncture analgesia (Lee et al., 2002). Studies of CCK-1 receptor modulation showed that EA at Zusanli reduced 30 and 60 min food intake in 48 h-fasted rats and these EA effects can be reversed by pretreating the lorglumide, a CCK-1 receptor antagonist (Kim et al., 2008). Further, EAs at Zusanli acupoint only induce satiety in LETO rats, but not in CCK-1 receptor knockout OLETF rats; whereas, satiety effect of EA can be attenuated by bilateral subdiaphragmatic vagotomy. These suggest that EA-induced satiety effect was mediated through endogenous CCK to activate the CCK-1 receptor signal pathway (Kim et al., 2008). Moreover, EA at Zusanli acupoints inhibited the pain-exciting neuron and potentiated the pain-inhibited neuron of the caudate nucleus, prolonging the TFL time induced by radiant-heat noxious stimulation in rats (Yang et al., 2010). These EA effects can be antagonized by an intracerebral ventricular injection of CCK-8, and the effect of CCK-8 that antagonizes EA analgesia can be reversed by injecting L-365, 260, an antagonist of the CCK-B receptor, into the caudate nucleus (Yang et al., 2010). In summary, those studies suggest that acupuncture analgesia is closely associated with the hypothalamic CCK expression.

4. ACUPUNCTURE AND DIABETES MELLITUS

Diabetes mellitus (DM) is a metabolic disease primarily characterized by hyperglycemia, resulting in defective insulin secretion or insulin action. DM is typically classified into type I diabetes because of an immune system attack or the destruction of insulin-producing β cells in the pancreas, called insulin-dependent diabetes or juvenile-onset diabetes, which occurs in approximately 5–10% of DM patients (American Diabetes Association, 2010). Type II DM is referred to as noninsulin-dependent diabetes or adult-onset diabetes and occurs in approximately 90–95% of DM patients (American Diabetes Association, 2010). Although the body may sufficiently produce insulin in type II DM, it cannot use the insulin due to insulin resistance. We focus on the effects of acupuncture in controlling the blood sugar and enhancing the insulin sensitivity of DM patients.

Cabioglu and Ergene (2006) investigated the effects of EA on serum insulin levels in 52 healthy women, EA stimulation at ear acupoints Hunger and Shen Men and at body acupoints LI4, LI11, ST36, and ST44, and produced a greater reduction of body weight and blood glucose in the EA group than in the control group. The serum insulin and C-peptide levels increased to a greater degree in the EA group than in the control group; so EA reduces blood sugar by increasing insulin and C-peptide levels in serum (Cabioglu & Ergene, 2006). In another study, EA at CV12 and CV4 reduced plasma glucose levels in type II (noninsulin-dependent) diabetic rats or normal rats, and the reduction of plasma glucose from EA can be abolished by naloxone, an opiate antagonist, suggesting that EA effect was mediated through inducing the secretion of β-endorphin, which decreased serum glucose levels (Chang, Lin, Chi, Liu, & Cheng, 1999). In another study, the effect of EA stimulation on the abdomen on glucose tolerance in the Goto–Kakizaki (GK) rat, a genetic model of type II diabetes, was investigated by Ishizaki, Okushi, Yano, and Yamamura (2009). It was shown that EA stimulation elicited a significant decrease in fasting blood glucose and increase in plasma insulin level and improvement in insulin sensitivity in GK rats (Ishizaki et al., 2009). These results indicate that EA improves hyperglycemia and increases insulin levels in both type I and type II diabetic models.

Previously, Shapira et al. (2000) showed that EA at the Zhongwan and Guanyuan acupoints produced a long-term hypoglycemia effect in diabetic *Psammomys obesus* (sand rats) compared to control group, but plasma insulin levels showed a nonsignificant difference between EA and sham EA groups, suggesting that EA enhanced insulin sensitivity (Shapira et al., 2000). Recently, it has been reported that EA applied at Zusanli and Guanyuan acupoints in genetic obese and type II diabetic db/db mice improved insulin sensitivity through partially activating sirtuin 1/PGC-1α in the skeletal muscle (Liang et al., 2011). A similar study found that EA at bilateral abdominal rectus muscle, soleus, and gastrocnemius enhanced insulin sensitivity through increasing the expression of glucose transport 4 in hydrotestosterone-induced POCS rats (Johansson et al., 2010). In another type II DM animal model, established by intraperitoneally injecting STZ (60 mg/kg) into rats, characterized by higher plasma blood glucose and lower plasma insulin levels, EA stimulation at bilateral Zusanli acupoints reduced the blood glucose level effect and increases the insulin level in type II diabetic as well as in normal rats (Pai et al., 2009). Tominaga, Ishizaki, Naruse, Kitakoji, and Yamamura (2011) reported that EA at bilateral Zusanli acupoints in rats with high-fructose diet (HFD)-induced insulin resistance increased the expression of AMP-activated protein kinase in the skeletal muscle, suggesting that EA improved HFD-induced insulin resistance (Tominaga et al., 2011). In summary, EA stimulation induced multiple effects that improve plasma glucose levels, which is beneficial to DM treatment. The effects include the release of endogenous β-endorphin, increased insulin levels, reduced insulin resistance, and increased insulin sensitivity.

5. ACUPUNCTURE STIMULATION AND FEMALE HORMONES

Polycystic ovarian syndrome (PCOS) is the most common endocrine disorder in women and is typically associated with high androgen levels, ovulatory dysfunction, and metabolism disturbances such as obesity, DM, hyperinsulinemia, insulin resistance, and cardiovascular disorder. Although the etiology of POCS is unclear, it is suggested that the primary defects of the HPAA resulting in increased sympathetic activity are one of the important factors (Stener-Victorin, Jedel, & Manneras, 2008). PCOS development may be caused by the insufficient inhibition of the central β-endorphin to GnRH release or the hyperactivity of peripheral sympathetic activity in the ovaries (Stener-Victorin et al., 2000). PCOSs are characterized by increase in GnRH secretion and in the luteinizing hormone (LH)/follicle-stimulating hormone (FSH) ratio (Hayes, Taylor, Martin, & Hall, 1998). Previous studies showed that EA stimulation at the biceps femoris and erector spinae, corresponding to ovary innervations, reduced enodothelin-1 and the levels of the nerve growth factor of ovaries in steroid-induced PCOS rats (Stener-Victorin et al., 2003). These results suggest that EA can modulate sympathetic nerve activities to modulate the neuroendocrinological function of ovaries. Further, acupuncture at the bilateral SP6, LR3, and CV3 acupoints significantly decreased the serum estradiol levels and endometrial leukaemia-inhibitor factor (LIF) and Osteopontin (OPN) expression and enhanced glandular development in the clomiphene citrate-induced rat model during the implantation period (Fu et al., 2011). Therefore, acupuncture improved endometrial receptivity, which is beneficial to female reproduction. In 5-αdihydrostesterone-induced PCOS rats, EA at the abdominal rectus muscle induced recovery of the estrous cycle and impeded increasing hypothalamic GnRH and AR expression, suggesting that EA has a beneficial modulation effect on PCOS (Feng et al., 2009). In another study, EA at the GV20, GV4, BL23, and K13 acupoints simultaneously downregulated the expression of neuropeptide Y (NPY) and GnRH in the hypothalamus in rats during early puberty (Zhaohui et al., 2012).

In a nonrandomized, longitudinal study, EA applied on BL23, BL28, and SP9 acupoints induced an increase in ovulation and decrease in the LH/FSH ratio, testosterone, and β-endorphin levels in women with PCOS (Stener-Victorin et al., 2000). A randomized, double-blind, sham-controlled clinical study in 96 women with PCOS was assigned to the acupuncture treatment group (46 patients) or the sham acupuncture group (50 patients) (Pastore, Williams, Jenkins, & Patrie, 2011). The results indicated a similar ovulation rate and a similar LH/FSH ratio decline compared to baseline between the two groups, but lower fasting insulin and free testosterone were correlated with the ovulation rate in the acupuncture group only, suggesting that acupuncture has a larger effect on women with less severe metabolic disturbances (Pastore et al., 2011). Very recently, the effect of acupuncture was shown on 11 patients with primary ovarian insufficient (POI) (Zhou, Jiang, Wu, & Liu, 2013). EA at BL33, CV4, ST29, and ST25 significantly increased estradiol 2 level at the end of treatment and remained at elevated levels during 3-month follow-up and markedly decreased FSH and LH at the end of treatment, which remained stable during the 3-month follow-up (Zhou et al., 2013). Ten of 11 patients also regained menstruation, and hot flashes and night sweating symptoms were improved, indicating that EA is beneficial in treating women with POI (Zhou et al., 2013). These studies suggest that acupuncture stimulation at specific acupoints modulates release of gonadotropin-releasing hormone. Overall, acupuncture stimulation at acupoints, corresponding to ovary innervations, produced both central and peripheral effects. The central effect is mediated by the release of the hypothalamic β-endorphin leading to inhibiting GnRF, resulting in a reduced LH/FSH ratio and estradiol and androgen levels of the ovaries. The

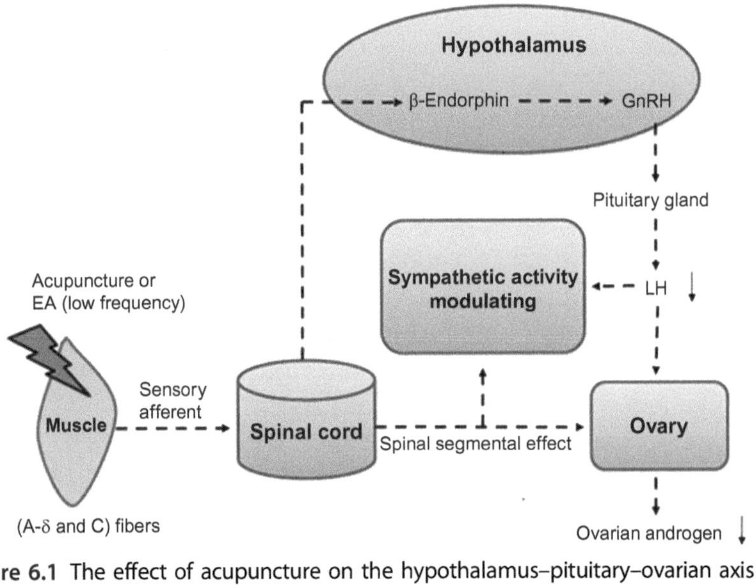

Figure 6.1 The effect of acupuncture on the hypothalamus–pituitary–ovarian axis and the sympathetic nervous system. Acupuncture needle insertion into the muscle layer followed by manual twisting (or low-frequency EA) produces obtaining qi with fish biting on bait feeling by the acupuncturist, and soreness, numbness or distension feeling by the subject, stimulating Aδ- and C-fibers leading to the generation of sensory afferent. This afferent impulse enhances the release of the hypothalamic β-endorphin, resulting in decreased gonadotropin-releasing hormone (GnRH) secretion, LH, and sympathetic nerve activities (central effect). The production of the spinal segmental effect inhibits the sympathetic nerve activity of ovaries (peripheral effect); thus, acupuncture stimulation may produce both central and peripheral effects; EA: electroacupuncture. For color version of this figure, the reader is referred to the online version of this chapter.

peripheral effect occurs through the spinal segmental effect. Finally, both the central and peripheral effects inhibit sympathetic nerve activities (Fig. 6.1).

6. ACUPUNCTURE STIMULATION AND ADRENAL GLANDS

Acupuncture has been used to treat numerous types of inflammatory disease, such as asthma and arthritis (Zijlstra, van den Berg-de Lange, Huygen, & Klein, 2003). Several studies have examined the negative feedback circuit between the autonomic nervous system (ANS) and innate immunity (Kavoussi & Ross, 2007). Stress can initiate the activity in HPAA (Ma & Morilak, 2005) and vagal efferent (Kavoussi & Ross, 2007). Acupuncture inhibits the production of proinflammatory cytokines, such as IL-1β and tumor necrosis factor-α through modulating cholinergic anti-inflammatory pathway, and plays a modulating role in ANS, neuroimmune system, and hormone regulation (Cho et al., 2006; Kavoussi & Ross, 2007). We previously showed that EA at bilateral GB30 acupoint reduced CFA-induced hind paw edema and increased the plasma levels of corticosterone; however, EA effects could be partially blocked by adrenalectomy (ADX), suggesting that EA exerted its anti-inflammatory action through activating HPAA (Zhang et al., 2005). Similarly, EA at bilateral Huantiao acupoint reduced edema and increased plasma corticosterone levels in the CFA-induced inflammation rat model (Li et al., 2007). However, this EA anti-edema effect can be blocked by ADX, suggesting that EA mediated adrenal gland activation and increased plasma corticosterone levels to improve edema (Li et al., 2007). Previous studies reported that acupuncture at GB20, Neiguan, and Zusanli acupoints increased the cortisol levels by 28% at 15 min and 50% at 45 min after acupuncture, respectively, in humans, indicating that cortisol or neurohormones may mediate the acupuncture effect (Lee, Yin, Lee, Tsai, & Sim, 1982). Further, acupuncture at Sanyinjiao acupoint inhibited leukocyte infiltration, neutrophils activity, and abdominal vascular permeability and increased the levels of peritoneal fluid in mice with carrageenan-induced peritonitis, whereas these acupuncture effects were reversed by ADX (da Silva et al., 2011).

Acupuncture stimulation at different frequencies result in eliciting different neural activities. For example, Kim et al. (2008) reported that, although 1 and 120 Hz EAs at bilateral Zusanli acupoint reduce edema, neutrophils, and thermal hyperalgesia in the carrageenan-induced paw inflammation mouse model, the anti-inflammatory effect of 1 Hz EA could be abolished by

6-hydroxydopamine treatment, whereas ADX abolished the anti-inflammatory effect of 120 Hz EA. In addition, propranolol a sympathetic nonselective beta-blocker completely abolishes the anti-inflammatory effect ofboth 1 and 120 Hz EAs. One hertz EA acts through the sympathetic post-ganglionic neuron, whereas 120 Hz EA acts through the sympathoadrenal medullary axis (Kim et al., 2008). Together, acupuncture on the adrenal cortex primarily induces immune modulation or anti-inflammatory effects involving the secretion of substances from the adrenal cortex including cortisol, corticosterone, and proinflammatory cytokines such as IL-1β and IL-10. On the other hand, acupuncture on the adrenal medulla modulates sympathetic nerve activity, including adrenalin and noradrenalin.

7. ACUPUNCTURE STIMULATION AND ANALGESIA

The acupuncture analgesia mechanism has been widely explored since the 1970s. These include acupuncture-induced inhibition of the release of histamine, noradrenalin, and dopamine in the PAG region and of endogenous opiates, including β-endorphin, enkephalin, endomorphin, and dynorphin (Murotani et al., 2010). The serotoninergic descending inhibitory pathway is considered to play an important role in acupuncture analgesia, which has become focused on the ANS and inflammatory reaction (Lin & Chen, 2008).

Endorphins are endogenous opioid peptides that functions as neurotransmitters. Different frequencies of EA stimulation could determine the release of different endogenous opiates in the CNS (Han, 2003). It has been reported that 2 Hz EA analgesia is mediated by μ- and δ-opioid receptors, whereas 100 Hz EA analgesia is mediated by the κ-opioid receptors (Chen & Han, 1992). EA stimulation at a frequency of 2 Hz induces the release of β-endorphin, enkephalin, and endomorphin within the CNS network, whereas 100 Hz EA stimulation causes the release of dynorphin (Chen & Han, 1992).

Serotonin (5-HT) and both 5-HT1A and 5-HT3 receptors mediate the EA-induced analgesic effect in the mice model with formalin-induced pain (Chang, Tsai, Yu, Yi, & Lin, 2004). The serotonergic neurons are major components of the raphe nuclei, and the rostral raphe nuclei project to the forebrain, whereas caudal raphe nuclei project to the brain stem and spinal cord (Hornung, 2003). Stamford (1995) reported that pain sensation can be modulated through the peripheral ascent to the cerebral cortex or the descent from the cerebral cortex, thalamus, and the brain stem, including the PAG, raphe nucleus, and the locus coeruleus, which is called the descending pain control system. The primary neurotransmitters of the descending pain control system include serotonin, noradrenalin, and endogenous opioids (Stamford, 1995). The serotonin pathway links the spinal cord and the brain stem, which plays an important role in mood and emotion (Suzuki, Rygh, & Dickenson, 2004). Acupuncture at Shenshu acupoint increased the release of serotonin from the nucleus accumbens in rats (Yoshimoto et al., 2006). Similar results have been observed in the patients with fibromyalgia. Acupuncture treatment over 6 weeks increased serum serotonin levels and decreased pain (Sprott, Franke, Kluge, & Hein, 1998). Low-frequency EA induced the arcuate nucleus of the hypothalamus to activate the PAG region through β-endorphin (Han, 2003). Serotonin and inhibitory interneurons modulated the expression of excitatory amino acids and their receptors in the spinal cord. These receptors are involved in nociception, and EA downregulated their expression (Choi, Kang, & Jo, 2005). EA at a frequency of 2–100 Hz at the Hegu and Sanyinjiao acupoints relieved labor pain through the release of β-endorphin and serotonin in the CNS, which affects the uterus directly (Qu & Zhou, 2007).

Oxytocin is a nonapeptide posterior pituitary hormone that is primarily synthesized in the hypothalamic PVN and the hypothalamic supraoptic nucleus. Intraventricular or intrathecal injection of oxytocin can enhance the analgesic effect of EA at Zusanli acupoint, whereas antioxytocin serum can weaken the analgesic effect of EA (Yang et al., 2007). However, the effect of EA is not affected by intravenous administration of oxytocin or antioxytocin, indicating that oxytocin acting as analgesia is central origin (Yang et al., 2007). Antinociceptive effect of oxytocin was made through activating oxytocin receptor and involved in μ-opioid receptor activity in the nucleus raphe magnus (Wang, Lundeberg, & Yu, 2003). Clinically, oxytocin plus 2–100 Hz EA at the bilateral Hegu acupoints enhanced uterine contraction and shortened the birth process in puerperant women with uterine inertia (Liu et al., 2008), suggesting a synergic effect of acupuncture and oxytocin.

Zhang, Zhang, Yung, and Zhang (2004) reported that EA at Zusanli and Sanyinjiao acupoints reduced edema and c-fos expression of the spinal cord in the ipsilateral carrageenan-induced edema model, and this EA effect was not affected by the opioid antagonist naloxone. Therefore, EA may inhibit inflammatory response and exerts its central neuronal activity through a nonopioid-dependent mechanism (Zhang et al., 2004). In healthy human subjects, EA at bilateral Hegu and Konqzui acupoints induced a normalized lymphocyte pattern and reduced heart rate resulting from parasympathetic nerve activity, suggesting that EA modulates immunologic response by regulating the ANS (Mori, Nishijo, Kawamura, & Abo, 2002). In another report, Son et al. (2002) found that acupuncture at HT8, BL66, and LR2 acupoints reduced hypothalamic mRNA levels of proinflammatory cytokine interleukin (IL)-6 and IL-1β induced by lipopolysaccharide (LPS) stimulation in rats; thus, acupuncture produced antipyretic effects through the downregulation of IL-6 and IL-1β cytokines (Son et al., 2002). The development of neuropathic pain and proinflammatory cytokine IL-6 and IL-1β levels increased in the peripheral nerve and the dorsal root ganglion (DRG) injury rat model. EA stimulation at Zusanli and Yinlingquan acupoints reduced neuropathic pain and levels of IL-6 and IL-1β of the peripheral nerve and DRG, suggesting anti-inflammation of EA might be partly mediated through inhibition of proinflammatory cytokines (Cha, Nam, Kwak, Lee, & Lee, 2012). Similar results have been shown that EA at Huantiao acupoint attenuated bone cancer cell-induced increase in thermal hyperalgesia and upregulation of IL-1β (Zhang et al., 2007). Together, those studies showed that the analgesic effect of acupuncture involved in modulating endogenous opiates, serotonin, and oxytocin release and regulating the functions of the autonomic and immune systems.

8. SUMMARY

Neuroendocrine regulation by acupuncture stimulation has been extensively investigated. Acupuncture can induce the releases of hormones, neurotransmitters, and neuromodulators in both central and peripheral nervous systems. Cumulative evidence suggests that acupuncture-induced neuroendocrine modulation affects many conditions, like analgesia, reproduction-related symptoms, drug addiction, and psychological disorders. However, the mechanisms underlying the regulation are not fully understood and thus are worthy of further investigation.

REFERENCES

American Diabetes Association (2010). Diagnosis and classification of diabetes mellitus. *Diabetes Care*, *33*(Suppl. 1), S62–S69. http://dx.doi.org/10.2337/dc10-S062.

Cabioglu, M. T., & Ergene, N. (2006). Changes in levels of serum insulin, C-peptide and glucose after electroacupuncture and diet therapy in obese women. *The American Journal of Chinese Medicine*, *34*(3), 367–376. http://dx.doi.org/10.1142/S0192415X06003904.

Cha, M. H., Nam, T. S., Kwak, Y., Lee, H., & Lee, B. H. (2012). Changes in cytokine expression after electroacupuncture in neuropathic rats. *Evidence-Based Complementary and Alternative Medicine*, *2012*, 792765. http://dx.doi.org/10.1155/2012/792765.

Chang, S. L., Lin, J. G., Chi, T. C., Liu, I. M., & Cheng, J. T. (1999). An insulin-dependent hypoglycaemia induced by electroacupuncture at the Zhongwan (CV12) acupoint in diabetic rats. *Diabetologia*, *42*(2), 250–255. http://dx.doi.org/10.1007/s001250051146.

Chang, F. C., Tsai, H. Y., Yu, M. C., Yi, P. L., & Lin, J. G. (2004). The central serotonergic system mediates the analgesic effect of electroacupuncture on ZUSANLI (ST36) acupoints. *Journal of Biomedical Science*, *11*(2), 179–185. http://dx.doi.org/10.1159/000076030.

Chen, X. H., & Han, J. S. (1992). Analgesia induced by electroacupuncture of different frequencies is mediated by different types of opioid receptors: Another cross-tolerance study. *Behavioural Brain Research*, *47*(2), 143–149.

Cho, Z. H., Hwang, S. C., Wong, E. K., Son, Y. D., Kang, C. K., Park, T. S., et al. (2006). Neural substrates, experimental evidences and functional hypothesis of acupuncture mechanisms. *Acta Neurologica Scandinavica*, *113*(6), 370–377. http://dx.doi.org/10.1111/j.1600-0404.2006.00600.x.

Choi, B. T., Kang, J., & Jo, U. B. (2005). Effects of electroacupuncture with different frequencies on spinal ionotropic glutamate receptor expression in complete Freund's adjuvant-injected rat. *Acta Histochemica*, *107*(1), 67–76. http://dx.doi.org/10.1016/j.acthis.2004.07.008.

Chrousos, G. P., Torpy, D. J., & Gold, P. W. (1998). Interactions between the hypothalamic-pituitary-adrenal axis and the female reproductive system: Clinical implications. *Annals of Internal Medicine*, *129*(3), 229–240.

da Silva, M. D., Guginski, G., Werner, M. F., Baggio, C. H., Marcon, R., & Santos, A. R. (2011). Involvement of interleukin-10 in the anti-inflammatory effect of Sanyinjiao (SP6) acupuncture in a mouse model of peritonitis. *Evidence-Based Complementary and Alternative Medicine*, *2011*, 217946. http://dx.doi.org/10.1093/ecam/neq036.

Feng, Y., Johansson, J., Shao, R., Manneras, L., Fernandez-Rodriguez, J., Billig, H., et al. (2009). Hypothalamic neuroendocrine functions in rats with dihydrotestosterone-induced polycystic ovary syndrome: Effects of low-frequency electro-acupuncture. *PLoS One*, *4*(8), e6638. http://dx.doi.org/10.1371/journal.pone.0006638.

Fu, H., He, Y., Gao, Y., Man, Y., Liu, W., & Hao, H. (2011). Acupuncture on the endometrial morphology, the serum estradiol and progesterone levels, and the expression of endometrial leukaemia-inhibitor factor and osteopontin in rats. *Evidence-Based Complementary and Alternative Medicine*, *2011*, 606514. http://dx.doi.org/10.1155/2011/606514.

Han, J. S. (2003). Acupuncture: Neuropeptide release produced by electrical stimulation of different frequencies. *Trends in Neurosciences*, *26*(1), 17–22.

Hayes, F. J., Taylor, A. E., Martin, K. A., & Hall, J. E. (1998). Use of a gonadotropin-releasing hormone antagonist as a physiologic probe in polycystic ovary syndrome: Assessment of neuroendocrine and androgen dynamics. *Journal of Clinical Endocrinology and Metabolism*, *83*(7), 2343–2349.

Hornung, J. P. (2003). The human raphe nuclei and the serotonergic system. *Journal of Chemical Neuroanatomy*, *26*(4), 331–343.

Hsieh, J. C., Tu, C. H., Chen, F. P., Chen, M. C., Yeh, T. C., Cheng, H. C., et al. (2001). Activation of the hypothalamus characterizes the acupuncture stimulation at the analgesic acupoint in human: A positron emission tomography study. *Neuroscience Letters*, *307*(2), 105–108.

Ishizaki, N., Okushi, N., Yano, T., & Yamamura, Y. (2009). Improvement in glucose tolerance as a result of enhanced insulin sensitivity during electroacupuncture in spontaneously diabetic Goto-Kakizaki rats. *Metabolism, 58*(10), 1372–1378. http://dx.doi.org/10.1016/j.metabol.2009.05.001.

Jan, Y. M., Li, T. C., & Hsieh, C. L. (2010). A segmental effect involved in the changes of skin blood flow induced by acupuncture in normal health human. *The American Journal of Chinese Medicine, 38*(3), 441–448. http://dx.doi.org/10.1142/S0192415X10007968.

Johansson, J., Feng, Y., Shao, R., Lonn, M., Billig, H., & Stener-Victorin, E. (2010). Intense electroacupuncture normalizes insulin sensitivity, increases muscle GLUT4 content, and improves lipid profile in a rat model of polycystic ovary syndrome. *American Journal of Physiology, Endocrinology and Metabolism, 299*(4), E551–E559. http://dx.doi.org/10.1152/ajpendo.00323.2010.

Kavoussi, B., & Ross, B. E. (2007). The neuroimmune basis of anti-inflammatory acupuncture. *Integrative Cancer Therapies, 6*(3), 251–257. http://dx.doi.org/10.1177/1534735407305892.

Kim, H. W., Uh, D. K., Yoon, S. Y., Roh, D. H., Kwon, Y. B., Han, H. J., et al. (2008). Low-frequency electroacupuncture suppresses carrageenan-induced paw inflammation in mice via sympathetic post-ganglionic neurons, while high-frequency EA suppression is mediated by the sympathoadrenal medullary axis. *Brain Research Bulletin, 75*(5), 698–705. http://dx.doi.org/10.1016/j.brainresbull.2007.11.015.

Ko, E. S., Kim, S. K., Kim, J. T., Lee, G., Han, J. B., Rho, S. W., et al. (2006). The difference in mRNA expressions of hypothalamic CCK and CCK-A and -B receptors between responder and non-responder rats to high frequency electroacupuncture analgesia. *Peptides, 27*(7), 1841–1845. http://dx.doi.org/10.1016/j.peptides.2006.01.002.

Kuo, Y. L., Chiu, J. H., Lin, J. G., Hsieh, C. L., & Wu, C. W. (2005). Localization of cholecystokinin and vasoactive intestinal peptide in lower biliary tract in cats following electroacupuncture on right Qimen (LR14) and Riyue (GB 24): An immunohistochemistry study. *Acupuncture & Electro-Therapeutics Research, 30*(1–2), 15–25.

Lee, G. S., Han, J. B., Shin, M. K., Hong, M. C., Kim, S. W., Min, B. I., et al. (2003). Enhancement of electroacupuncture-induced analgesic effect in cholecystokinin-A receptor deficient rats. *Brain Research Bulletin, 62*(2), 161–164.

Lee, G., Rho, S., Shin, M., Hong, M., Min, Bi, & Bae, H. (2002). The association of cholecystokinin-A receptor expression with the responsiveness of electroacupuncture analgesic effects in rat. *Neuroscience Letters, 325*(1), 17–20.

Lee, S. C., Yin, S. J., Lee, M. L., Tsai, W. J., & Sim, C. B. (1982). Effects of acupuncture on serum cortisol level and dopamine beta-hydroxylase activity in normal Chinese. *The American Journal of Chinese Medicine, 10*(1–4), 62–69. http://dx.doi.org/10.1142/S0192415X82000117.

Li, A., Lao, L., Wang, Y., Xin, J., Ren, K., Berman, B. M., et al. (2008). Electroacupuncture activates corticotrophin-releasing hormone-containing neurons in the paraventricular nucleus of the hypothalamus to alleviate edema in a rat model of inflammation. *BMC Complementary and Alternative Medicine, 8*, 20. http://dx.doi.org/10.1186/1472-6882-8-20.

Li, A., Zhang, R. X., Wang, Y., Zhang, H., Ren, K., Berman, B. M., et al. (2007). Corticosterone mediates electroacupuncture-produced anti-edema in a rat model of inflammation. *BMC Complementary and Alternative Medicine*, 7, 27. http://dx.doi.org/10.1186/1472-6882-7-27.

Liang, F., Chen, R., Nakagawa, A., Nishizawa, M., Tsuda, S., Wang, H., et al. (2011). Low-frequency electroacupuncture improves insulin sensitivity in obese diabetic mice through activation of SIRT1/PGC-1alpha in skeletal muscle. *Evidence-Based Complementary and Alternative Medicine, 2011*, 735297. http://dx.doi.org/10.1155/2011/735297.

Lin, J. G., & Chen, W. L. (2008). Acupuncture analgesia: A review of its mechanisms of actions. *The American Journal of Chinese Medicine, 36*(4), 635–645. http://dx.doi.org/10.1142/S0192415X08006107.

Liu, J., Han, Y., Zhang, N., Wang, B., Zhou, Y., Yang, D., et al. (2008). The safety of electroacupuncture at hegu (LI 4) plus oxytocin for hastening uterine contraction of puerperants—A randomized controlled clinical observation. *Journal of Traditional Chinese Medicine, 28*(3), 163–167.

Ma, S., & Morilak, D. A. (2005). Chronic intermittent cold stress sensitises the hypothalamic-pituitary-adrenal response to a novel acute stress by enhancing noradrenergic influence in the rat paraventricular nucleus. *Journal of Neuroendocrinology, 17*(11), 761–769.

Mori, H., Nishijo, K., Kawamura, H., & Abo, T. (2002). Unique immunomodulation by electro-acupuncture in humans possibly via stimulation of the autonomic nervous system. *Neuroscience Letters, 320*(1–2), 21–24.

Murotani, T., Ishizuka, T., Nakazawa, H., Wang, X., Mori, K., Sasaki, K., et al. (2010). Possible involvement of histamine, dopamine, and noradrenalin in the periaqueductal gray in electroacupuncture pain relief. *Brain Research, 1306*, 62–68. http://dx.doi.org/10.1016/j.brainres.2009.09.117.

N.I.H. Consensus Conference (1998). Acupuncture. *Journal of the American Medical Association, 280*(17), 1518–1524.

Pai, H. C., Tzeng, C. Y., Lee, Y. C., Chang, C. H., Lin, J. G., Cheng, J. T., et al. (2009). Increase in plasma glucose lowering action of rosiglitazone by electroacupuncture at bilateral Zusanli acupoints (ST.36) in rats. *Journal of Acupuncture and Meridian Studies, 2*(2), 147–151. http://dx.doi.org/10.1016/S2005-2901(09)60047-9.

Pan, B., Castro-Lopes, J. M., & Coimbra, A. (1996). Activation of anterior lobe corticotrophs by electroacupuncture or noxious stimulation in the anaesthetized rat, as shown by colocalization of Fos protein with ACTH and beta-endorphin and increased hormone release. *Brain Research Bulletin, 40*(3), 175–182.

Pan, B., Castro-Lopes, J. M., & Coimbra, A. (1997). Chemical sensory deafferentation abolishes hypothalamic pituitary activation induced by noxious stimulation or electroacupuncture but only decreases that caused by immobilization stress. A c-fos study. *Neuroscience, 78*(4), 1059–1068.

Pastore, L. M., Williams, C. D., Jenkins, J., & Patrie, J. T. (2011). True and sham acupuncture produced similar frequency of ovulation and improved LH to FSH ratios in women with polycystic ovary syndrome. *Journal of Clinical Endocrinology and Metabolism, 96*(10), 3143–3150. http://dx.doi.org/10.1210/jc.2011-1126.

Qu, F., & Zhou, J. (2007). Electro-acupuncture in relieving labor pain. *Evidence-Based Complementary and Alternative Medicine, 4*(1), 125–130. http://dx.doi.org/10.1093/ecam/nel053.

Shapira, M. Y., Appelbaum, E. Y., Hirshberg, B., Mizrahi, Y., Bar-On, H., & Ziv, E. (2000). A sustained, non-insulin related, hypoglycaemic effect of electroacupuncture in diabetic Psammomys obesus. *Diabetologia, 43*(6), 809–813. http://dx.doi.org/10.1007/s001250051379.

Son, Y. S., Park, H. J., Kwon, O. B., Jung, S. C., Shin, H. C., & Lim, S. (2002). Antipyretic effects of acupuncture on the lipopolysaccharide-induced fever and expression of interleukin-6 and interleukin-1 beta mRNAs in the hypothalamus of rats. *Neuroscience Letters, 319*(1), 45–48.

Sprott, H., Franke, S., Kluge, H., & Hein, G. (1998). Pain treatment of fibromyalgia by acupuncture. *Rheumatology International, 18*(1), 35–36.

Stamford, J. A. (1995). Descending control of pain. *British Journal of Anaesthesia, 75*(2), 217–227.

Stener-Victorin, E., Jedel, E., & Manneras, L. (2008). Acupuncture in polycystic ovary syndrome: Current experimental and clinical evidence. *Journal of Neuroendocrinology, 20*(3), 290–298. http://dx.doi.org/10.1111/j.1365-2826.2007.01634.x.

Stener-Victorin, E., Lundeberg, T., Cajander, S., Aloe, L., Manni, L., Waldenstrom, U., et al. (2003). Steroid-induced polycystic ovaries in rats: Effect of electro-acupuncture on concentrations of endothelin-1 and nerve growth factor (NGF), and expression of NGF mRNA in the ovaries, the adrenal glands, and the central nervous system. *Reproductive Biology and Endocrinology*, *1*, 33.

Stener-Victorin, E., Waldenstrom, U., Tagnfors, U., Lundeberg, T., Lindstedt, G., & Janson, P. O. (2000). Effects of electro-acupuncture on anovulation in women with polycystic ovary syndrome. *Acta Obstetricia et Gynecologica Scandinavica*, *79*(3), 180–188.

Suzuki, R., Rygh, L. J., & Dickenson, A. H. (2004). Bad news from the brain: Descending 5-HT pathways that control spinal pain processing. *Trends in Pharmacological Sciences*, *25*(12), 613–617. http://dx.doi.org/10.1016/j.tips.2004.10.002.

Tominaga, A., Ishizaki, N., Naruse, Y., Kitakoji, H., & Yamamura, Y. (2011). Repeated application of low-frequency electroacupuncture improves high-fructose diet-induced insulin resistance in rats. *Acupuncture in Medicine*, *29*(4), 276–283. http://dx.doi.org/10.1136/acupmed-2011-010006.

Wang, J. W., Lundeberg, T., & Yu, L. C. (2003). Antinociceptive role of oxytocin in the nucleus raphe magnus of rats, an involvement of mu-opioid receptor. *Regulatory Peptides*, *115*(3), 153–159.

Yang, C. X., Shi, T. F., Liang, Q. C., Yang, B. F., Jiao, R. S., Zhang, H., et al. (2010). Cholecystokinin-8 antagonizes electroacupuncture analgesia through its B receptor in the caudate nucleus. *Neuromodulation*, *13*(2), 93–98. http://dx.doi.org/10.1111/j.1525-1403.2009.00247.x.

Yang, J., Yang, Y., Chen, J. M., Liu, W. Y., Wang, C. H., & Lin, B. C. (2007). Effect of oxytocin on acupuncture analgesia in the rat. *Neuropeptides*, *41*(5), 285–292. http://dx.doi.org/10.1016/j.npep.2007.05.004.

Yoshimoto, K., Fukuda, F., Hori, M., Kato, B., Kato, H., Hattori, H., et al. (2006). Acupuncture stimulates the release of serotonin, but not dopamine, in the rat nucleus accumbens. *The Tohoku Journal of Experimental Medicine*, *208*(4), 321–326.

Zhang, R. X., Lao, L., Wang, X., Fan, A., Wang, L., Ren, K., et al. (2005). Electroacupuncture attenuates inflammation in a rat model. *Journal of Alternative and Complementary Medicine*, *11*(1), 135–142. http://dx.doi.org/10.1089/acm.2005.11.135.

Zhang, R. X., Li, A., Liu, B., Wang, L., Ren, K., Qiao, J. T., et al. (2007). Electroacupuncture attenuates bone cancer pain and inhibits spinal interleukin-1 beta expression in a rat model. *Anesthesia and Analgesia*, *105*(5), 1482–1488. http://dx.doi.org/10.1213/01.ane.0000284705.34629.c5.

Zhang, S. P., Zhang, J. S., Yung, K. K., & Zhang, H. Q. (2004). Non-opioid-dependent anti-inflammatory effects of low frequency electroacupuncture. *Brain Research Bulletin*, *62*(4), 327–334.

Zhao, H., Tian, Z. Z., & Chen, B. Y. (2003). Increased corticortropin-releasing hormone release in ovariectomized rats' paraventricular nucleus: Effects of electroacupuncture. *Neuroscience Letters*, *353*(1), 37–40.

Zhaohui, Z., Jingzhu, Z., Guipeng, D., Xuesong, W., Yuanming, Z., Yinping, W., et al. (2012). Role of neuropeptide Y in regulating hypothalamus-pituitary-gonad axis in the rats treated with electro-acupuncture. *Neuropeptides*, *46*(3), 133–139. http://dx.doi.org/10.1016/j.npep.2012.03.002.

Zhou, K., Jiang, J., Wu, J., & Liu, Z. (2013). Electroacupuncture modulates reproductive hormone levels in patients with primary ovarian insufficiency: Results from a prospective observational study. *Evidence-Based Complementary and Alternative Medicine*, *2013*, 657234. http://dx.doi.org/10.1155/2013/657234.

Zijlstra, F. J., van den Berg-de Lange, I., Huygen, F. J., & Klein, J. (2003). Anti-inflammatory actions of acupuncture. *Mediators of Inflammation*, *12*(2), 59–69. http://dx.doi.org/10.1080/0962935031000114943.

CHAPTER SEVEN

Current Development of Acupuncture Research in Parkinson's Disease

Bai-Yun Zeng[1], Sarah Salvage, Peter Jenner
Neurodegenerative Disease Research Group, Institute of Pharmaceutical Science, School of Biomedical Sciences, King's College London, London, United Kingdom
[1]Corresponding author: e-mail address: b.zeng@kcl.ac.uk

Contents

Abstract

Parkinson's disease is an age-related progressive neurodegenerative disease. The etiology and pathogenetic mechanisms that cause PD are still not fully understood. The available treatments to PD are only symptomatic relief. Acupuncture is used to treat many medical conditions for 1000 years in China and has gained wider and increasing acceptance within both public and medical profession because it has been a very safe and well-tolerated treatment. In this chapter, we reviewed relevant laboratory findings regarding acupuncture mechanism on Parkinson's. We showed that acupuncture stimulation in Parkinson's models had generated valuable mechanistic insight of Parkinson's and showed that acupuncture treatment is in fact a neuroprotective therapy that increase the release of various neuroprotective agents such as brain-derived neurotrophic factor, glial cell line-derived neurotrophic factor, and cyclophilin A. In addition, acupuncture therapy slows cell death process and attenuates oxidative stress to dopaminergic neurons in the substantia nigra. Further, acupuncture therapy modulates neuronal activity of the basal ganglia output structures. These results suggest that early application of acupuncture therapy to Parkinson's patients may be helpful for the best

International Review of Neurobiology, Volume 111
ISSN 0074-7742
http://dx.doi.org/10.1016/B978-0-12-411545-3.00007-9

efficacy of acupuncture treatment. It is hopeful that translation of achievement in acupuncture research in Parkinson's models will maximize the potentials of acupuncture treatment.

1. INTRODUCTION

Parkinson's disease (PD) is an age-related progressive neurodegenerative disorder that affects approximately 1 in 1000 population over the age of 60 years. The main cause of PD is the gradual loss of dopaminergic neurons in the substantia nigra pars compacta (SNpc) and subsequent dopamine depletion in the striatum. The main clinical symptoms include bradykinesia, resting tremor, rigidity, and postural instability, and many nonmotor symptoms frequently appear in PD (Chaudhuri, Odin, Antonini, & Martinez-Martin, 2011; Jankovic, 2008). The etiology and pathogenetic mechanisms that cause PD remain unclear, although some clues can be derived from the disease's neuropathologic features (Schapira & Jenner, 2011). Dopaminergic cell death is associated with increased oxidative stress, dysfunction of mitochondrial respiratory chain, apoptosis, excitotoxicity, loss of neurotrophic support, inflammation, and impairment of protein degradation pathway (Jenner, 2003; Sulzer, 2007). Classic description of PD mostly focuses on the progressive degeneration of dopaminergic neurons located in the SNpc. This pathology is thought to underlie several extrapyramidal motor symptoms, although it is believed that it does not explain other signs and symptoms that are associated with PD (Braak, Ghebremedhin, Rüb, Bratzke, & Del Tredici, 2004; Halliday, Del Tredici, & Braak, 2006). Pathology in other brainstem and subcortical and cortical structures is also prominent (Wolters & Braak, 2006). Recent clinical and *in vivo* studies demonstrating signs of parkinsonism-related degeneration in the spinal cord suggest much widespread pathological process of the disease (Knaryan, Samantaray, Le Gal, Ray, & Banik, 2011). In the clinic, treatment with levodopa, the precursor of dopamine, is only a symptomatic relief within limited time rather than a cure (Stocchi, Tagliati, & Olanow, 2008). Long-term treatment with levodopa induces adverse effects such as dyskinesia and motor fluctuation within approximately 50% of PD patients within 3–5 years following medication (Ahlskog & Muenster, 2001).

Acupuncture has been used to treat the medical problems for three thousand years in China (Wu, 1996). The concept of acupuncture is based on the premise that life energy, known as Qi, flows through channels or meridians throughout the body connected to all organs and to each other. In disease state, life energy is imbalanced within the meridians and acupuncture is believed to restore the healthy balance energy by stimulating the specific points along the meridians. Acupuncture, including electroacupuncture (EA), as part of complementary medicine, has gained increasing acceptance and popularity in recent years with both the public and physicians and is one of the most popular types of complementary medicines because it has been reported to be a very safe and well-tolerated treatment with only minor side effects (Rabinstein & Shulman, 2003; Shulman et al., 2002). Many patients with PD are reported using acupuncture as an alternative treatment during some points of their life. Indeed, it has been estimated that more than a quarter of patients with PD in the United States (40%), Britain (38.7%), Singapore (61%), and Argentina (25.7%) have used at least one form of complementary medicines for PD, while 7–49.4% of them have used acupuncture as an alternative therapy (Ferry et al., 2002; Pecci et al., 2010; Rajendran, Thompson, & Reich, 2001; Tan, Lau, Jamora, & Chan, 2006; Wang et al., 2013). In the patients who received the acupuncture treatment, the improvements were noted in motor symptoms and nonmotor symptoms (Cristian, Katz, Cutrone, & Walker, 2005; Eng, Lyons, Greene, & Pahwa, 2006; Shulman et al., 2002). A recent study reported that PD patients who underwent 8-week of twice-weekly acupuncture treatment showed significant improvement in motor symptoms (Unified Parkinson's Disease Rating Scale, UPDRS part III) and total UPDRS score compared to their baseline assessment (Cho et al., 2012). Although the potential benefits of acupuncture on patients with PD have gained increased attention, there is surprisingly little research on the mechanisms of acupuncture on the treatment of PD. By integrating the traditional Chinese medicine and modern biomedical science, many researchers have tried to reevaluate the effects of acupuncture on PD and find out the scientific explanations behind such effects. In this chapter, we have looked at the recent development of research studies of effects and mechanisms of acupuncture treatment on PD and, in particular, focused on effects of acupuncture and its underlying mechanisms on Parkinson's animal models.

2. NEUROPROTECTIVE EFFECT OF ACUPUNCTURE ON ANIMAL MODELS OF PD

The animal models used in acupuncture research are the same models of PD used in neuroscience research, including 6-hydroxydopamine

(6-OHDA)-lesioned rats, medial forebrain bundle (MFB)-axotomized rats, and 1-methyl-4-phenyl-1,2,3,6-tetrahydropyridine (MPTP)-treated mice (Betarbet, Sherer, & Greenamyre, 2002, Kang et al., 2007; Liang et al., 2002, 2003; Park et al., 2003). Those models have shown similar pathophysiology of human PD and proved highly valuable and effective in the search for clues to understand the underlying causes of illness (Dauer & Przedborski, 2003; Duty & Jenner, 2011).

In rats with 6-OHDA infusion into the striatum, acupuncture stimulation at acupoints GB34 and LR3 or at acupoints GV14 and GV20 for 14 days markedly reduced rotational behavioral and significantly reduced the loss of dopaminergic neurons in the SN ipsilateral to the lesioned side compared to lesioned rats with nonacupoint stimulation (Kim et al., 2005; Park et al., 2003); however, it only marginally increased tyrosine hydroxylase (TH)-immunoreactive fiber density in the lesioned striatum compared to lesioned group (Kim et al., 2005). Similarly, in rats with 6-OHDA infusion into the MFB, acupuncture stimulation at acupoints GB34, LR3, ST36, and SP10 for 14 days suppressed abnormal behavioral and protected nigral dopamine neurons from 6-OHDA toxicity (Yu et al., 2010).

To illustrate whether different frequencies of EA stimulation exert different effects in the models of PD, studies of EA at 2 and 100 Hz were carried out in MFB-axotomized rat. It was shown that both frequency stimulations at acupoints GV14 and GV20 significantly decreased rotational behavioral after 2-week treatment in unilateral MFB-axotomized rat. However, only 100 Hz stimulation showed significant rotational behavior reductions after 4-week treatment (Jia et al., 2009; Liang et al., 2003). Along with the behavioral recovery, acupuncture at 100 Hz significantly protected the degeneration of dopaminergic neurons in the SN (Jia et al., 2009; Liang et al., 2002) and HPLC analysis showed a trend of increase in dopamine content in the striatum (Jia et al., 2010; Liang et al., 2003), whereas 2 Hz stimulation did not show such a protection after 4-week treatment (Jia et al., 2010, 2009; Liang et al., 2002, 2003).

In MPTP-treated mice, acupuncture stimulation at acupoint GB34 for 12 days induced a significant motor function recovery by showing a lower overall rod performance compared to control group and almost completely rescued dopaminergic neurons from MPTP toxic insults; however, it did not restore striatal dopamine content (Jeon et al., 2008; Kim, Doo, et al., 2011). Very recently, 100 Hz EA stimulation at acupoints SP6 and ST36 in MPTP-treated mice showed that decreased the elevated total movements and the movement time, velocity, and distance was accompanied by increasing striatal dopamine content and decreasing striatal turnover compared to sham control group (Wang et al., 2013). In another study, acupuncture stimulation at GB34 and LR3 acupoints up to 7 days not only completely inhibited MPTP-induced loss of TH-immunoreactive dopaminergic neurons in the SNpc but also restored dopamine expression in the striatum, as shown by Western blot (Choi, Yeo, Hong, & Lim, 2011; Kang et al., 2007). Further, the expression of dopamine transporter in the striatum was restored by the acupuncture treatment at the same acupoints (Choi, Park, & Lim, 2009; Choi, Yeo, Hong, & Lim, 2011). Taken together, these studies clearly demonstrated that acupuncture treatment is effective to protect dopaminergic neurons against neurotoxin and mechanical lesion in nigrostriatal system.

3. NEUROPROTECTIVE MECHANISMS OF ACUPUNCTURE ON ANIMAL MODELS OF PD

3.1. Increased neurotrophic factor-related gene and protein expression in the basal ganglia

In 6-OHDA-lesioned and MFB-axotomized rats, acupuncture treatment at GV14 and GV20 or GB34 and LR3 markedly increased TrkB protein expression and mRNA expression of BDNF in the ipsilateral substantia nigra while improving motor behavioral and reducing the loss of dopaminergic neurons (Liang et al., 2002; Park et al., 2003). In a similar treatment regime, GDNF mRNA levels were increased in both sides of the globus pallidus (Liang et al., 2003). These studies demonstrated that acupuncture stimulation increases the expression of neurotrophic factors not only in the dopaminergic cell body but also in their terminal region. This suggests that the neuroprotection of acupuncture may be partly due to its stimulating expression of neurotrophic factors at multiple sites and at both protein and mRNA levels.

Using peptide fingerprinting mass spectroscopy to detect changes in proteins in the SN of MPTP-treated mice, Jeon et al. identified the changes in 22 proteins in MPTP-treated SN and nine of 22 proteins were normalized by 100 Hz EA at GB34 stimulation (Jeon et al., 2008). These acupuncture-normalized proteins were identified as cell death regulatory, inflammatory, or cell damage-restoring proteins, while acupuncture attenuated dopaminergic degeneration. Another finding in this study was that the levels of a natural neuroprotective agent, cyclophilin A, that was not changed in the SN in MPTP-treated alone mice were significantly elevated in the acupuncture-treated MPTP mice (Jeon et al., 2008). In other studies, cDNA microarray technology was applied to identify new biomarkers in response to acupuncture stimulation at GB34 in MPTP-treated mice, and it found that acupuncture stimulation suppressed MPTP-increased cytokine–cytokine receptor interaction pathway-related genes and oxidative phosphorylation pathway-related genes in the SN of MPTP-treated mice (Hong et al., 2010). These biomarkers could be potential targets for understanding the mechanism of acupuncture or other new therapy for PD. The same technique was used to identify gene profiles in the striatum and thalamus of MPTP-treated mice with acupuncture stimulation at GB34 and LR3 acupoints, and it was found that a proportion of probes upregulated by MPTP were attenuated by acupuncture and that acupuncture stimulation exerted an inhibitory effect on MPTP-induced striatal and thalamic degeneration (Choi, Yeo, Hong, & Lim, 2011; Yeo, Choi, Hong, & Lim, 2013). In another study, acupuncture stimulation at the same acupoints altered gene expression in the cervical spinal cord of MPTP-treated mice although it is unclear where the altered genes are located because the cervical spinal cord is not separated into dorsal and ventral roots in this study (Choi, Yeo, Hong, Kim, & Lim, 2011).

3.2. Antioxidative stress effect of acupuncture treatment

Oxidative stress has been considered to play an essential role in dopaminergic neuron death (Grunblatt, Mandel, & Youdim, 2000; Jenner, Dexter, Sian, Schapira, & Marsden, 1992). Accumulated evidence showed marked alterations in a range of oxidative-related markers such as superoxide dismutase (SOD), glutathione peroxidase (GSH-Px), catalase (CAT), reduced glutathione (GSH), and malondialdehyde (MDA) in the SN of postmortem brain of PD patients and PD models as well (Betarbet et al., 2002; Dauer & Przedborski, 2003; Javitch, D'Amato, Strittmatter, & Snyder, 1985; Perumal, Gopal, Tordzro, Cooper, & Cadet, 1992; Riederer et al., 1989; Sian et al., 1994). In 6-OHDA-lesioned rats, acupuncture stimulation at GB34, LR3, ST36, and SP10 acupoints for 14 days reversed 6-OHDA-induced changes in SOD, GSH, GSH-Px, CAT, and MDA in the lesioned SN and inhibited the production of hydrogen peroxide and MDA and increased GSH and SOD activity in the striatum of MPTP-treated mice while improving the motor function and protected dopaminergic neurons from MPTP toxicity (Wang et al., 2011; Yu et al., 2010). Further, acupuncture stimulation at the same acupoints previously mentioned counteracted MPTP-induced increase in oxidative stress-related proteins, such as cytosolic malate dehydrogenase and hydroxyacylglutathione hydrolase (Kim et al., 2010). This suggests that neuroprotective effect of acupuncture may be mediated, at least in part, by regulating antioxidant system.

Phosphatidylinositol-3-kinase/protein kinase B (PI3K/Akt) pathway is a key transducer of brain metabolic and mitogenic signals involved in neuronal survival, neurogenesis, and apoptosis (Engelman, Luo, & Cantley, 2006). A recent study confirmed DJ-1/Akt signaling pathway is necessary for neuroprotection against MPTP-induced oxidative stress (Aleyasin et al., 2010). Very recently, Kim et al. investigated the effect and underlying mechanism of acupuncture stimulation in MPTP-treated mice and they found that acupuncture effectively restored the MPTP-induced impairment of Akt activation in nigral dopaminergic neurons (Kim, Kim, Doo, Park, & Moon, 2011). Furthermore, acupuncture-induced dopaminergic neuron protection and motor function improvement were significantly blocked by administration of LY294002, a specific inhibitor of PI3K/Akt signaling pathway (Kim, Kim, et al., 2011). This study provided evidence that PI3K/Akt signaling pathway may play an important role in the mechanism underlying acupuncture-induced neuroprotection in mouse model of PD.

Properly regulated iron is necessary to maintain the vitality of the brain; however, excessive iron uptake can lead to cell damage and induce iron-mediated neuropathologic impairments (Burdo & Connor, 2003). Increased levels of iron were found in the SN of PD patients and animal models after treatment with MPTP or 6-OHDA, which indicates that iron metabolism is closely related to the pathogenesis of PD (Mandel, Maor, & Youdim, 2004; Sofic, Riederer, Heinsen, Beckmann, & Reynolds, 1988; Wang, Jiang, & Xie, 2004). In parkinsonian mice, ferric ion (Fe^{3+}) was markedly increased and ferritin heavy chain (FtH) was significantly decreased in the SN after chronic MPTP administration (Choi et al., 2009). Acupuncture stimulation at LR3 and GB34 acupoints for 15 days reversed the changes in ferric ion (Fe^{3+}) and ferritin heavy chain (FtH) induced by MPTP in those mice while protecting dopaminergic neurons in the SNpc and dopaminergic terminals and dopamine transporter in the striatum against MPTP toxicity (Choi et al., 2009). This suggests that acupuncture may inhibit iron-related oxidative damage and prevent the deleterious change in iron metabolism in PD animal model.

Overall, the accumulated evidence suggests that acupuncture simulation at certain acupoints in parkinsonian models may act as antioxidant agent to protect dopamine neurons against neurodegeneration in parkinsonian rodent models.

3.3. Anti-inflammation effect of acupuncture treatment

Studies at cellular and molecular levels suggested that activated microglia cells might contribute to dopaminergic cell death by releasing cytotoxic inflammatory compounds such as tumor necrosis factor (TNF)-α, IL-1β, and interferon-γ. Among these cytokines, TNF-α is believed to have a direct damaging effect on dopaminergic neurons by activating an intracellular death pathway coupled with TNF receptor 1 expressed on the cell surface of those neurons (Hirsch & Hunot, 2009). Pathways transduced by activation of TNF receptor 1 are linked to the induced expression of COX-2 within dopaminergic neurons (Hirsch & Hunot, 2009; Tansey, McCoy, & Frank-Cannon, 2007). The changes seen in the brains of patients with PD have been reproduced in several animal models, which provide strong evidence that neuroinflammatory processes are involved in the death of dopaminergic neurons (Akiyama & McGeer, 1989; Liberatore et al., 1999; Su et al., 2008). These studies suggest that microglial activation is a common feature of animal models in which parkinsonism is induced by neurotoxins or manipulations of genes involved in inherited forms of disease and that any approaches leading to the inhibition of microglial activation, COX-2, and iNOS expression and of the release of the proinflammatory cytokines will halt the inflammation process and stop or slow nigrostriatal degeneration (Wu et al., 2002a, 2002b).

Acupuncture has a marked anti-inflammatory effect on animal models of PD (Kang et al., 2007; Liu et al., 2004). In MFB-axotomized rats, acupuncture stimulation at acupoints Du14 and Du21 for 24 days significantly attenuated MFB axotomy-induced microglial activation and reversed the upregulation of TNF-α and IL-1β mRNA expression in the SNpc while reducing the loss of dopaminergic neurons from MFB-axotomized damage (Liu et al., 2004). Further, in animals that received acupuncture treatment after axotomy, the number of ED-1-positive macrophage cells decreased by 47% compared to axotomy-alone group (Liu et al., 2004). In MPTP-treated mice, acupuncture stimulation at acupoints GB34 and LR3 neutralized MPTP toxin-induced activation of microglia and reduced the increase in COX-2 and iNOS expression in the SNpc while not only protecting dopaminergic neurons from MPTP-induced neurodegeneration but also elevating the decreased dopamine levels in the striatum (Kang et al., 2007). Collectively, these studies suggest that anti-inflammatory effect of acupuncture at least in part acts as the inhibitors of COX-2 and iNOS expression in the SNpc in parkinsonian brain.

The MFB axotomy causes apoptosis of dopaminergic neurons in rat SN and ED-1-positive phagocytic microglia were apposed to axotomized dopaminergic neurons that were morphologically and biochemically intact 3 days after axotomy, indicating the ongoing microglial phagocytosis of degenerating dopamine neurons (Sugama et al., 2003). The occurrence of microglial phagocytosis at the early stage of apoptosis may indicate the evolution of apoptosis into an irreversible state. Thus, it is particularly important for neuroprotection to suppress early phagocytic action of microglia. In the studies by Liu et al. (2004), acupuncture was applied at the early time points after mechanical insults in order to obtain a meaning efficacy. This suggests that early acupuncture treatment may block phagocytic activity of microglia cells and attenuate the further loss of dopaminergic neurons by providing an opportunity for the axotomized dopaminergic neurons to recover from the injury in the presence of efficient neuroprotectants such as BDNF and GDNF simultaneously induced by acupuncture stimulation (Liang et al., 2002; Park et al., 2003).

4. MODULATORY EFFECT OF ACUPUNCTURE ON NEURONAL ACTIVITY IN THE BASAL GANGLIA

Acupuncture stimulation at acupoints not only induces the expression of neurotrophic factors and acts as an antioxidant and inflammatory inhibitor in parkinsonian rodent models but also may rebalance neuronal activity in the basal ganglia circuits. In MFB-axotomized rat, radioimmunoassay test showed that EA stimulation at acupoints GV14 and GV20 for 4 weeks from day 2 after MFB lesion reversed axotomy-induced decrease in substance P content in the ventral midbrain although it did not affect axotomy-induced increase in enkephalin in the globus pallidus (Jia et al., 2009). Further, in the same setting of another study, HPLC analysis showed EA attenuated MFB-induced increase in GABA content in the ventral midbrain but had no effect on GABA content in the globus pallidus (Jia et al., 2010). The level of glutamic acid decarboxylase (GAD), the rate-limiting enzyme for GABA synthesis, is altered in the parkinsonian condition; for example, the GAD67 protein and gene expression are increased in dopamine-depleted striatum, globus pallidus, and SN in rats (Segovia & Garcia-Munoz, 1987; Soghomonian & Chesselet, 1992). EA stimulation reversed MFB-induced increase in GAD67 mRNA levels in the ventral midbrain (Jia et al., 2009).

In both studies (Jia et al., 2010, 2009), EA treatment protected dopaminergic neurons against the mechanical-induced damage and improved motor coordination measured by the RotaRod treadmill test and reduced amphetamine-induced rotational behavioral although there is no significant increase in striatal dopamine level. This again suggests that modulation of neuronal activity by EA in the output nuclei of the basal ganglia such as SNpr even without improving striatal dopamine may still result in functional recovery and these studies provided the molecular basis to support this opinion. It has been reported that nigral dopaminergic neurons release dopamine not only from their striatal terminals but also from the dendrite network to the SNpr (Cheramy, Leviel, & Glowinski, 1981; Robertson & Robertson, 1989). So the basal ganglia output structure may play an important role in the improvement of motor function. It has been reported that intranigral grafts of either dopaminergic or GABAergic tissue or GDNF can improve the behavioral outcome in the Parkinson's models without affecting striatal neurotransmission (Gash et al., 1996; Tseng, Baetge, Zurn, & Aebischer, 1997; Winkler, Bentlage, Cenci, Nikkhah, & Björklund, 2003). Further, high-frequency stimulation of STN in rats reversed 6-OHDA-induced increase in GAD67 mRNA in the SN and partially antagonized in the entopeduncular nucleus but did not affect lesion-mediated changes in both enkephalin and substance P mRNA expression (Salin, Manrique, Forni, & Kerkerian-Le Goff, 2002). Overall, like STN stimulation and nigral grafting of dopaminergic tissue, acupuncture stimulation at GA14 and GV20 acupoints changes neuronal activity of the basal ganglia output structure and improves behavioral function in rat Parkinson's models.

Changes in GABAergic neurotransmission in PD condition suggest that PD motor symptom expression depends on increase in striatal GABAergic transmission. The effect of EA stimulation at acupoints GV14 and GV20 on striatal neurotransmitters was measured using *in vivo* microdialysis and HPLC in MFB-lesioned rats and EA stimulation significantly decreased lesion-induced increase in striatal glutamate and acetylcholine levels while rescuing dopamine neurons from mechanical lesion and improving behavioral function (Sun et al., 2012). Kim, Doo, et al. (2011), using HPLC, investigated the effect of acupuncture on the levels of striatal dopamine and its metabolites, DOPAC and HVA, and they found that although acupuncture stimulation at GB34 did not restore striatal dopamine content in parkinsonian animals, it markedly increased dopamine turnover ratio, which was corroborated by immunochemical study showing increased

TH expression in the striatum after acupuncture treatment (Kim, Doo, et al., 2011). Further, microdialysis-HPLC study showed that acupuncture significantly increased dopamine efflux in the striatum 20 min after stimulation and returned to its basal level 120 min later (Kim, Doo, et al., 2011). This suggests that acupuncture may induce an increased rate of dopamine metabolism in the protected dopaminergic neurons with an accompanying increase in dopamine release from dopaminergic terminals. It seems plausible that acupuncture-induced dopaminergic neuronal protection may lead to increase in dopamine efflux from surviving dopamine neurons, which in turn may improve motor function without necessitating restoration of striatal dopamine content. Taken together, acupuncture treatment not only modulates neurotransmitter activities in the striatum but also regulates the neurochemical modulations in the basal ganglia output structures leading to the functional improvement in PD rodent models.

5. NEURAL MECHANISM OF ACUPUNCTURE TREATMENT ON PD PATIENTS: fMRI STUDY

By safely monitoring brain activity, functional neuroimaging techniques such as functional magnetic resonance imaging (fMRI) and positron emission tomography have opened a window into the brain and made it possible to unravel the neurobiological mechanisms underlying acupuncture treatment to help map the neurophysiological correlation of acupuncture and better understand how multiple bodily functions can be modified simultaneously in humans (Dhond, Kettner, & Napadow, 2007; Ogawa et al., 1992). Further, with the development of these noninvasive techniques, studies on neurological effects of acupuncture have become more practical in clinical settings.

fMRI is very useful to monitor the brain neurotransmitter activity. In amphetamine-pretreated rats, fMRI showed an increase in cerebral blood volume (CBV) in the striatum, thalamus, medial prefrontal, and cingulated cortices, reflecting the elevated dopamine release in these brain regions, and microdialysis studies corroborated the increased release of dopamine in the striatum and further showed the decreased release of glutamate and GABA in the striatum (Chen, Ren, et al., 2008). Acupuncture stimulation at LI4 acupoint at 2 Hz for 10 min after amphetamine administration attenuated the increase in CBV in those brain regions and reversed release pattern of neurotransmitters in the striatum (Chen, Ren, et al., 2008). Further, 100 Hz stimulation at the LI4 acupoint induced a greater attenuation of CBV in those brain regions compared to 2 Hz stimulation (Chen, Wang, et al., 2008). In the striatal 6-OHDA-infused rat, a partial dopamine-depleted PD model, stimulation of acupoints LI4 and P7 at 100 Hz for 10 min significantly increased CBV levels in the striatum, nucleus accumbens, thalamus, and medial prefrontal cortex compared to controls (Chen, Ren, Kaptchuk, & Kwong, 2012). In a MPTP-induced dog model of PD, fMRI demonstrated that retained acupuncture at ST36 elicited a decrease in blood oxygenation level dependent signal intensity and affected preprioceptive brain activation compared to nontreatment group (Lee et al., 2013). These PD model fMRI studies demonstrated that even for a short period of acupuncture stimulation, it restored dopamine release in the brain and it is a step forward in unveiling how a simple external peripheral manipulation can effectively modulate the central nervous system to regulate dopamine release and activity in response to dopamine deficit.

In PD patients, fMRI studies demonstrated that brain activity associated with performance of motor tasks differs from that of healthy volunteers (Jahanshahi et al., 1995). Patients with PD showed a hyperactivation in the ipsilateral premotor area and a hypoactivation in structures of frontostriatal motor loop compared to healthy subjects (Jahanshahi et al., 1995; Mallol et al., 2007; Owen, Doyon, Dagher, Sadikot, & Evans, 1998; Rascol et al., 1997). Further, task performance in PD group was associated with relative decrease of activation in the striatum and thalamic area (Mallol et al., 2007). Acupuncture stimulation at acupoint GB34 on the left foot of 10 PD patients led to increased activation of the putamen and the primary motor cortex that correlated with the enhanced motor function of affected hand assessed by finger-tapping task following acupuncture treatment (Chae et al., 2009). This is consistent with studies showing that acupuncture stimulation at the same acupoint modulates motor cortical activities of somatomotor area in healthy volunteers (Jeun et al., 2005; Wu et al., 2002a, 2002b; Zhang et al., 2004). In another study, a new method called regional homogeneity was used to investigate PD-related modulations of neural activity in the resting stage and fMRI study showed that regional homogeneity activity was decreased in the putamen, thalamus, and supplementary motor area and increased in the cerebellum, primary sensorimotor cortex, and premotor area in the patients with PD group compared to control group (Wu et al., 2009). Acupuncture stimulation at GB34 acupoint modulated the brain resting-state network associated with

PD and increased neuronal responses in the SN, caudate, thalamus, and putamen of 12 patients with PD compared to healthy controls (Yeo et al., 2012). This confirms the results from acupuncture treatment on PD models that acupuncture may improve motor symptoms of PD patients by modulating the basal ganglia–thalamocortical circuit (Jia et al., 2010, 2009; Kim, Doo, et al., 2011; Kim, Kim, et al., 2011).

6. FUTURE ACUPUNCTURE RESEARCH IN PD

So far, the results from acupuncture studies in PD models suggest that acupuncture may act as neurotrophin release enhancer, antioxidant, and anti-inflammation agents and the modulator of neuronal activity in the basal ganglia circuits that ultimately protects dopaminergic neurons against degenerative process and modulates the outcome of neuronal activity of the basal ganglia circuit. Future studies needed to look other potential acupoints that modulate dopamine and other neurotransmitter transmissions in the basal ganglia to enhance the efficacy of acupuncture treatment. More importantly, we need to explore the acupoints that enhance releasing of neurotrophic factors after stimulation to slow down dopaminergic degeneration in particular at the early stage of disease. Further, it is important to select a group of principal acupoints that may improve motor function, enhance trophic factor releasing and modify basal ganglia activity and apply them into PD patients. It is hopeful that translation of achievement in acupuncture research in animal models into clinical treatment of Parkinson's will maximize the effectiveness of acupuncture treatment.

REFERENCES

Ahlskog, J. E., & Muenster, M. D. (2001). Frequency of levodopa-related dyskinesias and motor fluctuations as estimated from the cumulative literature. *Movement Disorders, 16*(3), 448–458.

Akiyama, H., & McGeer, P. L. (1989). Microglial response to 6-hydroxydopamine-induced substantia nigra lesions. *Brain Research, 489*, 247–253.

Aleyasin, H., Rousseaux, M. W., Marcogliese, P. C., Hewitt, S. J., Irrcher, I., Joselin, A. P., et al. (2010). DJ-1 protects the nigrostriatal axis from the neurotoxin MPTP by modulation of the AKT pathway. *Proceedings of the National Academy of Sciences of the United States of America, 107*(7), 3186–3191.

Betarbet, R., Sherer, T. B., & Greenamyre, J. T. (2002). Animal models of Parkinson's disease. *Bioessays, 24*(4), 308–318.

Braak, H., Ghebremedhin, E., Rüb, U., Bratzke, H., & Del Tredici, K. (2004). Stages in the development of Parkinson's disease-related pathology. *Cell and Tissue Research, 318*, 121–134.

Burdo, J. R., & Connor, J. R. (2003). Brain iron uptake and homeostatic mechanisms: An overview. *Biometals, 16*, 63–75.

Chae, Y., Lee, H., Kim, H., Kim, C. H., Chang, D. I., Kim, K. M., et al. (2009). Parsing brain activity associated with acupuncture treatment in Parkinson's diseases. *Movement Disorders*, *24*(12), 1794–1802.

Chaudhuri, K. R., Odin, P., Antonini, A., & Martinez-Martin, P. (2011). Parkinson's disease: The non-motor issues. *Parkinsonism and Related Disorders*, *17*(10), 717–723.

Chen, Y. I., Ren, J. Q., Kaptchuk, T. J., & Kwong, K. K. (2012). Restoring cerebral dopamine homeostasis by electrical forepaw stimulation: An FMRI study. *Synapse*, *66*(4), 331–339.

Chen, Y. I., Ren, J., Wang, F. N., Xu, H., Mandeville, J. B., Kim, Y., et al. (2008). Inhibition of stimulated dopamine release and hemodynamic response in the brain through electrical stimulation of rat forepaw. *Neuroscience Letters*, *431*(3), 231–235.

Chen, Y. I., Wang, F. N., Nelson, A. J., Xu, H., Kim, Y., Rosen, B. R., et al. (2008). Electrical stimulation modulates the amphetamine-induced hemodynamic changes: An fMRI study to compare the effect of stimulating locations and frequencies on rats. *Neuroscience Letters*, *444*(2), 117–121.

Cheramy, A., Leviel, V., & Glowinski, J. (1981). Dendritic release of dopamine in the substantia nigra. *Nature*, *289*, 537–542.

Cho, S. Y., Shim, S. R., Rhee, H. Y., Park, H. J., Jung, W. S., Moon, S. K., et al. (2012). Effectiveness of acupuncture and bee venom acupuncture in idiopathic Parkinson's disease. *Parkinsonism and Related Disorders*, *18*(8), 948–952.

Choi, Y. G., Park, J. H., & Lim, S. (2009). Acupuncture inhibits ferric iron deposition and ferritin-heavy chain reduction in an MPTP-induced parkinsonism model. *Neuroscience Letters*, *450*(2), 92–96.

Choi, Y. G., Yeo, S., Hong, Y. M., Kim, S. H., & Lim, S. (2011). Changes of gene expression profiles in the cervical spinal cord by acupuncture in an MPTP-intoxicated mouse model: microarray analysis. *Gene*, *481*(1), 7–16.

Choi, Y. G., Yeo, S., Hong, Y. M., & Lim, S. (2011). Neuroprotective changes of striatal degeneration-related gene expression by acupuncture in an MPTP mouse model of Parkinsonism: microarray analysis. *Cell and Molecular Neurobiology*, *31*(3), 377–391.

Cristian, A., Katz, M., Cutrone, E., & Walker, R. H. (2005). Evaluation of acupuncture in the treatment of Parkinson's disease: A double-blind pilot study. *Movement Disorders*, *20*, 1185–1188.

Dauer, W., & Przedborski, S. (2003). Parkinson's disease: Mechanisms and models. *Neuron*, *39*(6), 889–909.

Dhond, R. P., Kettner, N., & Napadow, V. (2007). Neuroimaging acupuncture effects in the human brain. *Journal of Alternative and Complementary Medicine*, *13*(6), 603–616.

Duty, S., & Jenner, P. (2011). Animal models of Parkinson's disease: A source of novel treatments and clues to the cause of the disease. *British Journal of Pharmacology*, *164*(4), 1357–1391.

Eng, M. L., Lyons, K. E., Greene, M. S., & Pahwa, R. (2006). Open-label trial regarding the use of acupuncture and yin tui na in Parkinson's disease outpatients: A pilot study on efficacy, tolerability, and quality of life. *Journal of Alternative and Complementary Medicine*, *12*, 395–399.

Engelman, J. A., Luo, J., & Cantley, L. C. (2006). The evolution of phosphatidylinositol 3-kinases as regulators of growth and metabolism. *Nature Reviews Genetics*, *7*, 606–619.

Ferry, P., Johnson, M., & Wallis, P. (2002). Use of complementary therapies and non-prescribed medication in patients with Parkinson's disease. *Postgraduate Medical Journal*, *78*, 612–614.

Gash, D. M., Zhang, Z., Ovadia, A., Cass, W. A., Yi, A., et al. (1996). Functional recovery in parkinsonian monkeys treated with GDNF. *Nature*, *380*, 252–255.

Grunblatt, E., Mandel, S., & Youdim, M. B. (2000). Neuroprotective strategies in Parkinson's disease using the models of 6-hydroxydopamine and MPTP. *Annals of the New York Academy of Sciences*, *899*, 262–273.

Halliday, G. M., Del Tredici, K., & Braak, H. (2006). Critical appraisal of brain pathology staging related to presymptomatic and symptomatic cases of sporadic Parkinson's disease. *Journal of Neural Transmission Supplement, 70*, 99–103.

Hirsch, E. C., & Hunot, S. (2009). Neuroinflammation in Parkinson's disease: A target for neuroprotection? *Lancet Neurology, 8*(4), 382–397.

Hong, M. S., Park, H. K., Yang, J. S., Park, H. J., Kim, S. T., Kim, S. N., et al. (2010). Gene expression profile of acupuncture treatment in 1-methyl-4-phenyl-1,2,3,6-tetrahydropyridine-induced Parkinson's disease model. *Neurology Research, 32*(Suppl. 1), 74–78.

Jahanshahi, M., Jenkins, I. H., Brown, R. G., Marsden, C. D., Passingham, R. E., & Brooks, D. J. (1995). Self-initiated versus externally triggered movements. I. An investigation using measurement of regional cerebral blood flow with PET and movement-related potentials in normal and Parkinson's disease subjects. *Brain, 118*(Pt. 4), 913–933.

Jankovic, J. (2008). Parkinson's disease: Clinical features and diagnosis. *Journal of Neurological Neurosurgery Psychiatry, 79*(4), 368–376.

Javitch, J. A., D'Amato, R. J., Strittmatter, S. M., & Snyder, S. H. (1985). Parkinsonism-inducing neurotoxin, N-methyl-4-phenyl-1,2,3,6-tetrahydropyridine: Uptake of the metabolite N-methyl-4-phenylpyridine by dopamine neurons explains selective toxicity. *Proceedings of the National Academy of Sciences of the United States of America, 82*, 2173–2177.

Jenner, P. (2003). Oxidative stress in Parkinson's disease. *Annals of Neurology, 53*(Suppl. 3), S26–S36.

Jenner, P., Dexter, D. T., Sian, J., Schapira, A. H., & Marsden, C. D. (1992). Oxidative stress as a cause of nigral cell death in Parkinson's disease and incidental Lewy body disease. The Royal Kings and Queens Parkinson's Disease Research Group. *Annals of Neurology, 32*(Suppl.), S82–S87.

Jeon, S., Kim, Y. J., Kim, S. T., Moon, W., Chae, Y., Kang, M., et al. (2008). Proteomic analysis of the neuroprotective mechanisms of acupuncture treatment in a Parkinson's disease mouse model. *Proteomics, 8*(22), 4822–4832.

Jeun, S. S., Kim, J. S., Kim, B. S., Park, S. D., Lim, E. C., Choi, G. S., et al. (2005). Acupuncture stimulation for motor cortex activities: A 3T fMRI study. *American Journal of Chinese Medicine, 33*, 573–578.

Jia, J., Li, B., Sun, Z. L., Yu, F., Wang, X., & Wang, X. M. (2010). Electro-acupuncture stimulation acts on the basal ganglia output pathway to ameliorate motor impairment in Parkinsonian model rats. *Behavioural Neuroscience, 124*(2), 305–310.

Jia, J., Sun, Z. L., Li, B., Pan, Y. L., Wang, H. M., Wang, X., et al. (2009). Electro-acupuncture stimulation improves motor disorders in Parkinsonian rats. *Behavioural Brain Research, 205*, 214–218.

Kang, J. M., Park, H. J., Choi, Y. G., Choe, I. H., Park, J. H., et al. (2007). Acupuncture inhibits microglial activation and inflammatory events in the MPTP-induced mouse model. *Brain Research, 1131*(1), 211–219.

Kim, S. N., Doo, A. R., Park, J. Y., Bae, H., Chae, Y., & Shim, I. (2011). Acupuncture enhances the synaptic dopamine availability to improve motor function in a mouse model of Parkinson's disease. *PLoS One, 6*(11), e27566.

Kim, S. N., Kim, S. T., Doo, A. R., Park, J. Y., & Moon, W. (2011). Phosphatidylinositol 3-kinase/Akt signaling pathway mediates acupuncture-induced dopaminergic neuron protection and motor function improvement in a mouse model of Parkinson's disease. *International Journal of Neuroscience, 121*(10), 562–569.

Kim, Y. K., Lim, H. H., Song, Y. K., Lee, H. H., Lim, S., et al. (2005). Effect of acupuncture on 6-hydroxydopamine-induced nigrostratal dopaminergic neuronal cell death in rats. *Neuroscience Letters, 384*(1–2), 133–138.

Kim, S. T., Moon, W., Chae, Y., Kim, Y. J., Lee, H., & Park, H. J. (2010). The effect of electroacupuncture for 1-methyl-4-phenyl-1,2,3,6-tetrahydropyridine-induced proteomic changes in the mouse striatum. *Journal of Physiological Sciences*, *60*(1), 27–34.

Knaryan, V. H., Samantaray, S., Le Gal, C., Ray, S. K., & Banik, N. L. (2011). Tracking extranigral degeneration in animal models of Parkinson's disease: Quest for effective therapeutic strategies. *Journal of Neurochemistry*, *118*(3), 326–338.

Lee, S. H., Jahng, G. H., Choe, I. H., Choi, C. B., Kim, D. H., & Kim, H. Y. (2013). Neural pathway interference by retained acupuncture: A functional MRI study of a dog model of Parkinson's disease. *CNS Neuroscience Therapy*, *19*, 585–595. http://dx.doi.org/10.1111/cns.12108.

Liang, X. B., Liu, X. Y., Li, F. Q., Luo, Y., Lu, J., Zhang, W. M., et al. (2002). Long-term high-frequency electro-acupuncture stimulation prevents neuronal degeneration and up-regulates BDNF mRNA in the substantia nigra and ventral tegmental area following medial forebrain bundle axotomy. *Brain Research Molecular Brain Research*, *108*(1–2), 51–59.

Liang, X. B., Luo, Y., Liu, X. Y., Lu, J., Li, F. Q., Wang, Q., et al. (2003). Electro-acupuncture improves behavior and upregulates GDNF mRNA in MFB transected rats. *Neuroreport*, *14*(8), 1177–1181.

Liberatore, G. T., Jackson-Lewis, V., Vukosavic, S., et al. (1999). Inducible nitric oxide synthase stimulates dopaminergic neurodegeneration in the MPTP model of Parkinson disease. *Nature Medicine*, *5*, 1403–1409.

Liu, X. Y., Zhou, H. F., Pan, Y. L., Liang, X. B., Niu, D. B., et al. (2004). Electro-acupuncture stimulation protects dopaminergic neurons from inflammation-mediated damage in medial forebrain bundle-transected rats. *Experimental Neurology*, *189*(1), 189–196.

Mallol, R., Barrós-Loscertales, A., López, M., Belloch, V., Parcet, M. A., & Avila, C. (2007). Compensatory cortical mechanisms in Parkinson's disease evidenced with fMRI during the performance of pre-learned sequential movements. *Brain Research*, *1147*, 265–271.

Mandel, S., Maor, G., & Youdim, M. B. (2004). Iron and alpha-synuclein in the substantia nigra of MPTP-treated mice: Effect of neuroprotective drugs R-apomorphine and green tea polyphenol (−)-epigallocatechin-3-gallate. *Journal of Molecular Neuroscience*, *24*, 401–416.

Ogawa, S., Tank, D., Menon, R., et al. (1992). Intrinsic signal changes accompanying sensory stimulation: Functional brain mapping with magnetic resonance imaging. *Proceedings of the National Academy of Sciences of the United States of America*, *89*, 5951–5955.

Owen, A. M., Doyon, J., Dagher, A., Sadikot, A., & Evans, A. C. (1998). Abnormal basal ganglia outflow in Parkinson's disease identified with PET. Implications for higher cortical functions. *Brain*, *121*, 949–965.

Park, H. J., Lim, S., Joo, W. S., Yin, C. S., Lee, H. S., et al. (2003). Acupuncture prevents 6-hydroxydopamine-induced neuronal death in the nigrostriatal dopaminergic system in the rat Parkinson's disease model. *Experimental Neurology*, *180*(1), 93–98.

Pecci, C., Rivas, M. J., Moretti, C. M., Raina, G., Ramirez, C. Z., et al. (2010). Use of complementary and alternative therapies in outpatients with Parkinson's disease in Argentina. *Movement Disorders*, *25*(13), 2094–2098.

Perumal, A. S., Gopal, V. B., Tordzro, W. K., Cooper, T. B., & Cadet, J. L. (1992). Vitamin E attenuates the toxic effects of 6-hydroxydopamine on free radical scavenging systems in rat brain. *Brain Research Bulletin*, *29*, 699–701.

Rabinstein, A. A., & Shulman, L. M. (2003). Acupuncture in clinical neurology. *The Neurologist*, *9*, 137–148.

Rajendran, P. R., Thompson, R. E., & Reich, S. G. (2001). The use of alternative therapies by patients with Parkinson's disease. *Neurology*, *57*, 790–794.

Rascol, O., Sabatini, U., Fabre, N., Brefel, C., Loubinoux, I., Celsis, P., et al. (1997). The ipsilateral cerebellar hemisphere is overactive during hand movements in akinetic parkinsonian patients. *Brain, 120*(Pt 1), 103–110.

Riederer, P., Sofic, E., Rausch, W. D., Schmidt, B., Reynolds, G. P., Jellinger, K., & Youdim, M. B. (1989). Transition metals, ferritin, glutathione, and ascorbic acid in parkinsonian brains. *Journal of Neurochemistry, 52*, 515–520.

Robertson, G. S., & Robertson, H. A. (1989). Evidence that L-dopa-induced rotational behavior is dependent on both striatal and nigral mechanisms. *Journal of Neuroscience, 9*, 3326–3331.

Salin, P., Manrique, C., Forni, C., & Kerkerian-Le Goff, L. (2002). High-frequency stimulation of the subthalamic nucleus selectively reverses dopamine denervation-induced cellular defects in the output structures of the basal ganglia in the rat. *Journal of Neuroscience, 22*(12), 5137–5148.

Schapira, A. H., & Jenner, P. (2011). Etiology and pathogenesis of Parkinson's disease. *Movement Disorders, 26*(6), 1049–1055.

Segovia, J., & Garcia-Munoz, M. (1987). Changes in the activity of GAD in the basal ganglia of the rat after striatal dopaminergic denervation. *Neuropharmacology, 26*(9), 1449–1451.

Shulman, L. M., Wen, X., Weiner, W. J., Bateman, D., Minagar, A., et al. (2002). Acupuncture therapy for the symptoms of Parkinson's disease. *Movement Disorders, 17*(4), 799–802.

Sian, J., Dexter, D. T., Lees, A. J., Daniel, S., Agid, Y., Javoy-Agid, F., Jenner, P., & Marsden, C. D. (1994). Alterations in glutathione levels in Parkinson's disease and other neurodegenerative disorders affecting basal ganglia. *Annals of Neurology, 36*, 348–355.

Sofic, E., Riederer, P., Heinsen, H., Beckmann, H., & Reynolds, G. P. (1988). Increased iron (III) and total iron content in post mortem substantia nigra of parkinsonian brain. *Journal of Neural Transmission, 74*, 199–205.

Soghomonian, J. J., & Chesselet, M. F. (1992). Effects of nigrostriatal lesions on the levels of messenger RNAs encoding two isoforms of glutamate decarboxylase in the globus pallidus and entopeduncular nucleus of the rat. *Synapse, 11*(2), 124–133.

Stocchi, F., Tagliati, M., & Olanow, C. W. (2008). Treatment of levodopa-induced motor complications. *Movement Disorders, 23*(Suppl. 3), S599–S612.

Su, X., Maguire-Zeiss, K. A., Giuliano, R., Prifti, L., Venkatesh, K., & Federo, H. J. (2008). Synuclein activates microglia in a model of Parkinson's disease. *Neurobiology Aging, 29*, 1690–1701.

Sugama, S., Cho, B. P., Degiorgio, L. A., Shimizu, Y., Kim, S. S., Kim, Y. S., et al. (2003). Temporal and sequential analysis of microglia in the substantia nigra following medial forebrain bundle axotomy in rat. *Neuroscience, 116*, 925–933.

Sulzer, D. (2007). Multiple hit hypotheses for dopamine neuron loss in Parkinson's disease. *Trends in Neuroscience, 30*(5), 244–250.

Sun, Z., Jia, J., Gong, X., Jia, Y., Deng, J., Wang, X., & Wang, X. (2012). Inhibition of glutamate and acetylcholine release in behavioral improvement induced by electroacupuncture in parkinsonian rats. *Neuroscience Letters, 520*(1), 32–37.

Tan, L. C., Lau, P. N., Jamora, R. D., & Chan, E. S. (2006). Use of complementary therapies in patients with Parkinson's disease in Singapore. *Movement Disorders, 21*(1), 86–89.

Tansey, M. G., McCoy, M. K., & Frank-Cannon, T. C. (2007). Neuroinflammatory mechanisms in Parkinson's disease: potential environmental triggers, pathways, and targets for early therapeutic intervention. *Experimental Neurology, 208*, 1–25.

Tseng, J. L., Baetge, E. E., Zurn, A. D., & Aebischer, P. (1997). GDNF reduces drug-induced rotational behavior after medial forebrain bundle transection by a mechanism not involving striatal dopamine. *Journal of Neuroscience, 17*, 325–333.

Wang, J., Jiang, H., & Xie, J. X. (2004). Time dependent effects of 6-OHDA lesions on iron level and neuronal loss in rat nigrostriatal system. *Neurochemistry Research*, *29*(12), 2239–2243.

Wang, H., Liang, X., Wang, X., Luo, D., Jia, J., & Wang, X. (2013). Electro-acupuncture stimulation improves spontaneous locomotor hyperactivity in MPTP intoxicated mice. *PLoS One*, *8*(5), 64403. http://dx.doi.org/10.1371/journal.pone.0064403.

Wang, H., Pan, Y., Xue, B., Wang, X., Zhao, F., et al. (2011). The antioxidative effect of electro-acupuncture in a mouse model of Parkinson's disease. *PLoS One*, *6*(5), e19790. http://dx.doi.org/10.1371/journal.pone.0019790.

Winkler, C., Bentlage, C., Cenci, M. A., Nikkhah, G., & Björklund, A. (2003). Regulation of neuropeptide mRNA expression in the basal ganglia by intrastriatal and intranigral transplants in the rat Parkinson model. *Neuroscience*, *118*(4), 1063–1077.

Wolters, E. C., & Braak, H. (2006). Parkinson's disease: Premotor clinico-pathological correlations. *Journal of Neural Transmission Supplement*, *70*, 309–319.

Wu, J.-N. (1996). A short history of acupuncture. *Journal of Alternative and Complementary Medicine*, *2*, 19–21.

Wu, D. C., Jackson-Lewis, V., Vila, M., et al. (2002a). Blockade of microglial activation is neuroprotective in the 1-methyl-4-phenyl-1,2,3,6-tetrahydropyridine mouse model of Parkinson disease. *Journal of Neuroscience*, *22*, 1763–1771.

Wu, T., Long, X., Zang, Y., Wang, L., Hallett, M., Li, K., & Chan, P. (2009). Regional homogeneity changes in patients with Parkinson's disease. *Human Brain Mapping*, *30*(5), 1502–1510.

Wu, M. T., Sheen, J. M., Chuang, K. H., et al. (2002b). Neuronal specificity of acupuncture response: A fMRI study with electroacupuncture. *Neuroimage*, *16*, 1028–1037.

Yeo, S., Choi, Y. G., Hong, Y. M., & Lim, S. (2013). Neuroprotective changes of thalamic degeneration-related gene expression by acupuncture in an MPTP mouse model of parkinsonism: Microarray analysis. *Gene*, *515*(2), 329–338. http://dx.doi.org/10.1016/j.gene.2012.12.002.

Yeo, S., Lim, S., Choe, I. H., Choi, Y. G., Chung, K. C., et al. (2012). Acupuncture stimulation on GB34 activates neural responses associated with Parkinson's disease. *CNS Neuroscience Therapy*, *18*(9), 781–790.

Yu, Y. P., Ju, W. P., Li, Z. G., Wang, D. Z., Wang, Y. C., & Xie, A. M. (2010). Acupuncture inhibits oxidative stress and rotational behavior in 6-hydroxydopamine lesioned rat. *Brain Research*, *1336*, 58–65.

Zhang, W. T., Jin, Z., Luo, F., Zhang, L., Zeng, Y. W., & Han, J. S. (2004). Evidence from brain imaging with fMRI supporting functional specificity of acupoints in humans. *Neuroscience Letters*, *354*, 50–53.

CHAPTER EIGHT

Acupuncture Therapy for Stroke Patients

Xin Li, Qiang Wang[1]

Department of Anesthesiology, Xijing Hospital, Fourth Military Medical University, Xi'an, China

[1]Corresponding author: e-mail address: wangqiang@fmmu.edu.cn

Contents

Abstract

Acupuncture is one of the most important parts of Traditional Chinese Medicine, has been used for more than 3000 years as prevention and treatment for various diseases in China as well as in adjacent regions, and is widely accepted in western countries in recent years. More and more clinical trials revealed that acupuncture shows positive

International Review of Neurobiology, Volume 111
ISSN 0074-7742
http://dx.doi.org/10.1016/B978-0-12-411545-3.00008-0

effect in stroke, not only as a complementary and alternative medicine for poststroke rehabilitation but also as a preventive strategy which could induce cerebral ischemic tolerance, especially when combined with modern electrotherapy. Acupuncture has some unique characteristics, which include acupoint specificity and parameter-dependent effect. It also involves complicated mechanism to exert the beneficial effect on stroke. Series of clinical trials have shown that acupuncture primarily regulates the release of neurochemicals, hemorheology, cerebral microcirculation, metabolism, neuronal activity, and the function of specific brain region. Animal studies showed that the effects of acupuncture therapy on stroke were possibly via inhibition of postischemic inflammatory reaction, stimulation of neurogenesis and angiogenesis, and influence on neural plasticity. Mechanisms for its preconditioning effect include activity enhancement of antioxidant, regulation of the endocannabinoid system, and inhibition of apoptosis. Although being controversial, acupuncture is a promising preventive and treatment strategy for stroke, but further high-quality clinical trials would be needed to provide more confirmative evidence.

1. INTRODUCTION

In recent years, the morbidity and mortality of stroke are dramatically increasing as a result of aging population, diet changes, and stress. Now, stroke is the second most common cause of death and the main cause of disability worldwide, which imposes a heavy burden on both family and community (Donnan, Fisher, Macleod, & Davis, 2008; Liu et al., 2007). Although many strategies have been used in clinical situations, such as angioplasty, stenting, thrombolysis, and many neuroprotective agents, their safety and efficacy remain controversial after a systematic review and meta-analysis of the clinical trials (Lansberg, Bluhmki, & Thijs, 2009; Toyoda et al., 2004). Moreover, researchers developed plenty of new methods either to prevent or to treat stroke in recent years, and most of them have been proved to reduce infarction volume and improve neurological recovery with animal models, but few of them show certain effects in clinical trials (Lees et al., 2006). Therefore, it is still a huge and urgent medical challenge to develop novel and rational ways aimed at preventing the occurrence of stroke or reducing impairments caused by stroke.

Acupuncture is a kind of traditional Chinese medicine, which has been used for more than 3000 years as prevention and treatment for many diseases. It is a procedure in which a fine needle is inserted into the skin at discrete acupoints and is manipulated manually, electrically, by heat or some drug injections. Although it was first developed in China, it has spread worldwide and is popularly used. In 2007, there were approximately 3 millions of adults who received acupuncture in the United States, 1 million more than that in 2002 (Barnes, Bloom, & Nahin, 2008). Since the acupuncture stimulation is supposed to be tightly associated with the nervous system, acupuncture is expected to improve the neurological function after stroke. So, acupuncture is most commonly used in the rehabilitation of stroke patients (Rabinstein & Shulman, 2003). Actually, 46% of the stroke survivors in the United States were engaged in some form of complementary and alternative medicine (CAM) therapy, in which acupuncture was the most frequently used CAM therapy among stroke survivors (Shah, Engelhardt, & Ovbiagele, 2008). A survey for stroke patients in Australia showed that almost all respondents (98%) wanted to know more about acupuncture in stroke rehabilitation and 87% were willing to consider acupuncture as a treatment option (Yam & Wilkinson, 2010). Although several systematic reviews indicated no strong evidence to support the positive effect of acupuncture on functional recovery after stroke (Kong, Lee, Shin, Song, & Ernst, 2010; Long & Wu, 2012; Park, Hopwood, White, & Ernst, 2001; Sze, Wong, Or, Lau, & Woo, 2002; Wu et al., 2006; Zhang, Liu, Asplund, & Li, 2005; Zhao, Du, Liu, & Wang, 2012), an increasing number of high-quality randomized controlled trials (RCTs) are in process, which might shed lights on the application of acupuncture therapy on stroke patients.

2. EFFICACY OF ACUPUNCTURE THERAPY FOR STROKE

Acupuncture is commonly used in rehabilitation of poststroke patients. A number of clinical trials have proved that acupuncture stimulation improved the balance function (Liu et al., 2009) and spastic states (Zhao et al., 2009), increased the muscle strength (Yan & Hui-Chan, 2009), reduced the muscle spasticity, and improved the motor function for chronic stroke survivors with moderate or severe muscle spasticity (Liu, Mukherjee, Sun, Liu, & McPeak, 2008). Acupuncture also improved the function of the affected upper limb in chronic hemiparetic stroke patients by increasing activity in the ipsilesional motor cortex (Schaechter et al., 2007). It can also improve insomnia in stroke patients by reducing the sympathetic nervous activity (Lee et al., 2009). The results of one clinical trial showed a clinically relevant decrease of relapse in the patients treated with resuscitating acupuncture intervention by the end of 6 months. The resuscitating acupuncture intervention could also improve self-care ability and quality of life (Shen, Kong, et al., 2012). A systematic review for seven RCTs concluded that acupuncture combined with exercise was effective for shoulder pain after stroke (Lee et al., 2012). With bee venom acupoint injection, acupuncture could significantly relieve the central poststroke pain (Cho, Park, et al., 2013). It is also an effective therapy on aphasia after stroke (Sun, Xue, & Zuo, 2012). Acupuncture plus conventional care was similar in effectiveness to physiotherapy treatment plus conventional care for poststroke rehabilitation without any obvious adverse events, which suggested that it might represent an additional treatment option for the stroke patients (Zhuangl et al., 2012).

3. ELECTROACUPUNCTURE THERAPY FOR STROKE

Integrated with electrotherapy, electroacupuncture (EA) is traditional acupuncture connected with electric stimulation apparatus, which is conducted by inserting acupuncture needles into the acupoints and then changing electric stimulation parameters, including the stimulation frequency, current intensity, pulse width, and pulse interval. Except for a small current passing through a pair of needles, EA is similar to the regular acupuncture. Therefore, EA not only inherits the benefits of the traditional acupuncture but also integrates with the physiological effects of electric stimulation (Napadow et al., 2005).

Compared with regular acupuncture, there are two advantages for EA. On the one hand, the acupoint of EA is not as precise as regular acupuncture, since current delivered by needles stimulates a larger area than that of the needle itself. On the other hand, there is an alternative technique for EA, which is called transcutaneous electrical nerve stimulation (TENS). TENS uses electrodes that are taped to the surface of the skin other than needles being inserted, so it can be applied in condition that patients deny insertion of needle or cannot be needled. Based on these two advantages, a growing number of basic researches and clinical trials are preceded in investigating the neuroprotective effect of EA and mechanism of this effect. A large number of animal studies have shown that EA could reduce neural apoptosis, promote cell proliferation, increase cerebral blood flow (CBF), and improve neurological function after stroke (Du et al., 2011; Liu, Zou, Du, & Wong, 2010; Tao et al., 2010; Wang et al., 2002). These results provide some evidence for further translational studies. So far, some clinical trials have focused on the effects of EA in stroke patients, but the results are ambiguous. Although many investigations with small enrollments support that EA treatment has positive effects on the motor function and quality of life (Chou, Chu, & Lin, 2009; Hsieh, Wang, & Lee, 2007), two RCTs including more patients find that there is no significant difference between EA and control groups in the improvement of functional outcome and life satisfaction (Gosman-Hedstrom et al., 1998; Johansson et al., 2001). In respect of this contradiction, the new method of EA treatment is required for further clinical application.

4. ACUPUNCTURE PRECONDITION FOR STROKE

The phenomenon of precondition-induced ischemic tolerance provides a new idea for the prevention of stroke injury. The preconditioning effect is that a brief exposure to sublethal or noninjurious stimuli can increase resistance to the subsequent, prolonged and lethal damage (Murry, Jennings, & Reimer, 1986). There are various kinds of precondition measures, such as ischemia (regional or remote), hypoxia, endotoxin, cytokines, and anesthetics. Accumulating preclinical evidences have demonstrated that these pretreatment methods, especially ischemia pretreatment, could induce neuroprotection and myocardial protection against ischemia and reperfusion (I/R) injury (Perez-Pinzon, Xu, Dietrich, Rosenthal, & Sick, 1997; Przyklenk, Bauer, Ovize, Kloner, & Whittaker, 1993). But from the clinical point of view, all the mentioned pretreatment ways have limitations and adverse effects to be applied in patients, especially the patients with severe illness.

As an indispensable part of traditional Chinese medicine, acupuncture has played an important role in prevention and treatment of diseases throughout the history. "Treating before sickness" is the plain idea of preventive medicine in traditional Chinese medicine. "Preventive acupuncture" is an approach using acupuncture to "treat before sick," namely, applying acupuncture in healthy or mildly sick patients, to stimulate the meridians of the body and enhance the body's resistance to disease, in order to prevent disease or to reduce the extent of damage following disease (Wang & Liang, 2008). In general, most of preclinical studies and clinical trials on acupuncture have been focused on its therapeutic role after stroke. However, prevention is definitely superior to treatment. Since acupuncture is economical, easily operated and has fewer negative side effects than the other preventative methods (e.g., pharmacological, ischemic, etc.), it should be more valuable and advantageous in preventing ischemic cerebral vascular disease, especially on patients with high risk of ischemic injury.

In 2003, Xiong and colleagues first defined the concept of EA pretreatment. They reported that repeated EA stimulation at the Baihui acupoint (GV20) before cerebral ischemia in rats could significantly reduce the infarct volume caused by transient middle cerebral artery occlusion (MCAO) as well as improve the neurological outcomes (Xiong et al., 2003). The results of previously published studies have shown that EA stimulation before ischemia could produce an effect similar to that of the ischemia pretreatment and could also induce ischemic tolerance. Later, this group showed that pretreatment with a single EA session could also induce tolerance to focal cerebral ischemia in rats (Wang, Xiong, Chen, Liu, & Zhu, 2005). At the same time, another group found that EA pretreatment at Hegu (LI4), a well-known acupoint commonly used in Oriental medicine for the treatment of neuronal injury resulting from hypoxia-ischemia, could induce neuroprotective effect in neonatal hypoxic–ischemic rat brains (Jiang, Zhang, & Shui, 2003).

Just like ischemia pretreatment, EA pretreatment has potential protective effects on the mammalian brain. Furthermore, both ischemia pretreatment and EA pretreatment can produce the acute and delayed neuroprotection. A single EA stimulation at Baihui acupoint for 30 min could induce biphasic tolerance against focal cerebral ischemia: the acute phase occurred 2 h after EA pretreatment, while the delayed ischemic tolerance was observed 24 h after the stimulus (Wang et al., 2005).

In addition, EA pretreatment was also observed in the heart. Researchers applied EA to bilateral Neiguan (PC6) before or during myocardial I/R induced by ligating and reperfusing the left anterior descending coronary artery. The results showed that there were significant reductions in cardiac enzymes, the duration of arrhythmia, and mortality rate in rats that were either preconditioned or treated with EA on PC6, as compared with those that did not underwent EA (Tsou, Huang, & Chiu, 2004).

The research for the EA precondition does not just stay in the labs. Two latest clinical trials provided some evidence for the effectiveness of EA pretreatment in the patients. Lu et al. enrolled 32 patients, requiring selective craniocerebral tumor resection to study the neuroprotection after EA pretreatment. The results showed that the serum levels of S100 calcium-binding protein β (S-100β) and neuron-specific enolase (NSE) in the EA group were significantly lower than that of the control group at the end of the surgery and 24 h postsurgery (Lu et al., 2010). In the same year, Yang and colleagues also designed an RCT, enrolling 60 patients to investigate the cardioprotective effect of EA pretreatment in patients undergoing heart valve replacement surgery. EA or sham stimuli were applied at bilateral Neiguan (PC6), Lieque (LU7), and Yunmen (LU2) for 30 min each day for 5 consecutive days before surgery. The level of serum cardiac troponin I was significantly decreased in the EA group at 6, 12, and 24 h after aortic cross-clamp removal. Meanwhile, EA pretreatment also reduced the inotrope use at 12, 24, and 48 h after the intensive care unit arrival and shortened intensive care unit stay time (Yang et al., 2010). These two clinical trials have indicated that EA pretreatment might have beneficial effects on patients undergoing surgery. But, this evidence was limited since the number of enrolled patients was small and both of these trials were conducted in a single center. Thus, multicenter RCTs would be needed to provide further evidence on EA pretreatment.

5. ACUPOINT OPTION OF ACUPUNCTURE THERAPY FOR STROKE

Based on meridian theory, an acupoint is relatively specific to certain functions or certain organs, and different effects occur when different acupoints are stimulated. The neuronal specificity of acupoints has been tested by functional magnetic resonance imaging (fMRI), providing neurobiological evidence for the existence of acupoint specific (Na et al., 2009; Wu et al., 2002). Lu et al. found that EA pretreatment of the Baihui acupoint could induce more robust neuroprotection against cerebral I/R injury as compared to stimulation 1 cm lateral to the Baihui acupoint or nonmeridian points of the distal limbs (Lu, Xiong, Zhu, Wang, & Cheng, 2002). The Baihui acupoint was chosen because the theory of meridians in the traditional Chinese medicine indicated that the Du meridian was closely related to the brain and spinal cord and Baihui is one of the acupoints of the Du meridian. At the same time, according to the neuroanatomy of western medicine, Baihui acupoint (GV20) is in the projection area of the motor and sensory cortex, as well as in the projection area of the anterior cerebral artery. Therefore, Baihui is probably an important acupoint in preventing and treating cerebral diseases, and this is why Baihui seems to be the mostly used acupoint in lab to study acupuncture effect for stroke. Similarly, EA pretreatment at Weizhong acupoint (BL40) was more beneficial for the spinal cord I/R injury in rabbits as compared to that of the Zusanli acupoint (ST36) (Lei et al., 2003). Hegu (LI4) was chosen for the treatment of neuronal injury resulting from hypoxia–ischemia (Jiang et al., 2003), and Neiguan (PC6) was preferred in EA pretreatment-induced cardioprotection for its effect on heart disease (Tsou et al., 2004).

The option for acupoint is much more important in clinical trials. A recent fMRI study found that acupuncture on Quchi (LI11) and Zusanli (ST36) induced activity of different brain regions, and the responses were also different between the healthy and stroke patients (Cho, Kim, et al., 2013). Sanjiao (SJ8) is a language-implicated acupoint. Acupuncture on SJ8 of the poststroke aphasia patients could induce a significant activation in opercular, triangular, and insula via fMRI, which demonstrated that language-deficit-implicated acupoint stimulation could selectively activate the language-associated brain region on the lesion side in the poststroke aphasia patients (Chau, Fai Cheung, Jiang, Au-Yeung, & Li, 2010; Li & Yang, 2011). Acupuncture stimulation at somatosensory-implicated acupoints (LI4, LI11, GB34) could induce a greater activation in the somatosensory cortex in stroke patients as compared to controls (Jeun et al., 2005; Li, Jack, & Yang, 2006). Qian et al. compared the effects of acupuncture at "Shuigou" (GV26), "Neiguan" (PC6), "Chize" (LU5), "Sanyinjiao" (SP6), and "Weizhong" (BL40) on MCAO rats. The results showed that "Shuigou" (GV26) and "Neiguan" (PC6) had a more obvious effect in the improvement of CBF (Qian et al., 2009). Wu and colleagues used the data mining technology to explore the characteristics of meridian points in the treatment of poststroke disorder with acupuncture. They found that the acupoints of the Yang meridians were the first option, which mainly distributed on the limbs, and the combination of Yangming and Shaoyang meridians was the most common (Wu, Li, & Ren, 2013).

6. PARAMETERS OF ACUPUNCTURE THERAPY FOR STROKE

Since the acupuncture needs to be manually manipulated after the needle was inserted into the skin of acupoints, different manipulation methods and time could significantly influence the effect of the acupuncture. He et al. observed the effect of different needle-retaining duration on hemorheology in stroke patients. The results showed that the effective rate of the 60-min group was significantly higher than those of the 20- and 40-min groups, and the improvement of the hemorheology parameters in the 60-min group was also remarkably higher than that of the 20- and 40-min groups (He et al., 2007).

The parameters are especially important in EA. Studies have shown that electric stimulation of different parameters might have different effects on these functions. Nested study design (Yang et al., 2004) was adopted to identify the influence of different parameters and their combination on EA preconditioning induced cerebral ischemic tolerance in rats. This study confirmed the optimal electrical stimulation parameters of the EA preconditioning to induce cerebral ischemic tolerance in rats. The results showed that varying frequency and waveform of the stimulus could produce different protective effects, but the current did not matter much. The density-sparse wave had the most obvious neuroprotective effect, followed by the intermittent wave, and the continuous wave's neuroprotective effect was relatively poor. The reason for this observation could be that the continuous wave tended to induce the tolerance of the electric stimulus, whereas the density-sparse wave could stimulate the release of different types of neurochemicals by transformation between low-, medium-, and high-frequency stimuli. Therefore, the density-sparse wave EA stimulation could generate neuroprotective effect at different targets via the activation of different signaling pathways. Two other groups did similar work and figured out the optimal parameters of EA of their labs to treat stroke (Wei, Fan, Wang, Yang, & Shi, 2010; Zhou, Guo, Cheng, Wu, & Xia, 2011).

7. MECHANISMS OF ACUPUNCTURE THERAPY FOR STROKE

Acupuncture is pleiotropic and could have multiple complicated influences on the physiology of brain and on the pathophysiology of stroke. Although effective mechanisms of acupuncture for stroke are largely unknown, a series of clinical trials have shown that acupuncture primarily regulates the release of neurochemicals, hemorheology, cerebral microcirculation, metabolism, neuronal activity, and the function of specific brain region. Animal studies showed that the effects of acupuncture therapy on stroke were possibly the inhibition of postischemic inflammatory reaction, cell apoptosis, stimulation of neurogenesis, angiogenesis, and influence on neural plasticity.

8. MECHANISMS OF ACUPUNCTURE THERAPY FOR POSTSTROKE PATIENTS

8.1. Acupuncture regulated the release of neurochemicals

The motor functions of the limbs and the activities of daily living in hemiplegic patients caused by acute stroke were significantly improved after the treatment with EA. This improvement was associated with reduced serum levels of NSE, S-100B, and endothelin (Zhang, Kang, Li, & Zhang, 2013). Proteomic analysis showed that serpin G1 protein expression in serum was downregulated, while the expression of gelsolin, complement component I, C3, C4B, and beta-2-glycoprotein I proteins were upregulated in the post-stroke patients after the EA stimulation, which indicated that EA appears to be effective in regulating the differential expression of multiple serum proteins involved in stroke (Pan et al., 2011).

8.2. Acupuncture-regulated hemorheology and cerebral microcirculation

Acupuncture could induce changes in the cerebral microvascular blood flow perfusion, that cause cardiovascular regulatory activities and changes of blood pressure, which in turn affect the cerebral circulation (Hsiu, Hsu, Chen, Hsu, & Lin, 2013). Using laser Doppler flowmetry, Hsiu et al. noted significant bilateral differences in patients' parameters following the acupuncture stimulation, with an increased pulsatile component of the microcirculatory blood flow (MBF), decreased blood flow resistance, and decreased MBF variability in the vascular beds on the stroke-affected side. Spectral analysis revealed that the vasodilation on the stroke-affected side could be partly attributed to decreased sympathetic neural activity (Hsiu, Huang, Chen, Hsu, & Hsu, 2011).

8.3. Acupuncture influenced the metabolism and activity of neurons

Acupuncture regulates the glucose metabolism and activates the cerebral structures and plasticity in the cerebral functional regions in chronic stage ischemic stroke patients (Fang, Ning, Xiong, & Shulin, 2012; Huang et al., 2012). It is effective for protecting neurons and facilitating the recovery proved by higher apparent diffusion coefficient and fractional anisotropy as compared to the control group, which correlates with the patient recovery and reveals the progress of secondary degeneration (Shen, Li, Wei, & Lou, 2012).

8.4. Acupuncture influenced the local function of brain and muscles

Manual acupuncture provides sufficient neuromuscular stimuli to promote immediate changes in the motor unit gross recruitment without repercussion in maximal force output in the healthy subjects. However, poststroke patients did not exhibit a significant reduction on the myoelectric activity and maximal force output (Fragoso & Ferreira, 2012). Enduring motor cortex functional changes were observed after acupuncture treatment, in terms of cortical excitability and output mapping using transcranial magnetic stimulation (Lo, Cui, & Fook-Chong, 2005).

9. MECHANISMS OF ACUPUNCTURE FOR ANIMAL STROKE MODELS

9.1. Acupuncture regulated oxidative stress and inhibited inflammation and neuronal apoptosis

In animal stroke models, acupuncture exerts anti-inflammatory effects via suppression of the TLR4/NF-kB pathway, which could ameliorate the cognitive impairment induced by stroke (Feng et al., 2013; Lan et al., 2013). Furthermore, it promotes neurological functional recovery via the retinoic acid signaling pathway (Hong, Wu, Zou, Tao, & Chen, 2013). Acute EA stimulation after a moderate focal cerebral ischemia improves the tissue and functional recovery. The Ach/eNOS-mediated perfusion augmentation might be related to these beneficial effects (Kim et al., 2013). Liu et al. acquired 3-D fluorodeoxyglucose-microPET images after using acupuncture in stroke mice, which demonstrated that acupuncture reduced the injury volume by improving the metabolic recovery after the stroke (Liu et al., 2013). EA can alleviate the cerebral edema after ischemic stroke (Zhang, Wu, & Jia, 2011), attenuate the extracellular glutamate level in the ischemic cerebral tissues (Lee et al., 2010), and repair the cells under stress undergoing apoptosis via activation of heat-shock proteins (Cakmak, 2009). It could activate PI3K/Akt signaling in the ischemic cerebral tissue, which resulted in the inhibition of cerebral cell apoptosis, and increased the serum secretion levels of the PI3K activators, brain-derived neurotrophic factor (BDNF) and glial cell-derived neurotrophic factor (GDNF), as well as upregulated the antiapoptotic *Bcl-2/Bax* ratio in the ischemic cerebrum (Chen et al., 2012). Acupuncture stimulation is responsible for the potential protection of neurons through suppression of CBF response in the increased plasma osmolarity and extracellular glutamine release in diabetic rats under ischemic conditions (Choi et al., 2010).

9.2. Acupuncture stimulates neurogenesis and angiogenesis, facilitates plasticity

Acupuncture has been reported to improve the neuronal regeneration at the edge of the ischemic lesions (Wu et al., 2012) and to increase cell proliferation in dentate gyrus (Kim et al., 2001). Angiogenesis and improved CBF in the ischemic boundary area were detected after the EA treatment in the rats with ischemic stroke (Du et al., 2011). The acupuncture prevented the impairments of the spike encoding and synaptic transmission at the GABAergic neurons from ischemia. This prevention was associated with the resistance of these cells to ischemia-induced changes in the spike threshold potentials and refractory period (Zhang, Li, et al., 2011).

10. MECHANISMS OF ACUPUNCTURE PRECONDITIONING FOR STROKE

10.1. EA pretreatment regulates oxidative stress, maintains the integrity of BBB, and inhibits apoptosis

Abrupt reperfusion after ischemia results in an overproduction of reactive oxygen species (ROS), which can lead to the brain injury (Chan, 1996). EA pretreatment enhances the activity of mitochondrial respiratory enzymes, attenuates the lipid peroxidation, and reduces the production of ROS, which consequently improves the function of the respiratory chain and antioxidant capacity in the ischemic penumbra (Siu, Lo, & Leung, 2004; Zhong, Li, Huan, & Chen, 2009). In addition, it increases the levels of antiapoptotic genes like *Bcl-2* while decreasing the levels of proapoptotic genes such as *c-Jun* and *c-Fos*, which inhibits the subsequent apoptotic cascades (Jiang et al., 2003; Jiang, Zhao, Shui, & Xia, 2004). Jiang et al. also found that the antagonizing effect of EA pretreatment on the cerebral hypoxic/ischemic injury might be related to its activation of K_{ATP}, which inhibits the neuronal apoptosis induced by the immediate genes, *c-Fos* and *c-Jun*, at the early injury stages (Jiang et al., 2004).

Moreover, the blood-brain barrier (BBB) integration and stress reactions are involved in the neuroprotection after EA pretreatment. The BBB integration is disrupted by cerebral ischemia, resulting in the brain edema. Matrix metalloproteinases (MMPs) are neutral proteases that disrupt the BBB and are associated with subcortical ischemic vascular disease (Candelario-Jalil et al., 2011). The expression and activity of the matrix metalloproteinases-9 (MMP-9), one of MMPs, are decreased after the EA pretreatment, and subsequently the brain edema and BBB damage are significantly alleviated (Dong et al., 2009). This phenomenon has also been observed in another experiment, which indicates that extracellular signal-regulated kinases (ERK) pathway is involved in this process (Chaudhry et al., 2010).

10.2. Endocannabinoid system contributes to the neuroprotective effects of the EA pretreatment

Recent investigations have shown that the endocannabinoid system might be a new mechanism of EA pretreatment-induced neuroprotection. EA pretreatment increases the release of 2-arachidonylglycerol (2-AG) as well as *N*-arach-idonoylethanolamine-anandamide (AEA), two endocannabinoids, and upregulates the expression of cannabinod CB1 receptor in the brain. Selective CB1 antagonist AM251 or CB1 short-interfering RNA (siRNA) blocked the neuroprotective effects of the EA pretreatment. Meanwhile, pretreatment with 2-AG and AEA also reduced the infarct size and improved the neurological outcomes (Wang et al., 2009). Moreover, further studies (Ma et al., 2011) showed that both acute and delayed ischemic tolerance were associated with endocannabinoid system: the acute phase, which occurred in 2 h after the EA pretreatment, was mediated by CB1, whereas the delayed phase occurred in 24 h after the EA pretreatment via CB2 (Ma et al., 2011). These findings indicated that the endocannabinoid system might play an important role in the neuroprotective effect of EA pretreatment.

Activation of the CB1 receptor triggers the signaling transduction events that can influence the ischemic compensatory responses. The cellular responses that elicit neuroprotection might involve the CB1 receptors and their link to a variety of signaling elements, including the Gi/Go family of G-proteins, mitogen-activated protein kinase, MAP Kinase kinase (MEK1/2), and its substrate, ERK1/2. Further studies have demonstrated that the neuroprotection of EA pretreatment could be abolished by U0126 (a specific inhibitor of the MEK1/2) or TAT-εV1–2 (an εPKC-selective peptide inhibitor). The blockade of the CB1 receptor by a CB1 receptor antagonist, AM251, reversed the activation of the ERK1/2 and εPKC as a result of EA pretreatment. These findings suggested that the ERK1/2 and εPKC pathways might be involved in the EA pretreatment-induced cerebral ischemic tolerance via the cannabinoid CB1 receptor (Du et al., 2010; Wang et al., 2011).

10.3. EA pretreatment attenuates glutamate excitotoxicity via NMDAR

Cerebral ischemia has been reported to induce excessive glutamate release and excitotoxicity (Lei, Berthet, Hirt, & Gruetter, 2009). Transient increase of CBF during reperfusion (hyperemia) would aggravate the brain injury induced by excitotoxicity. Pre-, intra-, or posttreatment of EA could rescue hippocampal neurons from the ischemic insults via decreasing the production of glutamate and reducing hyperemia (Meng, Sun, Liu, & Yan, 2008; Pang, Itano, Sumitani, Negi, & Miyamoto, 2003). Previous studies indicated that NMDARs were responsible for the glutamate-induced excitotoxicity in the postischemic brain (Briz, Galofre, & Sunol, 2010; Lei et al., 2009). EA pretreatment suppressed the expression of NR1, a subunit of the NMDARs, which might contribute to its effect in reducing apoptosis and protecting the cerebral neurons (Meng et al., 2008). Further study suggested that the reduced NR1 expression could be reverted by specific inhibitors of the PI3K pathway, but the inhibition of the ERK pathway did not show the same effects (Sun et al., 2005). Therefore, EA pretreatment attenuated glutamate excitotoxicity by modulating the PI3K pathway.

11. PROSPECTS OF ACUPUNCTURE THERAPY FOR STROKE

Available data indicate that acupuncture might have a positive effect on the stroke prevention and rehabilitation, but the evidence is not strong enough. Although large numbers of clinical trials were conducted, most of them were small-sampled, single-centered, and with low quality. There are some difficult to conduct acupuncture studies in relation to stroke. First, unlike other treatments, acupuncture has acupoints specificity and optimal parameters. Therefore, it needs a lot of practice and training before application. But there are identified differences in practice and training between acupuncture practitioners in different countries (Robinson et al., 2012). This is probably the main reason for the different results among these clinical trials. Second, most of the trials were conducted in China and were published in Chinese. Doctors and scientists in the field of traditional Chinese medicine usually lack the experience to communicate with foreign peers and have less knowledge for the standards of RCT, which is designed for modern western medicine. Similarly, researchers interested in acupuncture but from different cultures might be confronted with the language obstacle to acquire enough information. Finally, acupuncture is a procedure from which the patients could have strong feelings. So it is very hard to use placebo procedures as control. Researchers often give sham acupuncture at nonacupoint area or do not give electric stimulation when conducted with EA. But the patients can easily tell whether they receive the treatment in this situation, and the emotion of happiness or disappointment will seriously influence the results of the study.

However, current studies intend to confirm whether the acupuncture has a positive effect on stroke and just provide limited information about the mechanisms of acupuncture therapy for stroke. More studies need to be conducted to explain how the acupuncture works, which would provide further reasoning as to advocate for its effects among general public in order for them to accept it as a method of treatment.

In conclusion, there have been some proof for the efficiency of acupuncture therapy for stroke, but more large, collaborative, innovative, and high-quality trials are still needed to provide more confirmative and conceivable evidence.

REFERENCES

Barnes, P. M., Bloom, B., & Nahin, R. L. (2008). Complementary and alternative medicine use among adults and children: United States, 2007. *National Health Statistics Reports*, *12*, 1–23.

Briz, V., Galofre, M., & Sunol, C. (2010). Reduction of glutamatergic neurotransmission by prolonged exposure to dieldrin involves NMDA receptor internalization and metabotropic glutamate receptor 5 downregulation. *Toxicological Sciences*, *113*, 138–149.

Cakmak, Y. O. (2009). A review of the potential effect of electroacupuncture and moxibustion on cell repair and survival: The role of heat shock proteins. *Acupuncture in Medicine*, *27*, 183–186.

Candelario-Jalil, E., Thompson, J., Taheri, S., Grossetete, M., Adair, J. C., Edmonds, E., et al. (2011). Matrix metalloproteinases are associated with increased blood-brain barrier opening in vascular cognitive impairment. *Stroke*, *42*, 1345–1350.

Chan, P. H. (1996). Role of oxidants in ischemic brain damage. *Stroke*, *27*, 1124–1129.

Chau, A. C., Fai Cheung, R. T., Jiang, X., Au-Yeung, P. K., & Li, L. S. (2010). An fMRI study showing the effect of acupuncture in chronic stage stroke patients with aphasia. *Journal of Acupuncture and Meridian Studies*, *3*, 53–57.

Chaudhry, K., Rogers, R., Guo, M., Lai, Q., Goel, G., Liebelt, B., et al. (2010). Matrix metalloproteinase-9 (MMP-9) expression and extracellular signal-regulated kinase 1 and 2 (ERK1/2) activation in exercise-reduced neuronal apoptosis after stroke. *Neuroscience Letters*, *474*, 109–114.

Chen, A., Lin, Z., Lan, L., Xie, G., Huang, J., Lin, J., et al. (2012). Electroacupuncture at the Quchi and Zusanli acupoints exerts neuroprotective role in cerebral ischemia-reperfusion injured rats via activation of the PI3K/Akt pathway. *International Journal of Molecular Medicine*, *30*, 791–796.

Cho, S. Y., Kim, M., Sun, J. J., Jahng, G. H., Kim, H. J., Park, S. U., et al. (2013). A comparison of brain activity between healthy subjects and stroke patients on fMRI by acupuncture stimulation. *Chinese Journal of Integrative Medicine*, *19*, 269–276.

Cho, S. Y., Park, J. Y., Jung, W. S., Moon, S. K., Park, J. M., Ko, C. N., et al. (2013). Bee venom acupuncture point injection for central post stroke pain: A preliminary single-blind randomized controlled trial. *Complementary Therapies in Medicine*, *21*, 155–157.

Choi, S., Lee, G. J., Chae, S. J., Kang, S. W., Yin, C. S., Lee, S. H., et al. (2010). Potential neuroprotective effects of acupuncture stimulation on diabetes mellitus in a global ischemic rat model. *Physiological Measurement*, *31*, 633–647.

Chou, P., Chu, H., & Lin, J. G. (2009). Effects of electroacupuncture treatment on impaired cognition and quality of life in Taiwanese stroke patients. *Journal of Alternative and Complementary Medicine, 15*, 1067–1073.

Dong, H., Fan, Y. H., Zhang, W., Wang, Q., Yang, Q. Z., & Xiong, L. Z. (2009). Repeated electroacupuncture preconditioning attenuates matrix metalloproteinase-9 expression and activity after focal cerebral ischemia in rats. *Neurological Research, 31*, 853–858.

Donnan, G. A., Fisher, M., Macleod, M., & Davis, S. M. (2008). Stroke. *Lancet, 371*, 1612–1623.

Du, Y., Shi, L., Li, J., Xiong, J., Li, B., & Fan, X. (2011). Angiogenesis and improved cerebral blood flow in the ischemic boundary area were detected after electroacupuncture treatment to rats with ischemic stroke. *Neurological Research, 33*, 101–107.

Du, J., Wang, Q., Hu, B., Peng, Z., Zhao, Y., Ma, L., et al. (2010). Involvement of ERK 1/2 activation in electroacupuncture pretreatment via cannabinoid CB1 receptor in rats. *Brain Research, 1360*, 1–7.

Fang, Z., Ning, J., Xiong, C., & Shulin, Y. (2012). Effects of electroacupuncture at head points on the function of cerebral motor areas in stroke patients: A PET study. *Evidence-Based Complementary and Alternative Medicine, 2012*, 902413.

Feng, X., Yang, S., Liu, J., Huang, J., Peng, J., Lin, J., et al. (2013). Electroacupuncture ameliorates cognitive impairment through inhibition of NF-kappaB-mediated neuronal cell apoptosis in cerebral ischemia-reperfusion injured rats. *Molecular Medicine Reports, 7*, 1516–1522.

Fragoso, A. P., & Ferreira, A. S. (2012). Immediate effects of acupuncture on biceps brachii muscle function in healthy and post-stroke subjects. *Chinese Medicine, 7*, 7.

Gosman-Hedstrom, G., Claesson, L., Klingenstierna, U., Carlsson, J., Olausson, B., Frizell, M., et al. (1998). Effects of acupuncture treatment on daily life activities and quality of life: A controlled, prospective, and randomized study of acute stroke patients. *Stroke, 29*, 2100–2108.

He, Y. Z., Han, B., Zheng, S. F., Wang, L. N., Chen, Z. M., Hu, J., et al. (2007). Effect of different acupuncture needle-retaining time on hemorheology in ischemic stroke patients. *Zhen Ci Yan Jiu, 32*, 338–341.

Hong, J., Wu, G., Zou, Y., Tao, J., & Chen, L. (2013). Electroacupuncture promotes neurological functional recovery via the retinoic acid signaling pathway in rats following cerebral ischemia-reperfusion injury. *International Journal of Molecular Medicine, 31*, 225–231.

Hsieh, R. L., Wang, L. Y., & Lee, W. C. (2007). Additional therapeutic effects of electroacupuncture in conjunction with conventional rehabilitation for patients with first-ever ischaemic stroke. *Journal of Rehabilitation Medicine, 39*, 205–211.

Hsiu, H., Hsu, C. L., Chen, C. T., Hsu, W. C., & Lin, F. C. (2013). Effects of acupuncture on the harmonic components of the radial arterial blood-pressure waveform in stroke patients. *Biorheology, 50*, 69–81.

Hsiu, H., Huang, S. M., Chen, C. T., Hsu, C. L., & Hsu, W. C. (2011). Acupuncture stimulation causes bilaterally different microcirculatory effects in stroke patients. *Microvascular Research, 81*, 289–294.

Huang, Y., Tang, C., Wang, S., Lu, Y., Shen, W., Yang, J., et al. (2012). Acupuncture regulates the glucose metabolism in cerebral functional regions in chronic stage ischemic stroke patients—A PET-CT cerebral functional imaging study. *BMC Neuroscience, 13*, 75.

Jeun, S. S., Kim, J. S., Kim, B. S., Park, S. D., Lim, E. C., Choi, G. S., et al. (2005). Acupuncture stimulation for motor cortex activities: A 3T fMRI study. *The American Journal of Chinese Medicine, 33*, 573–578.

Jiang, K. W., Zhang, Y., & Shui, Q. X. (2003). Possible mechanism of electroacupuncture preconditioning for hypoxia/ischemic brain injury protection effect in neonatal rats. *Zhongguo Zhong Xi Yi Jie He Za Zhi, 23*, 914–917.

Jiang, K., Zhao, Z., Shui, Q., & Xia, Z. (2004). Electro-acupuncture preconditioning abrogates the elevation of c-Fos and c-Jun expression in neonatal hypoxic-ischemic rat brains induced by glibenclamide, an ATP-sensitive potassium channel blocker. *Brain Research, 998*, 13–19.

Johansson, B. B., Haker, E., von Arbin, M., Britton, M., Langstrom, G., & Terent, A. (2001). Acupuncture and transcutaneous nerve stimulation in stroke rehabilitation: A randomized, controlled trial. *Stroke, 32*, 707–713.

Kim, J. H., Choi, K. H., Jang, Y. J., Bae, S. S., Shin, B. C., Choi, B. T., et al. (2013). Electroacupuncture acutely improves cerebral blood flow and attenuates moderate ischemic injury via an endothelial mechanism in mice. *PLoS One, 8*, e56736.

Kim, E. H., Kim, Y. J., Lee, H. J., Huh, Y., Chung, J. H., Seo, J. C., et al. (2001). Acupuncture increases cell proliferation in dentate gyrus after transient global ischemia in gerbils. *Neuroscience Letters, 297*, 21–24.

Kong, J. C., Lee, M. S., Shin, B. C., Song, Y. S., & Ernst, E. (2010). Acupuncture for functional recovery after stroke: A systematic review of sham-controlled randomized clinical trials. *CMAJ, 182*, 1723–1729.

Lan, L., Tao, J., Chen, A., Xie, G., Huang, J., Lin, J., et al. (2013). Electroacupuncture exerts anti-inflammatory effects in cerebral ischemia-reperfusion injured rats via suppression of the TLR4/NF-kappaB pathway. *International Journal of Molecular Medicine, 31*, 75–80.

Lansberg, M. G., Bluhmki, E., & Thijs, V. N. (2009). Efficacy and safety of tissue plasminogen activator 3 to 4.5 hours after acute ischemic stroke: A metaanalysis. *Stroke, 40*, 2438–2441.

Lee, S. Y., Baek, Y. H., Park, S. U., Moon, S. K., Park, J. M., Kim, Y. S., et al. (2009). Intradermal acupuncture on shen-men and nei-kuan acupoints improves insomnia in stroke patients by reducing the sympathetic nervous activity: A randomized clinical trial. *The American Journal of Chinese Medicine, 37*, 1013–1021.

Lee, J. A., Park, S. W., Hwang, P. W., Lim, S. M., Kook, S., Choi, K. I., et al. (2012). Acupuncture for shoulder pain after stroke: A systematic review. *Journal of Alternative and Complementary Medicine, 18*, 818–823.

Lee, G. J., Yin, C. S., Choi, S. K., Choi, S., Yang, J. S., Lee, H., et al. (2010). Acupuncture attenuates extracellular glutamate level in global ischemia model of rat. *Neurological Research, 32*(Suppl 1), 79–83.

Lees, K. R., Zivin, J. A., Ashwood, T., Davalos, A., Davis, S. M., Diener, H. C., et al. (2006). NXY-059 for the treatment of acute ischemic stroke. *The New England Journal of Medicine, 354*, 588–600.

Lei, H., Berthet, C., Hirt, L., & Gruetter, R. (2009). Evolution of the neurochemical profile after transient focal cerebral ischemia in the mouse brain. *Journal of Cerebral Blood Flow and Metabolism, 29*, 811–819.

Lei, Y., Xiong, L., Zeng, Y., Sun, Y., Chen, S., Wang, Q., et al. (2003). The compare of preconditioning with repeated electroacupuncture on different acupoints to induce neuroprotection against spinal cord ischemia-reperfusion injury in rabbits. *Chinese Journal of Clinical Rehabilitation, 7*, 3.

Li, G., Jack, C. R., Jr., & Yang, E. S. (2006). An fMRI study of somatosensory-implicated acupuncture points in stable somatosensory stroke patients. *Journal of Magnetic Resonance Imaging, 24*, 1018–1024.

Li, G., & Yang, E. S. (2011). An fMRI study of acupuncture-induced brain activation of aphasia stroke patients. *Complementary Therapies in Medicine, 19*(Suppl. 1), S49–S59.

Liu, S. Y., Hsieh, C. L., Wei, T. S., Liu, P. T., Chang, Y. J., & Li, T. C. (2009). Acupuncture stimulation improves balance function in stroke patients: A single-blinded controlled, randomized study. *The American Journal of Chinese Medicine, 37*, 483–494.

Liu, W., Mukherjee, M., Sun, C., Liu, H., & McPeak, L. K. (2008). Electroacupuncture may help motor recovery in chronic stroke survivors: A pilot study. *Journal of Rehabilitation Research and Development, 45*, 587–595.

Liu, H., Shen, X., Tang, H., Li, J., Xiang, T., & Yu, W. (2013). Using microPET imaging in quantitative verification of the acupuncture effect in ischemia stroke treatment. *Scientific Reports, 3*, 1070.

Liu, M., Wu, B., Wang, W. Z., Lee, L. M., Zhang, S. H., & Kong, L. Z. (2007). Stroke in China: Epidemiology, prevention, and management strategies. *Lancet Neurology, 6*, 456–464.

Liu, Y., Zou, L. P., Du, J. B., & Wong, V. (2010). Electro-acupuncture protects against hypoxic-ischemic brain-damaged immature rat via hydrogen sulfide as a possible mediator. *Neuroscience Letters, 485*, 74–78.

Lo, Y. L., Cui, S. L., & Fook-Chong, S. (2005). The effect of acupuncture on motor cortex excitability and plasticity. *Neuroscience Letters, 384*, 145–149.

Long, Y. B., & Wu, X. P. (2012). A meta-analysis of the efficacy of acupuncture in treating dysphagia in patients with a stroke. *Acupuncture in Medicine, 30*, 291–297.

Lu, Z. H., Bai, X. G., Xiong, L. Z., Wang, Y. H., Wang, Y., & Wang, Q. (2010). Effect of electroacupuncture preconditioning on serum S100beta and NSE in patients undergoing craniocerebral tumor resection. *Chinese Journal of Integrative Medicine, 16*, 229–233.

Lu, Z. H., Xiong, L. Z., Zhu, Z. H., Wang, Q., & Cheng, M. (2002). Acupoint specificity of electroacupuncture preconditioning effect on cerebral ischemia injury in rats. *Zhongguo Zhen Jiu, 22*, 3.

Ma, L., Zhu, Z., Zhao, Y., Hou, L., Wang, Q., Xiong, L., et al. (2011). Cannabinoid receptor type 2 activation yields delayed tolerance to focal cerebral ischemia. *Current Neurovascular Research, 8*, 145–152.

Meng, P. Y., Sun, G. J., Liu, S. H., & Yan, H. M. (2008). Effect of electroacupuncture pretreatment on glutamate-NMDAR signal pathway in hippocampal neurons of vascular dementia rats. *Zhen Ci Yan Jiu, 33*, 103–106.

Murry, C. E., Jennings, R. B., & Reimer, K. A. (1986). Preconditioning with ischemia: A delay of lethal cell injury in ischemic myocardium. *Circulation, 74*, 1124–1136.

Na, B. J., Jahng, G. H., Park, S. U., Jung, W. S., Moon, S. K., Park, J. M., et al. (2009). An fMRI study of neuronal specificity of an acupoint: Electroacupuncture stimulation of Yanglingquan (GB34) and its sham point. *Neuroscience Letters, 464*, 1–5.

Napadow, V., Makris, N., Liu, J., Kettner, N. W., Kwong, K. K., & Hui, K. K. (2005). Effects of electroacupuncture versus manual acupuncture on the human brain as measured by fMRI. *Human Brain Mapping, 24*, 193–205.

Pan, S., Zhan, X., Su, X., Guo, L., Lv, L., & Su, B. (2011). Proteomic analysis of serum proteins in acute ischemic stroke patients treated with acupuncture. *Experimental Biology and Medicine (Maywood), 236*, 325–333.

Pang, J., Itano, T., Sumitani, K., Negi, T., & Miyamoto, O. (2003). Electroacupuncture attenuates both glutamate release and hyperemia after transient ischemia in gerbils. *The American Journal of Chinese Medicine, 31*, 295–303.

Park, J., Hopwood, V., White, A. R., & Ernst, E. (2001). Effectiveness of acupuncture for stroke: A systematic review. *Journal of Neurology, 248*, 558–563.

Perez-Pinzon, M. A., Xu, G. P., Dietrich, W. D., Rosenthal, M., & Sick, T. J. (1997). Rapid preconditioning protects rats against ischemic neuronal damage after 3 but not 7 days of reperfusion following global cerebral ischemia. *Journal of Cerebral Blood Flow and Metabolism, 17*, 175–182.

Przyklenk, K., Bauer, B., Ovize, M., Kloner, R. A., & Whittaker, P. (1993). Regional ischemic 'preconditioning' protects remote virgin myocardium from subsequent sustained coronary occlusion. *Circulation, 87*, 893–899.

Qian, Y. F., Fan, X. N., Li, Y. J., Zhang, Y. N., Wei, Y. Y., Zhang, X., et al. (2009). Effects of acupuncture at different acupoints on cerebral blood flow in cerebral ischemia model rats. *Zhongguo Zhen Jiu, 29*, 213–216.

Rabinstein, A. A., & Shulman, L. M. (2003). Acupuncture in clinical neurology. *The Neurologist, 9*, 137–148.

Robinson, N., Lorenc, A., Ding, W., Jia, J., Bovey, M., & Wang, X. M. (2012). Exploring practice characteristics and research priorities of practitioners of traditional acupuncture in China and the EU—A survey. *Journal of Ethnopharmacology, 140*, 604–613.

Schaechter, J. D., Connell, B. D., Stason, W. B., Kaptchuk, T. J., Krebs, D. E., Macklin, E. A., et al. (2007). Correlated change in upper limb function and motor cortex activation after verum and sham acupuncture in patients with chronic stroke. *Journal of Alternative and Complementary Medicine, 13*, 527–532.

Shah, S. H., Engelhardt, R., & Ovbiagele, B. (2008). Patterns of complementary and alternative medicine use among United States stroke survivors. *Journal of Neurological Sciences, 271*, 180–185.

Shen, P. F., Kong, L., Ni, L. W., Guo, H. L., Yang, S., Zhang, L. L., et al. (2012). Acupuncture intervention in ischemic stroke: A randomized controlled prospective study. *The American Journal of Chinese Medicine, 40*, 685–693.

Shen, Y., Li, M., Wei, R., & Lou, M. (2012). Effect of acupuncture therapy for postponing Wallerian degeneration of cerebral infarction as shown by diffusion tensor imaging. *Journal of Alternative and Complementary Medicine, 18*, 1154–1160.

Siu, F. K., Lo, S. C., & Leung, M. C. (2004). Effectiveness of multiple pre-ischemia electro-acupuncture on attenuating lipid peroxidation induced by cerebral ischemia in adult rats. *Life Sciences, 75*, 1323–1332.

Sun, Y., Xue, S. A., & Zuo, Z. (2012). Acupuncture therapy on apoplectic aphasia rehabilitation. *Journal of Traditional Chinese Medicine, 32*, 314–321.

Sun, N., Zou, X., Shi, J., Liu, X., Li, L., & Zhao, L. (2005). Electroacupuncture regulates NMDA receptor NR1 subunit expression via PI3-K pathway in a rat model of cerebral ischemia-reperfusion. *Brain Research, 1064*, 98–107.

Sze, F. K., Wong, E., Or, K. K., Lau, J., & Woo, J. (2002). Does acupuncture improve motor recovery after stroke? A meta-analysis of randomized controlled trials. *Stroke, 33*, 2604–2619.

Tao, J., Xue, X. H., Chen, L. D., Yang, S. L., Jiang, M., Gao, Y. L., et al. (2010). Electroacupuncture improves neurological deficits and enhances proliferation and differentiation of endogenous nerve stem cells in rats with focal cerebral ischemia. *Neurological Research, 32*, 198–204.

Toyoda, K., Fujii, K., Kamouchi, M., Nakane, H., Arihiro, S., Okada, Y., et al. (2004). Free radical scavenger, edaravone, in stroke with internal carotid artery occlusion. *Journal of Neurological Sciences, 221*, 11–17.

Tsou, M. T., Huang, C. H., & Chiu, J. H. (2004). Electroacupuncture on PC6 (Neiguan) attenuates ischemia/reperfusion injury in rat hearts. *The American Journal of Chinese Medicine, 32*, 951–965.

Wang, Q., Li, X., Chen, Y., Wang, F., Yang, Q., Chen, S., et al. (2011). Activation of epsilon protein kinase C-mediated anti-apoptosis is involved in rapid tolerance induced by electroacupuncture pretreatment through cannabinoid receptor type 1. *Stroke, 42*, 389–396.

Wang, H., & Liang, F. X. (2008). Analysis on the features of preventive treatment with acupuncture and moxibustion. *Journal of Traditional Chinese Medicine, 28*, 281–285.

Wang, S. J., Omori, N., Li, F., Jin, G., Zhang, W. R., Hamakawa, Y., et al. (2002). Potentiation of Akt and suppression of caspase-9 activations by electroacupuncture after transient middle cerebral artery occlusion in rats. *Neuroscience Letters, 331*, 115–118.

Wang, Q., Peng, Y., Chen, S., Gou, X., Hu, B., Du, J., et al. (2009). Pretreatment with electroacupuncture induces rapid tolerance to focal cerebral ischemia through regulation of endocannabinoid system. *Stroke, 40*, 2157–2164.

Wang, Q., Xiong, L., Chen, S., Liu, Y., & Zhu, X. (2005). Rapid tolerance to focal cerebral ischemia in rats is induced by preconditioning with electroacupuncture: Window of protection and the role of adenosine. *Neuroscience Letters, 381*, 158–162.

Wei, Y. Y., Fan, X. N., Wang, S., Yang, S., & Shi, X. M. (2010). Specificity effect of acupuncture at "Shuigou" (GV 26) on brain infarction area in MCAO rats and the influence of acupuncture parameter. *Zhongguo Zhen Jiu, 30*, 221–225.

Wu, Z., Hu, J., Du, F., Zhou, X., Xiang, Q., & Miao, F. (2012). Long-term changes of diffusion tensor imaging and behavioural status after acupuncture treatment in rats with transient focal cerebral ischaemia. *Acupuncture in Medicine, 30*, 331–338.

Wu, L. T., Li, Y., & Ren, Y. L. (2013). Exploration on the characteristics of meridian points in the treatment of post-stroke disorder with acupuncture and moxibustion based on the data mining technology. *Zhongguo Zhen Jiu, 33*, 125–130.

Wu, M. T., Sheen, J. M., Chuang, K. H., Yang, P., Chin, S. L., Tsai, C. Y., et al. (2002). Neuronal specificity of acupuncture response: A fMRI study with electroacupuncture. *NeuroImage, 16*, 1028–1037.

Wu, H. M., Tang, J. L., Lin, X. P., Lau, J., Leung, P. C., Woo, J., et al. (2006). Acupuncture for stroke rehabilitation. *The Cochrane Database of Systematic Reviews*, CD004131.

Xiong, L., Lu, Z., Hou, L., Zheng, H., Zhu, Z., Wang, Q., et al. (2003). Pretreatment with repeated electroacupuncture attenuates transient focal cerebral ischemic injury in rats. *Chinese Medical Journal (English), 116*, 108–111.

Yam, W., & Wilkinson, J. M. (2010). Is acupuncture an acceptable option in stroke rehabilitation? A survey of stroke patients. *Complementary Therapies in Medicine, 18*, 143–149.

Yan, T., & Hui-Chan, C. W. (2009). Transcutaneous electrical stimulation on acupuncture points improves muscle function in subjects after acute stroke: A randomized controlled trial. *Journal of Rehabilitation Medicine, 41*, 312–316.

Yang, J., Xiong, L., Wang, Q., Liu, Y., Chen, S., & Xu, N. (2004). Effects of different stimulating parameters and their various combinations on electroacupuncture induced cerebral ischemic tolerance in rats. *Zhongguo Zhen Jiu, 24*, 5.

Yang, L., Yang, J., Wang, Q., Chen, M., Lu, Z., Chen, S., et al. (2010). Cardioprotective effects of electroacupuncture pretreatment on patients undergoing heart valve replacement surgery: A randomized controlled trial. *The Annals of Thoracic Surgery, 89*, 781–786.

Zhang, H., Kang, T., Li, L., & Zhang, J. (2013). Electroacupuncture reduces hemiplegia following acute middle cerebral artery infarction with alteration of serum NSE, S-100B and endothelin. *Current Neurovascular Research, 10*, 216–221.

Zhang, S., Li, G., Xu, X., Chang, M., Zhang, C., & Sun, F. (2011). Acupuncture to point Baihui prevents ischemia-induced functional impairment of cortical GABAergic neurons. *Journal of Neurological Sciences, 307*, 139–143.

Zhang, S. H., Liu, M., Asplund, K., & Li, L. (2005). Acupuncture for acute stroke. *The Cochrane Database of Systematic Reviews*, CD003317.

Zhang, F., Wu, Y., & Jia, J. (2011). Electro-acupuncture can alleviate the cerebral oedema of rat after ischemia. *Brain Injury, 25*, 895–900.

Zhao, J. G., Cao, C. H., Liu, C. Z., Han, B. J., Zhang, J., Li, Z. G., et al. (2009). Effect of acupuncture treatment on spastic states of stroke patients. *Journal of Neurological Sciences, 276*, 143–147.

Zhao, X. F., Du, Y., Liu, P. G., & Wang, S. (2012). Acupuncture for stroke: Evidence of effectiveness, safety, and cost from systematic reviews. *Topics in Stroke Rehabilitation, 19*, 226–233.

Zhong, S., Li, Z., Huan, L., & Chen, B. Y. (2009). Neurochemical mechanism of electroacupuncture: Anti-injury effect on cerebral function after focal cerebral ischemia in rats. *Evidence-Based Complementary and Alternative Medicine, 6*, 51–56.

Zhou, F., Guo, J., Cheng, J., Wu, G., & Xia, Y. (2011). Electroacupuncture increased cerebral blood flow and reduced ischemic brain injury: Dependence on stimulation intensity and frequency. *Journal of Applied Physiology*, *111*, 1877–1887.

Zhuangl, L. X., Xu, S. F., D'Adamo, C. R., Jia, C., He, J., Han, D. X., et al. (2012). An effectiveness study comparing acupuncture, physiotherapy, and their combination in poststroke rehabilitation: A multicentered, randomized, controlled clinical trial. *Alternative Therapies in Health and Medicine*, *18*, 8–14.

CHAPTER NINE

Effect and Mechanism of Acupuncture on Alzheimer's Disease

Bai-Yun Zeng[1], Sarah Salvage, Peter Jenner
Neurodegenerative Disease Research Group, Institute of Pharmaceutical Science, School of Biomedical Sciences, King's College London, London, United Kingdom
[1]Corresponding author: e-mail address: b.zeng@kcl.ac.uk

Contents

Abstract

Alzheimer's disease is the most common form of dementia diagnosed in the aging population worldwide. The cause of Alzheimer's is still not clear. There is no cure for the disease and current treatments are only symptomatic relieve. The search for new treatment is made ever more urgent due to increasing population aging. Acupuncture has been in practice in China for more than 3000 years and used to treat a wide variety of conditions including cardiovascular and psychiatric diseases, acute, and chronic pain. In this chapter, we review recent development on the effects and mechanisms of acupuncture on Alzheimer's disease. In Alzheimer's animal models, acupuncture stimulation at acupoints enhances cholinergic neurotransmission, trophic factor releasing, reduces apoptotic and oxidative damages, improves synaptic plasticity and decreases the levels of Aβ proteins in the hippocampus and relevant brain regions. The biochemical modulations by acupuncture in the brains of Alzheimer's models are correlated with the cognitive improvement. In Alzheimer's patients, functional brain images demonstrated that acupuncture increased in the activity in the temporal lobe and prefrontal lobe which are related to the memory and cognitive function. Although only a few acupuncture clinical studies with a small number of participants are reported, they represent an important step forward in the research of both acupuncture and Alzheimer's. Translation of acupuncture research in animal model studies into the human subjects

International Review of Neurobiology, Volume 111
ISSN 0074-7742
http://dx.doi.org/10.1016/B978-0-12-411545-3.00009-2

will undoubtedly enhance acupuncture efficacy in clinical study and treatment which could eventually lead to a safer, well-tolerated and inexpensive form of care for Alzheimer's patients.

1. INTRODUCTION

Alzheimer's disease (AD) is the most common cause of dementia. AD is now recognized to progress through three stages: preclinical, mild cognitive impairment (MCI), and dementia (Albert et al., 2011; McKhann et al., 2011; Sperling et al., 2011). In preclinical AD, biomarkers for AD are present, but symptoms have not yet appeared (Sperling et al., 2011). In MCI, patients have cognitive deficits, where the memory of recent facts and spatial orientation is affected but not functional impairment (Albert et al., 2011). In AD dementia, a decline in two or more cognitive domains has gradually progressed to the point that functioning at work or daily activities such as speech is impaired (McKhann et al., 2011). Pathological changes in the brain, which define the disease, are abundant extracellular amyloid β (Aβ) plaques and intracellular neurofibrillary tangles, accompanied by synaptic and neuronal loss in the cerebral cortex and certain subcortical regions (Alonso, Zaidi, Novak, Grundke-Iqbal, & Iqbal, 2001; Goedert, Spillantini, & Crowther, 1991; Hardy & Allsop, 1991; Selkoe, 1986), where reduced synthesis of the neurotransmitter acetylcholine is observed and implicated as one of the causes of AD (Geula & Mesulam, 1995). The pathological changes result in gross atrophy of affected regions, including temporal lobe and parietal lobe, and parts of frontal cortex and cingulate gyrus (Wenk, 2003). Although the accumulated knowledge from research substantiates our understanding of AD, the cause and progression of the disease are still not clear. Intracellular Aβ oligomers are known to initiate a cascade of pathological events ranging from mitochondria dysfunction, abnormal synaptic transmission, oxidative stress, and loss of calcium regulation to inflammation (Delacourte, 2006; Hardy & Higgins, 1992; Leuner, Müller, Reichert, 2012; Yankner, Lu, & Loerch, 2008). On the other hand, hyperphosphorylation and aggregation of tau proteins can destabilize microtubules and impair axonal transport, compromising synaptic function and giving rise to neurofibrillary tangles (Chun & Johnson, 2007; Iqbal, 2005). The two neurodegenerative processes are considered to play an important role in the progressive loss of neurons, which undergoes a progressive spreading along a pathway from the entorhinal and hippocampal formation toward polymodal association and then primary brain regions (Braak & Braak, 1996). Current treatments for AD are mainly acetylcholinesterase inhibitors, for example, rivastigmine and donepezil, which is based on the proposal that AD is caused by reduced synthesis of the neurotransmitter acetylcholine. These drugs can only offer relatively small symptom benefit but remain palliative in nature, and there are no available treatments to stop or reverse the progression of the disease.

Acupuncture has been in practice in China for more than 3000 years. Acupuncture is one of the most popular forms of complementary medicine available, and it is used by many populations for a variety of conditions (Barnes, Bloom, & Nahin, 2008). Modern acupuncture can be defined as the insertion of sharp, thin needle into the specific point on the body with mechanical, electrical, or other physical manipulations, which stimulate nerve receptors both directly and indirectly through mechanical coupling via the connective tissue surrounding the needle (Dong et al., 2002; Langevin et al., 2002). In general, the acupuncture stimulation, through the local reflex and central nervous system, induces endocrine, autonomic, and systemic behavioral responses. This suggests that acupuncture therapy beneficially affects a whole body even if it stimulates only limited sites of body using fine needles (Kim et al., 2009; Samuels, Gropp, Singer, & Oberbaum, 2008; Sandberg, Lundeberg, Lindberg, & Gerdle, 2003). Within the past few decades, acupuncture is becoming increasingly popular with both public and physicians, and it has been reported to be a very safe and well-tolerated therapy with only minor side effects (Posadzki, Alotaibi, & Ernst, 2012; Rabinstein & Shulman, 2003; Shulman et al., 2002). It has been reported to effectively treat a wide variety of disorders including cardiovascular and psychiatric diseases and acute and chronic pain and neurological disorders (Longhurst, 2010; Pittler & Ernst, 2008; Rabinstein & Shulman, 2003; Schroer & Adamson, 2011). Recently, acupuncture has been reported to treat Alzheimer's and dementia and showed to be effective in improving intelligence and ameliorating depression and anxiety (Lee, Shin, & Ernst, 2009; Schroer & Adamson, 2011). In this review, we summarize the recent research development of acupuncture treatment in AD.

2. PRECLINICAL STUDY

The hippocampus, which plays a central role in learning and memory, demonstrates a higher degree of structural plasticity (Olson, Eadie, Ernst, &

Christie, 2006; Squire, 1993). The dentate gyrus is part of hippocampal formation and continues to develop through adulthood. Progenitor cells in the subgranular zone of the dentate gyrus (SZDG) continuously generate granule cells, as they fully integrate into hippocampal network (van Praag et al., 2002). Damage to the granule cells produces cognitive impairment (Moreira, Moreira, Bueno, & Xavier, 1997) as the hippocampus regulates learning and memory. Electroacupuncture stimulation at acupoints ST36 and GV20 in normal rats once a day for 3 weeks induce cell proliferation and differentiation demonstrated by specific markers, Ki67 and doublecortin, in the SZDG (Hwang et al., 2010a). Electro acupuncture (EA) significantly increased the number of phosphorylated cyclic AMP response element-binding (pCREB) protein immunoreactive cells and markedly elevated the levels of brain-derived neurotrophic factor (BDNF) and pCREB protein in the dentate gyrus (Hwang et al., 2010b). Further, administration of BDNF inhibitor, K252a, significantly reduced cell proliferation in the SZDG (Hwang et al., 2010b). This suggests that EA significantly increased neuroblast plasticity via BDNF and pCREB activation in the dentate gyrus.

2.1. Mouse model study

Mouse models are one of the most important research tools for evaluating new therapeutic strategies and understanding pathological mechanisms involved in the disease process in AD. Many mouse models are based on disease-causing mutations such as amyloid precursor protein (APP) transgenic mice (Sturchler-Pierrat et al., 1997); others are derived from selective breeding such as the senescence-accelerated mouse prone (SAMP), showing deficits in learning and memory (Takeda et al., 1981). Recently, AD mouse models have been used to explore the effect and mechanism of acupuncture treatment on AD.

2.1.1 APP transgenic mice

APP transgenic mice expressing a mutant form of the human APP gene develop fibrillar amyloid plaques and Alzheimer's-like brain pathology with spatial learning deficits behavioral (Games, 1995; Hsiao, 1996). EA stimulation at GV20 and KI1 of APP transgenic mice at 2 Hz/100 Hz, 3–4 mA for 15 min once a day for 3 months showed significant improvement in Lashley III water maze test and minimized neuronal mitochondrial damages in hippocampal CA1 region compared to control group (Xue, Ge, Zhang, Xu, & Bai, 2009). EA markedly reduced the expression of Aβ precursor protein and Aβ protein in the cerebral cortex and hippocampal CA1 region and significantly increased the levels of choline acetyltransferase (ChAT) in those brain areas compared to control group (Xue, Zhang, Bai, Xu, & Wu, 2009). Low-density lipoprotein receptor-related protein-1 (LRP-1) is involved in the regulating brain apolioprotein E and cholesterol metabolism, and LRP1 dysfunction is implicated in the Aβ aggregation in AD brain (Liu et al., 2007; Shibata et al., 2000). Following the acupuncture stimulation at acupoints GV20 and KI1, the expression of LRP1 was significantly elevated in the hippocampal sulcus microvessels in APP transgenic mice compared to controls (Xue, Zhang, Xu, Wu, & Bai, 2011). These serial studies suggest that EA stimulation improves learning-memory capacity by preserving neuronal mitochondrial integration, promoting APP normal function, reducing Aβ plaques, and increasing acetylcholine transmissions in the hippocampus of APP mutant mice.

2.1.2 SAMP mice

The SAMP is an accelerated aging model that was established through phenotypic selection from a common genetic pool of AKR/J strain mice including nine major SAMP substrains (Takeda et al., 1981). SAMP8 mice is characterized with the deposition of Aβ protein in cerebral cortex and hippocampus and the number of deposition increased with age, which is similar to those of AD (Takeda et al., 1997; Takemura et al., 1993). While SAMP8 is an age-related deficit of memory and learning model, SAMP10 is characterized by spontaneously occurring age-related brain atrophy with deficits in learning and memory and emotional disorders (Takeda, 2009). These models are useful in acupuncture search to investigate the mechanism of its brain modulatory effects.

In SAMP8 mice, acupuncture stimulation at acupoints CV17, ST36, CV12, CV6, and SP10 once a day for 15 days showed a markedly reduction of latency to find the hidden platform in the Morris water maze task test in 4-month old SAMP8 mice compared to control groups (Cheng et al., 2008). Further, Bromodeoxyuridine (Brdu) immunohistochemistry demonstrated that cell proliferation in dentate gyrus and ventricular/subventricular zones was significantly enhanced, with the cell proliferation pattern liking a stream-like distribution of newly proliferated cells presented alone alveus hippocampi extending from lateral ventricle to corpus callosum, in acupuncture-treated SAMP8 mice compared to control groups (Cheng et al., 2008; Li et al., 2012). In SAMP10 mice, acupuncture was performed at the acupoints CV17, CV12, CV6, SP10, and ST36 once a day for 15 days to investigate the effects of acupuncture on the expression of genes related to aging in the cerebellum of mice (Ding, Yu, Yu, Fu, & Han, 2006). DNA microarray technique showed that acupoint stimulation completely or partly reversed age-related gene expression profiles in the hippocampus of SAMP10 mice and subsequent RT-PCR and Northern blotting studies revealed expression of three genes, such as Hsp84, Hsp86, and YB-1, which are closely involved in oxidative stress-induced damage, are reversed in acupuncture-treated group (Ding et al., 2006). The data indicates that acupuncture stimulation at the specific acupoints improves the cognitive function of SAMP mice and this may related to its stimulation of neurogenesis potential in the learning and memory affected brain regions and reversal of age-related gene expression profiles.

2.2. Rat model study

In AD research, the rats have for decades been a very important model; for example, the acute neurodegenerative effect of Aβ and amyloid cores from the brains of AD patients was demonstrated *in vivo* in 1991, when these substances were injected into the brains of two different rat models (Frautschy, Baird, & Cole, 1991; Kowall, Beal, Busciglio, Duffy, & Yankner, 1991). In both cases, a significant induction of abnormal tau phosphorylation was observed in the immediate vicinity of the Aβ immunoreactive site. Subsequently, many studies reported the AD-like astrogliosis, tau hyperphosphorylation and memory decline in intracerebral Aβ-injected rat models (Harkany et al., 1998; Nakamura, Murayama, Noshita, Annoura, & Ohno, 2001; Sweeney, Luedtke, McDonald, & Overmier, 1997; Weldon et al., 1998). These models recently have been used to test the protective effects of ginseng, ginkgo biloba extracts, and other neuroprotective agents (Nguyen et al., 2007; Tang, Nag, Shiu, & Pang, 2002; Wang, Wang, Ng, & Lee, 2006; Wang, Zang, et al., 2006). Very recently, Aβ rat models are utilized to explore the mechanisms and effects of acupuncture treatment in AD. In a serial studies, EA was performed at acupoints Xiusanzhen including bilateral LI20 and EX-HN3 at 80–100 Hz, 1–3 mA for 10 min, once daily (excluding Saturday and Sunday) for 6 weeks in rats which received intrahippocampal injection of Aβ and developed learning and memory deficits (Liu, Niu, Yang, Niu, & Wang, 2009, Liu, Niu, Yang, Niu, & Yuan, 2011, Yang, Liu, Niu, & Niu, 2011). Morris water maze tests showed that acupoint stimulation significantly shortened the average escape latency and increased target-platform crossing times compared to control groups (Liu et al., 2009). Biochemical studies revealed that the activity of ChAT and AChE of hippocampal tissues was markedly decreased in Aβ-alone rat group and acupuncture stimulation reversed the decrease in both ChAT and AChE (Liu et al., 2009). Immunohistochemistry showed that acupuncture significantly increased Bcl-2, an antiappototic protein, and decreased Bax, a proappototic protein, expression in hippocampus compared to control groups (Liu et al., 2011). The muscarinic acetylcholine receptor (mAChR) density and its B_{max} were elevated, and disassociation constant (K_d) of mAChR was significantly reduced in hippocampus in acupuncture-treated Aβ rats compared to nontreated Aβ rat groups (Yang et al., 2011). In Aβ rat AD model, the step-down test showed that the error number and total error number were significantly increased, and electrophysiological recording showed that long-term potential (LTP) of hippocampus was markedly decreased (Shen, Tang, Li, & Ma, 2010). EA stimulation at acupoints GV20, GV14 (Dazhui), bilateral BL23 (Shenshu), and bilateral KI1 for 30 min, once daily for 7 days, significantly improved step-down test and increased hippocampal LTP in acupuncture-treated Aβ rat compared to nontreatment group (Shen et al., 2010). Following EA stimulation at acupoints GV20, CV14, KI3, BL23, and ST36, once daily, 7 days for a course, and lasted for four courses in Aβ rat model of AD, Aβ protein expression was significantly decreased and superoxide dismutases (SOD) was markedly elevated in the hippocampus of acupuncture-treated Aβ rats compared to nontreatment group (Zhang, Guan, & Jiang, 2010). These studies suggest that acupuncture improved learning and memory deficits in Aβ rat model of AD and its beneficial effects may be related to its ability to reverse the decreased acetylcholine neurotransmission, improve synaptic transmission, and decrease oxidative damage in the hippocampus and related brain regions in Aβ rats.

3. CLINICAL STUDY

There are no many reports of acupuncture treatment with AD patients. In an earlier study, eight patients diagnosed with mild-to-moderate AD were treated with acupuncture (Kao, 2000). Acupuncture stimulation was performed at acupoints HT7 and KI3 for 30 min each session, a 7-day treatment cycle with 3-day break in between for a total of 30 days. Patients were graded using the mini-mental state examination (MMSE) to measure their levels of orientation, memory, attention, and the ability to name an object, follow verbal and written commands, and write a sentence spontaneously. At the end of treatment, patients showed an obvious improvement on measures of verbal orientation and motor coordination and had higher overall MMSE scores (Kao, 2000). However, the study did not have a control group and the number of patients treated was too small to show any significant meaning.

Aβ-peptide is heavily deposited in the brains of AD patients, and free radical oxidative stress, particularly manifested by lipid peroxidation and protein oxidation, is extensive (Butterfield & Lauderback, 2002). Aβ causes lipid peroxidation in brain cell membranes that is inhibited by antioxidants (Avdulov et al., 1997). Among other markers of oxidative damage, iPF2α, a sensitive and specific marker of lipid peroxidation, was elevated and correlated with the cognitive and functional impairment in patients with AD (Praticò et al., 2000, 2004). To explore the therapeutic effects and its mechanism on AD, 20 patients with AD were treated with acupuncture (Zhu et al., 2010). Acupuncture was performed at acupoints GV20, BL23, SP10, and BL17 in AD patients for 12 weeks, and it significantly reduced the cognitive impairment in AD patients assessed by Alzheimer's Disease Assessment Scale-Cognitive Section (ADAS-Cog) and decreased in levels of iPF2α, in the cerebrospinal fluid, blood, and urine in patients with AD. This suggests that acupuncture could act as antioxidant by reducing oxidative damage in the brain of AD patients (Zhu et al., 2010).

4. BRAIN IMAGING STUDIES

In the past decades, brain imaging techniques allow us to study the changes in anatomical structures and physiological function. In particular, functional magnetic resonance imaging (fMRI) is used to investigate the neural underpinnings of higher cognitive functions by measuring regional hemodynamic changes, which are thought to be closely linked to underlying cellular activity (Logothetis, Pauls, Augath, Trinath, & Oeltermann, 2001; Shmuel, Augath, Oeltermann, & Logothetis, 2006). In its most widely used form, fMRI is used to infer the regional increases in activity based on increases in the blood-oxygen-level-dependent (BOLD) contrast, while decreases in the BOLD indicate the deactivation during performance of specific cognitive tasks (Shmuel et al., 2006). More specifically, BOLD signal changes detected by fMRI are thought to represent integrated synaptic activity by measuring the changes in blood flow, blood volume, and blood oxyhemoglobin/deoxyhemoglobin ratio underlying such synaptic activity (Gusnard & Raichle, 2001). The fMRI BOLD signals are observed in responses to acupuncture stimulation and have been applied to investigate the mechanisms of acupuncture in the brain.

fMRI studies showed that acupuncture is able to modulate a widely distributed brain network. It has been reported that increased functional correlations of throughout entire brain following acupuncture stimulation at acupoints ST36 in healthy subjects were primarily related with limbic/paralimbic and subcortical regions, whereas decreased correlations were mainly related with the sensory and frontal cortex (Feng et al., 2012). Very recently, the fractional amplitude of low-frequency fluctuations (fALFF), an index of regional brain activity, is used to investigate the effects of acupuncture on PC6 on spontaneous fluctuation in the resting-state BOLD fMRI signal in health adult subjects (Zhang et al., 2012). fMRI showed acupuncture stimulation at acupoint PC6 altered the amplitude of the intrinsic cortical activity and increased fALFF amplitudes in the anterior cingulate cortex, PCC, and cerebellum during acupoints-stimulating scan (Zhang et al., 2012). The fALFF was still significantly higher in the PCC during postacupoint-stimulating resting-state scan than during preacupoint-stimulating resting-state scan, while fALFF decreased in other brain regions during postacupoint-stimulating resting-state scan (Zhang et al., 2012). In AD brains, SPECT studies showed that the hypoperfusion and hypometabolism in the PCC and fMRI reported the reduced activity in this region (Matsuda, 2007; Rombouts, Barkhof, Goekoop, Stam, Scheltens, 2005). In MCI patients, fMRI displayed a decreased functional connectivity between PCC and a set of brain regions (Allen et al., 2007; Wang et al., 2012 Wang, Wang, et al., 2006; Wang, Zang, et al., 2006). So, the long-lasting increased activity in the PCC induced by PC6 stimulation indicates that acupuncture on PC6 may have therapeutic modulating effect on cognitive dysfunction.

Recently, fMRI is used to verify the mechanism of acupuncture in treating MCI and AD. In one study, 26 patients with mild-to-moderate AD underwent fMRI while undergoing acupuncture treatment at acupoint HT7, ST36, ST40, and KI3 on the left side of body for 6 min (Zhou & Jin, 2008). fMRI showed that acupuncture stimulation induced activity in the temporal lobe (including hippocampus), some regions of parietal lobe, and cerebellum in AD patients (Zhou & Jin, 2008). In another study, acupoint thread-embedding technique was used at HT7, ST40, KI3, and ST36 acupoints on 13 AD patients, once every month for six times in total, while sham operation was performed at the same acupoints of 13 AD patients of control group (Zhou, Han, & Jia, 2008). Cognitive changes were assessed by MMSE and ADAS-Cog scoring, and fMRI was monitored on two-time points, that is, 1 week before starting treatment and 1 week post-acupuncture treatment (Zhou et al., 2008). Acupoint thread-embedding stimulation significantly improved both MMSE and ADAS-Cog scoring in acupuncture-treated group compared to sham-treated group, and this is correlated with the changes in activity in the frontal and temporal lobes and marginal system in acupuncture-treated group (Zhou et al., 2008). However, both studies did not include healthy control group. In a better designed study,14 patients with AD, 8 patients with MCI, and 14 healthy subjects were treated with acupuncture at acupoints LV3 and LI4 bilaterally to investigate regional brain activity using fMRI in MCI patients, AD patients, and control subjects under three conditions including resting state, acupuncture state, and poststimulus resting state (Wang et al., 2012). The brain activities in MCI and AD patients were different from those of control subjects during resting state; during acupuncture stimulation, MCI and AD patients showed activities in the regions consistent with impaired brain function, and poststimulus resting state showed the increased or decreased activities in several regions in MCI and AD patients compared to control subjects. Most of the regions are involved in the temporal lobe and the frontal lobe, which are closely related to the memory and cognition (Wang et al., 2012). In a similar study, fMRI is used to investigate effect of acupuncture at acupoint KI3 on the functional connectivity throughout the entire brains in12 MCI patients and 12 healthy subjects, and fMRI showed abnormal functional connectivity during the resting state in MCI patients compared to control subject (Feng et al., 2012). Acupuncture stimulation significantly altered functional connectivity within abnormal regions, and the correlations related with temporal lobe were enhanced during poststimulus in MCI patients compared to controls. Moreover, deep acupuncture (with needling depth usually 1–2 cm) stimulation significantly increased the correlations related with temporal lobe compared to superficial acupuncture (with needling depth usually 1–2 mm) in MCI patients (Feng et al., 2012). The results indicate that acupuncture, in particular, deep stimulation can activate certain memory and cognition-related regions in AD and MCI. Further studies needed to look whether the activated brain regional activities by acupuncture correlate with symptomatic improvement in AD patients.

5. CONCLUSIONS

In summary, the data from acupuncture research in AD show that acupuncture or EA improves some of cognitive function by enhancing cholinergic neurotransmission and trophic factor releasing, protecting cerebral neurons from apoptosis and oxidative stress, improving synaptic plasticity, and reducing levels of abnormal proteins such as Aβ. Although all clinical studies mentioned above have a small number of participants and some studies even did not have normal controls, they, as the pioneer studies, represent an important step forward in the research of both acupuncture and AD. They could help lay the groundwork for larger, controlled investigations to determine how acupuncture combats Alzheimer's, which could eventually lead to safer, inexpensive forms of care for the millions of patients with AD.

REFERENCES

Albert, M. S., DeKosky, S. T., Dickson, D., Dubois, B., Feldman, H. H., Fox, N. C., et al. (2011). The diagnosis of mild cognitive impairment due to Alzheimer's disease: Recommendations from the National Institute on Aging-Alzheimer's Association workgroups on diagnostic guidelines for Alzheimer's disease. *Alzheimers Dementia*, 7, 270–279.

Allen, G., Barnard, H., McColl, R., Hester, A. L., Fields, J. A., Weiner, M. F., et al. (2007). Reduced hippocampal functional connectivity in Alzheimer disease. *Archives of Neurology*, *64*, 1482–1487.

Alonso, A., Zaidi, T., Novak, M., Grundke-Iqbal, I., & Iqbal, K. (2001). Hyperphosphorylation induces self-assembly of tau into tangles of paired helical filaments/straight filaments. *Proceedings of the National Academy of Sciences of the United States of America*, *98*, 6923–6928.

Avdulov, N. A., Chochina, S. V., Igbavboa, U., O'Hare, O., Schroeder, F., Cleary, J. P., et al. (1997). Amyloid beta-peptides increase annular and bulk fluidity and induce lipid peroxidation in brain synaptic plasma membranes. *Journal of Neurochemistry*, *68*, 2086–2091.

Barnes, P. M., Bloom, B., & Nahin, R. L. (2008). Complementary and alternative medicine use among adults and children: United States, 2007. In *National Health Statistics Reports; No 12*. Hyattsville, MD: National Center for Health Statistics.

Braak, H., & Braak, E. (1996). Development of Alzheimer-related neurofibrillary changes in the neocortex inversely recapitulates cortical myelogenesis. *Acta Neuropathologica (Berlin)*, *92*, 197–201.

Butterfield, D. A., & Lauderback, C. M. (2002). Lipid peroxidation and protein oxidation in Alzheimer's disease brain: Potential causes and consequences involving amyloid beta-peptide-associated free radical oxidative stress. *Free Radical Biology and Medicine*, *32*, 1050–1060.

Cheng, H., Yu, J., Jiang, Z., Zhang, X., Liu, C., Peng, Y., et al. (2008). Acupuncture improves cognitive deficits and regulates the brain cell proliferation of SAMP8 mice. *Neuroscience Letters*, *432*, 111–116.

Chun, W., & Johnson, G. V. (2007). The role of tau phosphorylation and cleavage in neuronal cell death. *Frontiers in Bioscience*, *12*, 733–756.

Delacourte, A. (2006). The natural and molecular history of Alzheimer's disease. *Journal of Alzheimer's Disease*, *9*(3 Suppl.), 187–194.

Ding, X., Yu, J., Yu, T., Fu, Y., & Han, J. (2006). Acupuncture regulates the aging-related changes in gene profile expression of the hippocampus in senescence-accelerated mouse (SAMP10). *Neuroscience Letters*, *399*, 11–16.

Dong, Q., Dong, X., Li, H., Sun, G., Ren, J., Sun, Q., et al. (2002). The influence of electroacupuncture on the activity of deep body receptors. *Shanghai Journal of Acupuncture, 21*, 38–40.

Feng, Y., Bai, L., Ren, Y., Chen, S., Wang, H., Zhang, W., et al. (2012). FMRI connectivity analysis of acupuncture effects on the whole brain network in mild cognitive impairment patients. *Magnetic Resonance Imaging, 30*, 672–682.

Frautschy, S. A., Baird, A., & Cole, G. M. (1991). Effects of injected Alzheimer beta-amyloid cores in rat brain. *Proceedings of the National Academy of Sciences of the United States of America, 88*, 8362–8366.

Games, D. (1995). Alzheimer-type neuropathology in transgenic mice overexpressing V717F beta-amyloid precursor protein. *Nature, 373*, 523–527.

Geula, C., & Mesulam, M. M. (1995). Cholinesterases and the pathology of Alzheimer disease. *Alzheimer Disease and Associated Disorders, 9*(Suppl. 2), 23–28.

Goedert, M., Spillantini, M. G., & Crowther, R. A. (1991). Tau proteins and neurofibrillary degeneration. *Brain Pathology, 1*, 279–286.

Gusnard, D. A., & Raichle, M. E. (2001). Searching for a baseline: Functional imaging and the resting human brain. *Nature Reviews Neuroscience, 2*, 685–694.

Hardy, J., & Allsop, D. (1991). Amyloid deposition as the central event in the aetiology of Alzheimer's disease. *Trends in Pharmacological Sciences, 12*, 383–388.

Hardy, J. A., & Higgins, G. A. (1992). Alzheimer's disease: The amyloidβ cascade hypothesis. *Science, 256*, 184–185.

Harkany, T., O'Mahony, S., Kelly, J. P., Soós, K., Törõ, I., & Penke, B. (1998). Beta-amyloid (Phe(SO3H)24)25–35 in rat nucleus basalis induces behavioral dysfunctions, impairs learning and memory and disrupts cortical cholinergic innervation. *Behavioural Brain Research, 90*, 133–145.

Hsiao, K. (1996). Correlative memory deficits, Abeta elevation, and amyloid plaques in transgenic mice. *Science, 274*, 99–102.

Hwang, I. K., Chung, J. Y., Yoo, D. Y., Yi, S. S., Youn, H. Y., Seong, J. K., et al. (2010a). Effects of electroacupuncture at Zusanli and Baihui on brain-derived neurotrophic factor and cyclic AMP response element-binding protein in the hippocampal dentate gyrus. *Journal of Veterinary Medical Science, 72*, 1431–1436.

Hwang, I. K., Chung, J. Y., Yoo, D. Y., Yi, S. S., Youn, H. Y., Seong, J. K., et al. (2010b). Comparing the effects of acupuncture and electroacupuncture at Zusanli and Baihui on cell proliferation and neuroblast differentiation in the rat hippocampus. *Journal of Veterinary Medical Science, 72*, 279–284.

Iqbal, K. (2005). Tau pathology in Alzheimer disease and other tauopathies. *Biochimica et Biophysica Acta, 1739*, 198–210.

Kao, H. (2000). Acupuncture enhancement in clinical symptoms and cognitive-motor abilities of the Alzheimer's disease patients. In: *Presented at the World Alzheimer's Conference, Washington, DC, July 9–18.*

Kim, H., Park, H. J., Han, S. M., Hahm, D. H., Lee, H. J., Kim, K. S., et al. (2009). The effects of acupuncture stimulation at PC6 (Neiguan) on chronic mild stress-induced biochemical and behavioral responses. *Neuroscience Letters, 460*, 56–60.

Kowall, N. W., Beal, M. F., Busciglio, J., Duffy, L. K., & Yankner, B. A. (1991). An in vivo model for the neurodegenerative effects of beta amyloid and protection by substance P. *Proceedings of the National Academy of Sciences of the United States of America, 88*, 7247–7251.

Langevin, H. M., Churchill, D. L., Wu, J., Badger, G. J., Yandow, J. A., Fox, J. R., et al. (2002). Evidence of connective tissue involvement in acupuncture. *FASEB Journal, 16*, 872–874.

Lee, M. S., Shin, B. C., & Ernst, E. (2009). Acupuncture for Alzheimer's disease: A systematic review. *International Journal of Clinical Practice, 63*, 874–879.

Leuner, K., Müller, W. E., & Reichert, A. S. (2012). From mitochondrial dysfunction to amyloid beta formation: Novel insights into the pathogenesis of Alzheimer's disease. *Molecular Neurobiology, 46*, 186–193.

Li, G., Zhang, X., Cheng, H., Shang, X., Xie, H., Zhang, X., et al. (2012). Acupuncture improves cognitive deficits and increases neuron density of the hippocampus in middle-aged SAMP8 mice. *Acupuncture in Medicine, 30*, 339–345.

Liu, Z. B., Niu, W. M., Yang, X. H., Niu, X. M., & Wang, Y. (2009). Effect of "Xiusanzhen" on learning-memory ability and hippocampal ChAT and AChE activity in Alzheimer disease rats. *Zhen Ci Yan Jiu, 34*, 48–51.

Liu, Z. B., Niu, W. M., Yang, X. H., Niu, X. M., & Yuan, W. (2011). Effect of "Xiusanzhen" on expression of hippocampal Bcl-2 and Bax proteins in Alzheimer disease rats. *Zhen Ci Yan Jiu, 36*, 7–11.

Liu, Q., Zerbinatti, C. V., Zhang, J., Hoe, H. S., Wang, B., Cole, S. L., et al. (2007). Amyloid precursor protein regulates brain apolipoprotein E and cholesterol metabolism through lipoprotein receptor LRP1. *Neuron, 56*, 66–78.

Logothetis, N. K., Pauls, J., Augath, M., Trinath, T., & Oeltermann, A. (2001). Neurophysiological investigation of the basis of the fMRI signal. *Nature, 412*, 150–157.

Longhurst, J. C. (2010). Defining meridians: A modern basis of understanding. *Journal of Acupuncture and Meridian Studies, 3*, 67–74.

Matsuda, H. (2007). Role of neuroimaging in Alzheimer's disease, with emphasis on brain perfusion SPECT. *Journal of Nuclear Medicine, 48*, 1289–1300.

McKhann, G. M., Knopman, D. S., Chertkow, H., Hyman, B. T., Jack, C. R., Kawas, C. H., Jr., et al. (2011). The diagnosis of dementia due to Alzheimer's disease: Recommendations from the National Institute on Aging-Alzheimer's Association workgroups on diagnostic guidelines for Alzheimer's disease. *Alzheimer's & Dementia, 7*, 263–269.

Moreira, R. C., Moreira, M. V., Bueno, J. L., & Xavier, G. F. (1997). Hippocampal lesions induced by ionizing radiation: A parametric study. *Journal of Neuroscience Methods, 75*, 41–47.

Nakamura, S., Murayama, N., Noshita, T., Annoura, H., & Ohno, T. (2001). Progressive brain dysfunction following intracerebroventricular infusion of beta(1–42)-amyloid peptide. *Brain Research, 912*, 128–136.

Nguyen, P. T., Kimura, T., Ho, S. A., Tran, A. H., Ono, T., & Nishijo, H. (2007). Ameliorative effects of a neuroprotective agent, T-817MA, on place learning deficits induced by continuous infusion of amyloid-beta peptide (1–40) in rats. *Hippocampus, 17*, 443–455.

Olson, A. K., Eadie, B. D., Ernst, C., & Christie, B. R. (2006). Environmental enrichment and voluntary exercise massively increase neurogenesis in the adult hippocampus via dissociable pathways. *Hippocampus, 16*, 250–260.

Pittler, M. H., & Ernst, E. (2008). Complementary therapies for neuropathic and neuralgic pain: Systematic review. *Clinical Journal of Pain, 24*, 731–733.

Posadzki, P., Alotaibi, A., & Ernst, E. (2012). Prevalence of use of complementary and alternative medicine (CAM) by physicians in the UK: A systematic review of surveys. *Clinical Medicine, 12*, 505–512.

Praticò, D., Clark, C. M., Lee, V. M., Trojanowski, J. Q., Rokach, J., & FitzGerald, G. A. (2000). Increased 8,12-iso-iPF2alpha-VI in Alzheimer's disease: Correlation of a noninvasive index of lipid peroxidation with disease severity. *Annals of Neurology, 48*, 809–812.

Praticò, D., Yao, Y., Rokach, J., Mayo, M., Silverberg, G. G., & McGuire, D. (2004). Reduction of brain lipid peroxidation by CSF drainage in Alzheimer's disease patients. *Journal of Alzheimer's Disease, 6*, 385–389.

Rabinstein, A. A., & Shulman, L. M. (2003). Acupuncture in clinical neurology. *The Neurologist, 9*, 137–148.

Rombouts, S. A., Barkhof, F., Goekoop, R., Stam, C. J., & Scheltens, P. (2005). Altered resting state networks in mild cognitive impairment and mild Alzheimer's disease: An fMRI study. *Human Brain Mapping, 26*, 231–239.
Samuels, N., Gropp, C., Singer, S. R., & Oberbaum, M. (2008). Acupuncture for psychiatric illness: A literature review. *Behavioral Medicine, 34*, 55–64.
Sandberg, M., Lundeberg, T., Lindberg, L. G., & Gerdle, B. (2003). Effects of acupuncture on skin and muscle blood flow in healthy subjects. *European Journal of Applied Physiology, 90*, 114–119.
Schroer, S., & Adamson, J. (2011). Acupuncture for depression: A critique of the evidence base. *CNS Neuroscience & Therapeutics, 17*, 398–410.
Selkoe, D. J. (1986). Altered structural proteins in plaques and tangles: What do they tell us about the biology of Alzheimer's disease? *Neurobiology of Aging, 7*, 425–432.
Shen, M. H., Tang, Q. Q., Li, Z. R., & Ma, C. (2010). Effect of electroacupuncture on hippocampal LTP in Alzheimer's disease rats induced by Abeta(25-35). *Zhen Ci Yan Jiu, 35*, 3–7.
Shibata, M., Yamada, S., Kumar, S. R., Calero, M., Bading, J., & Frangione, B. (2000). Clearance of Alzheimer's amyloid-ss(1-40) peptide from brain by LDL receptor-related protein-1 at the blood-brain barrier. *Journal of Clinical Investigation, 106*, 1489–1499.
Shmuel, A., Augath, M., Oeltermann, A., & Logothetis, N. K. (2006). Negative functional MRI response correlates with decreases in neuronal activity in monkey visual area V1. *Nature Neuroscience, 9*, 569–577.
Shulman, L. M., Wen, X., Weiner, W. J., Bateman, D., Minagar, A., Duncan, R., et al. (2002). Acupuncture therapy for the symptoms of Parkinson's disease. *Movement Disorders, 17*, 799–802.
Sperling, R. A., Aisen, P. S., Beckett, L. A., Bennett, D. A., Craft, S., Fagan, A. M., et al. (2011). Toward defining the preclinical stages of Alzheimer's disease: Recommendations from the National Institute on Aging-Alzheimer's Association workgroups on diagnostic guidelines for Alzheimer's disease. *Alzheimer's Dementia, 7*, 280–292.
Squire, L. R. (1993). The hippocampus and spatial memory. *Trends in Neuroscience, 16*, 56–57.
Sturchler-Pierrat, C., Abramowski, D., Duke, M., Wiederhold, K. H., Mistl, C., Rothacher, S., et al. (1997). Two amyloid precursor protein transgenic mouse models with Alzheimer disease-like pathology. *Proceedings of the National Academy of Sciences of the United States of America, 94*, 13287–13292.
Sweeney, W. A., Luedtke, J., McDonald, M. P., & Overmier, J. B. (1997). Intrahippocampal injections of exogenous beta-amyloid induce postdelay errors in an eight-arm radial maze. *Neurobiology of Learning and Memory, 68*, 97–101.
Takeda, T. (2009). Senescence-accelerated mouse (SAM) with special references to neurodegeneration models, SAMP8 and SAMP10 mice. *Neurochemical Research, 34*, 639–659.
Takeda, T., Hosokawa, M., Takeshita, S., Irino, M., Higuchi, K., Matsushita, T., et al. (1981). A new murine model of accelerated senescence. *Mechanisms of Ageing and Development, 17*, 183–194.
Takeda, T., Matsushita, T., Kurozumi, M., Takemura, K., Higuchi, K., & Hosokawa, M. (1997). Pathobiology of the senescence-accelerated mouse (SAM). *Experimental Gerontology, 32*, 117–127.
Takemura, M., Nakamura, S., Akiguchi, I., Ueno, M., Oka, N., Ishikawa, S., et al. (1993). b/A4 proteinlike granular structures in the brain of senescence-accelerated mouse. *American Journal of Pathology, 142*, 1887–1897.
Tang, F., Nag, S., Shiu, S. Y., & Pang, S. F. (2002). The effects of melatonin and Ginkgo biloba extract on memory loss and choline acetyltransferase activities in the brain of rats infused intracerebroventricularly with beta-amyloid 1–40. *Life Science, 71*, 2625–2631.

van Praag, H., Schinder, A. F., Christie, B. R., Toni, N., Palmer, T. D., & Gage, F. H. (2002). Functional neurogenesis in the adult hippocampus. *Nature, 415*, 1030–1034.
Wang, Z., Liang, P., Jia, X., Jin, G., Song, H., et al. (2012). The baseline and longitudinal changes of PCC connectivity in mild cognitive impairment: A combined structure and resting-state fMRI study. *PLoS One*, 7, e36838.
Wang, L. C., Wang, B., Ng, S. Y., & Lee, T. F. (2006). Effects of ginseng saponins on beta amyloid-induced amnesia in rats. *Journal of Ethnopharmacology, 103*, 103–108.
Wang, L., Zang, Y., He, Y., Liang, M., Zhang, X., Tian, L., et al. (2006). Changes in hippocampal connectivity in the early stages of Alzheimer's disease: Evidence from resting state fMRI. *NeuroImage, 31*, 496–504.
Weldon, D. T., Rogers, S. D., Ghilardi, J. R., Finke, M. P., Cleary, J. P., O'Hare, E., et al. (1998). Fibrillar beta amyloid induces microglial phagocytosis, expression of inducible nitric oxide synthase, and loss of a select population of neurons in the rat CNS in vivo. *Journal of Neuroscience, 18*, 2161–2173.
Wenk, G. L. (2003). Neuropathologic changes in Alzheimer's disease. *Journal of Clinical Psychiatry, 64*(Suppl. 9), 7–10.
Xue, W. G., Ge, G. L., Zhang, Z., Xu, H., & Bai, L. M. (2009). Effect of electroacupuncture on the behavior and hippocampal ultrastructure in APP 695 V 717 I transgenic mice. *Zhen Ci Yan Jiu, 34*, 309–314.
Xue, W. G., Zhang, Z., Bai, L. M., Xu, H., & Wu, H. X. (2009). Effect of electroacupuncture on the behavior and the expression of amyloid beta-protein, amyloid precursor protein and ChAT in APP 695 V 717 I transgenic mice. *Zhen Ci Yan Jiu, 34*, 152–158.
Xue, W. G., Zhang, Z., Xu, H., Wu, H. X., & Bai, L. M. (2011). Effect of electroacupuncture on learning-memory ability, and Abeta and LRP1 immunoactivity in hippocampal sulcus microvessels in APP transgenic mice. *Zhen Ci Yan Jiu, 36*, 95–100.
Yang, X. H., Liu, Z. B., Niu, W. M., & Niu, X. M. (2011). Effect of "Xiusanzhen" on hippocampal muscarinic cholinergic receptor activity in Alzheimer disease rats. *Zhen Ci Yan Jiu, 36*, 90–94.
Yankner, B. A., Lu, T., & Loerch, P. (2008). The aging brain. *Annual Review of Phytopathology, 3*, 41–66.
Zhang, P., Guan, S. S., & Jiang, G. H. (2010). Effects of electroacupuncture on expression of Abeta positive cells of the hippocampus and SOD activity in rats with streptozocin-Alzheimer's disease. *Zhongguo Zhen Jiu, 30*, 1007–1010.
Zhang, G., Yin, H., Zhou, Y. L., Han, H. Y., Wu, Y. H., et al. (2012). Capturing amplitude changes of low-frequency fluctuations in functional magnetic resonance imaging signal: A pilot acupuncture study on NeiGuan (PC6). *Journal of Alternative and Complementary Medicine, 18*, 387–393.
Zhou, Y. L., Han, H. Y., & Jia, J. P. (2008). Correlation analysis on changes between cognitive ability and brain fMRI after acupoint thread embedding in Alzheimer's disease patients. *Zhongguo Zhong Xi Yi Jie He Za Zhi, 28*, 689–693.
Zhou, Y., & Jin, J. (2008). Effect of acupuncture given at the HT 7, ST 36, ST 40 and KI 3 acupoints on various parts of the brains of Alzheimer's disease patients. *Acupuncture & Electro-Therapeutics Research, 33*, 9–17.
Zhu, H., Dong, K. L., Wu, Y., Zhang, T., Li, R. M., Dai, S. S., et al. (2010). Influence of acupuncture on isoprostane in patients with Alzheimer's disease. *Zhongguo Zhen Jiu, 30*, 18–21.

CHAPTER TEN

Acupuncture Therapy for Psychiatric Illness

Karen Pilkington[1]
Faculty of Science and Technology, University of Westminster, London, United Kingdom
[1]Corresponding author: e-mail address: k.pilkington@westminster.ac.uk

Contents

Abstract

Acupuncture has traditionally been used for problems including anxiety, insomnia, stress, and depression in China and other East Asian countries. A range of different neurobiological responses to acupuncture have been investigated including modulation of serotonergic, noradrenergic, and dopaminergic systems; effects on GABA and the hypothalamic–pituitary–adrenal axis; and inflammatory responses. Interpretation of the findings is challenging because the neurobiology of psychiatric disorders has yet to be fully elucidated. Limitations also arise from the use of animal models and the selection of appropriate control treatments. Further complexity is added by

International Review of Neurobiology, Volume 111
ISSN 0074-7742
http://dx.doi.org/10.1016/B978-0-12-411545-3.00010-9

acupuncture treatment being nonstandardized with acupuncture points often selected on the basis on traditional practice and theory. Potentially promising findings require further investigation and substantiation.

1. INTRODUCTION

1.1. Prevalence of acupuncture use in psychiatric illness

Acupuncture is one of a number of complementary and alternative therapies used by those with psychiatric illness. Many such patients remain untreated but a large-scale survey in the United States of over 9000 people with DSM-IV anxiety, mood, impulse control, and substance disorders indicated that approximately 7% had been treated by complementary practitioners in the previous 12 months (Wang et al., 2005). Patterns of use appear to vary with the psychiatric illness, and those with adjustment disorders and stress, anxiety, and depression appeared to be more frequent users than those without these problems (Druss & Rosenheck, 2000; Unutzer et al., 2000).

Acupuncture has traditionally been widely used for problems including anxiety, stress, insomnia, and depression in countries such as China, Japan, and Korea (Mukaino, Park, White, & Ernst, 2005), but there is significant use in Western populations. Eleven percent of visits to US acupuncturists were specifically for a mental health problem and, in up to 23% of visits, acute stress, anxiety, depression, insomnia, substance abuse, and other psychological problems emerged during the consultation (Simon et al., 2004). Similar figures were revealed in a survey of UK acupuncturists: 11% of consultations with over 9000 patients were for psychological problems (MacPherson, Sinclair-Lian, & Thomas, 2006). Conversely, research in Australia revealed frequent use of complementary therapies for depression but not use of acupuncture specifically (Jorm, Griffiths, Christensen, Parslow, & Rogers, 2004).

1.2. Acupuncture: Philosophy and practice

Acupuncture is an important component of traditional Chinese and other East Asian medical systems. As such, it remains one of the oldest therapeutic approaches still in contemporary use. The term "acupuncture" refers to puncturing the skin using fine metal needles at various points on the body known as acupuncture points or "acupoints." According to traditional theory, energy or "qi" flows along energy channels or "meridians" that run through the body, which, when disrupted or "blocked," can be released by stimulating specific points along each channel (Wang, Kain, & White, 2008). Correction of the "flow" is perceived as necessary for ensuring maintenance or restoration of health.

More recently, different methods of stimulating the acupoints have been introduced including electrical stimulation (electroacupuncture), laser acupuncture (use of low-level laser therapy at points), and acupressure (applying pressure at the points). Acupuncture is often part of a complex intervention that includes use of herbal mixtures, moxibustion (burning of herbs), and dietary and lifestyle advice. The search for physiological or histological evidence of the existence of qi, the meridians, and specific acupuncture points has been the focus of considerable effort on the part of researchers working in this area (Ahn et al., 2008). To date, conclusive proof has not been obtained.

In contrast to the traditional approach to acupuncture, "Western" or "medical acupuncture," a practice adapted from the traditional approach, is considered to be based on contemporary neurophysiology and anatomy. Conventional diagnoses are used and, particularly for pain-related conditions, a combination of trigger points, tender points, and segmental points is employed (Filshie & Cummings, 1999).

Within traditional Chinese medicine philosophy, there is no clear distinction between physical and psychological problems. Low mood and anxiety are considered part of an overall pattern of signs and symptoms and an indication of disrupted flow of energy. Thus, selection of acupuncture points depends on this overall pattern and has little to do with conventional physiology and anatomy. Even within the Western medical acupuncture approach, nonpainful, generalized conditions such as depression are likely to be treated using a number of well-known "traditional points" (Filshie & Cummings, 1999). The choice of specific traditional acupuncture points may, however, draw on a range of theoretical and diagnostic frameworks (MacPherson & Schroer, 2007).

1.3. Overview of potentially relevant neurobiological effects of acupuncture

Manual acupuncture and electroacupuncture have been reported to result in measurable responses in the nervous system, including modulation of neurotransmitters and other endogenous substances (Ma, 2004). Investigation of a physiological mechanism of action of acupuncture has focused primarily on its analgesic effects. Earlier studies suggested that acupuncture analgesia is mediated by endogenous opioid peptides, with several studies demonstrating a block of the analgesic effect by naloxone, an opioid antagonist (Lin & Chen, 2008). Various endogenous substances have been suggested as mediators of the analgesic effects apparently elicited by acupuncture. These include serotonin (5-hydroxytryptamine), catecholamines, and gamma-aminobutyric acid (GABA) (Ma, 2004). Opioid peptides at mu-, delta-, and kappa-receptors are considered to play a pivotal role, but glutamate and cholecystokinin octapeptide are also thought to be involved (Zhao, 2008). Further evidence for the complexity of the effects of acupuncture was provided by a recent neuroimaging study, which indicated that multiple responses continue after needling has been terminated (Bai et al., 2010). The precise role of each of these substances in generating the effects of acupuncture is still to be fully elucidated and the implications for psychiatric disorders are less well defined and understood.

1.4. Acupuncture and neurobiological responses relevant to psychiatric illness

Neurobiological responses result from complex interactions between the various mediators and regulatory processes. For example, chronic stress affects activity of the hypothalamus and amygdala with increased sympathetic tone and cytokine release that ultimately causes further neuroendocrine disruption (Maletic et al., 2007). Each of the potentially relevant mediators and mechanisms will be considered in turn with examples of studies illustrating the range of research that has been conducted in this area.

2. METHODS

Relevant studies were initially identified by searches of PubMed, and articles retrieved were screened for studies relevant to each topic. *In vivo* and neuroimaging studies for the section on specific psychiatric disorders were retrieved through searches of AMED (Allied and Complementary Medicine), Embase, Ovid MEDLINE, and PsycINFO using a search strategy containing relevant neurobiological terms identified from recent reviews and a neuroscience monograph. This was combined with condition-specific strategies. Clinical trials and reviews were identified by searching the Cochrane Library, PubMed, and PsycINFO. All searches were conducted in May 2013.

2.1. Modulation of serotonin and norepinephrine

The response to acupuncture treatment that is possibly of most relevance to psychiatric illness, and particularly to depressive disorders, is the reported effect on synthesis and release of serotonin and norepinephrine in the central nervous system (Han, 1986). Low levels of both these neurotransmitters have been reported in depression, and antidepressant therapy has focused on increasing levels or prolonging their effects. The tricyclic antidepressants primarily inhibit serotonin or norepinephrine reuptake transporters (Nestler et al., 2002). The lack of specificity of these agents led to the introduction of the selective serotonin reuptake inhibitors (SSRIs) and serotonin–norepinephrine reuptake inhibitors (SNRIs) (Feighner, 1999). Albeit more selective, serotonin is linked to a range of biological substrates, receptors, and pathways, which may explain why these agents have proved at least partially effective in other psychiatric conditions including obsessive–compulsive disorder, panic disorder, and eating disorders (Flament, Bissada, & Spettigue, 2012; Mochcovitch & Nardi, 2010; Soomro, Altman, Rajagopal, & Oakley-Browne, 2008).

The earliest indication of a possible effect of acupuncture on serotonin and norepinephrine was from series of animal studies, which suggested that acupuncture or electroacupuncture accelerated the synthesis and release of both neurotransmitters in the central nervous system (Han, 1986). This was followed by further *in vivo* investigations of this effect. In one such study, freely moving rats received acupuncture for 60 min at BL23 or a point not considered to be "active" (Yoshimoto et al., 2006). BL23, an acupuncture point located on either side of the lower back, is traditionally used to treat psychiatric disorders and considered to elicit sedative effects. Increases in serotonin were observed in microdialysates in the nucleus accumbens from rats treated at BL23. These observations led the researchers to suggest that acupuncture caused activation of neurons in the dorsal raphe nuclei in the midbrain via spinoreticular tract neurons, increasing serotonin release in the nucleus accumbens. The differential effects at the two points were not fully explained.

The effect of acupuncture on serotonin has also been investigated in maternally separated rat pups, an animal model for depression (Park et al., 2012). The point HT7, located at the ulnar end of the wrist crease, was chosen because of previous use in clinical studies of depression, while ST36 located near the knee was chosen as the "sham" point due to its traditional use for gastrointestinal rather than psychological conditions. Acupuncture was applied once daily for 30 s for 7 days. High-performance liquid chromatography was used to measure serotonin (5-HT) and a serotonin metabolite, 5-hydroxyindole-3-acetic acid (5-HIAA), in tissue from the prefrontal cortex (PFC). Acupuncture stimulation at HT7 was reported to alleviate a behavioral change (immobility during a tail suspension test) and to regulate the 5-HIAA/5-HT ratio and serotonin transporter expression, while stimulation at ST36 had no significant effect.

2.2. Modulation of dopaminergic systems

Changes in the dopamine system are also implicated in several psychiatric disorders, and the dopamine transporter is a major target for drug therapy (Haenisch & Bonisch, 2011). Depletion studies demonstrated a decrease in mood in those with a family history of depression and in remission from depression but not in healthy individuals (Ruhe, Mason, & Schene, 2007). Schizophrenia has also been linked with dopamine dysregulation but, in this case, with increased central dopaminergic neurotransmission (Farde, 1997). In substance abuse, Cocaine and amphetamines increase dopamine levels in the synapse via inhibition of reuptake or increased release (Kreek et al., 2012), while heroin (diamorphine) acts on opioid receptors to indirectly activate dopaminergic systems (Johnson & North, 1992).

As described earlier, changes in serotonin but not dopamine levels in response to acupuncture treatment were reported in one animal study (Yoshimoto et al., 2006). Conversely, a second report of the 2012 study by Park and colleagues reported that significantly lower levels of dopamine measured after maternal separation of rats were alleviated by acupuncture stimulation (Kwon, Kim, et al., 2012). Similarly, acupuncture treatment was reported to normalize dopamine levels during withdrawal from ethanol (Zhao et al., 2006). A review of the neurochemical and behavioral evidence concluded that acupuncture may suppress the reinforcing effects of drugs of abuse through modulation of the mesolimbic dopamine neurons (Yang, Lee, & Sohn, 2008). There is, however, further evidence of inconsistent effects from animal studies, for example, in one study, dopamine response to electroacupuncture stimulation was reported to depend on point location and stimulation frequency (Shen & Lai, 2007).

2.3. GABA and glutamate

GABA, an inhibitory neurotransmitter, has a role in regulating dopaminergic neurotransmission and has also been implicated in the neurobiology of psychiatric disorders. In substance abuse, heroin metabolites bind to opioid receptors and relieve the inhibitory effects of GABA on dopaminergic neurons (Kreek et al., 2012). Glutamate is a major excitatory neurotransmitter in the central nervous system and a precursor of GABA. Alterations in glutaminergic neurotransmission have been reported in relation to schizophrenia (Falkai & Moller, 2012).

Acupuncture at HT7, but not at the control point LI5 (on the radial side of the wrist), significantly decreased morphine self-administration in rats, and this effect was blocked by two GABA antagonists (Yoon et al., 2010). A second study reported similar results: acupuncture at HT7 significantly decreased symptoms of morphine withdrawal with these effects blocked by pretreatment with GABA antagonists (Lee et al., 2011). Studies of the effect on glutamate include one in which scalp acupuncture was applied to immature rats (Franco-Santana et al., 2013). GABA in the amygdala and hippocampus was reported to have increased while glutamate decreased in all regions in acupuncture-treated compared with untreated animals.

2.4. Modulation of the HPA axis

One of the main responses to acute or chronic stress is activation of the hypothalamic–pituitary–adrenal (HPA) axis (Nestler et al., 2002). Corticotropin-releasing factor (CRF) secreted by the hypothalamus stimulates the anterior pituitary to synthesize and release adrenocorticotropic hormone (ACTH). This stimulates the synthesis and release of glucocorticoids from the adrenal cortex, with cortisol being released in humans, while corticosterone is released in rodents. Excessive activation of the HPA axis has been reported in depressed individuals with increased production of cortisol and hypersecretion of CRF being observed (Nestler et al., 2002). Chronic stress and hypersecretion of cortisol are implicated in the pathophysiology of anxiety through a series of changes involving the serotonergic system (Ionescu, Niciu, Mathews, Richards, & Zarate, 2013). Insomnia, often comorbid with depression, also appears related to hyperactivity of the HPA axis with elevated cortisol levels reported in some studies although others report conflicting results (Wollweber & Wetter, 2011). CRF may also play a role in alcohol dependence (Gilpin & Koob, 2008).

A number of studies have investigated the action of acupuncture on the HPA axis. Pretreatment with electroacupuncture at ST36 resulted in significantly reduced peripheral ACTH and corticosterone in animals with chronic cold-induced stress compared with those that were untreated or treated at a "sham" point (Eshkevari, Permaul, & Mulroney, 2013). Similar effects were seen centrally with CRF levels significantly reduced compared with control animals. Acupuncture at HT7 was also shown to decrease corticosterone and ACTH plasma levels as well as anxiety-related behaviors in maternally separated rats (Park, Chae, Kim, Lee, & Chung, 2011). The elevated plus maze model and forced swimming test were used to investigate the effect of acupuncture at the PC6 point on dysregulation of the HPA axis and depression-like behavior induced by high doses of corticosteroid in rats (Lee, Han, & Shim, 2009). Acupuncture prior to corticosteroid injections was found to reduce this behavior and increase expression of neuropeptide Y.

2.5. Neuropeptide Y

Neuropeptide Y, a peptide neurotransmitter, is the most abundant neuropeptide in the brain. It has an important role in the stimulation of the appetite and is relevant to the pathophysiology of anxiety and depression (Ionescu et al., 2013). Neuropeptide Y, along with another neuropeptide, leptin, is also implicated in eating disorders with levels altered in anorexic patients although this may be a consequence rather than causative (Kaye, 2008). It may also play a role in the etiology and pathophysiology of posttraumatic stress disorder (Wu et al., 2011).

Several studies using the maternal separation animal model of anxiety and depression have reported that acupuncture modulates neuropeptide Y expression. Investigators in one such study reported that acupuncture treatment at HT7 increased expression of neuropeptide Y and alleviation of depression-like behavior (Lim, Ryu, Kim, Hong, & Park, 2003). A subsequent study by researchers from this group reported a difference between stimulation at two different acupuncture points, with reduced anxiety-like behavior and increased numbers of neuropeptide Y-immunoreactive cells in the amygdala in response to HT7 but not to ST36 stimulation (Park et al., 2005).

2.6. Inflammation and the role of cytokines

The role of inflammatory cytokines in depression has received much attention (Felger & Lotrich, 2013). Chronic raised levels can lead to psychiatric disorders including depression, the mechanism linked to changes in monoamine, glutamate, and neuropeptide systems. A decrease in growth factors such as brain-derived neurotrophic factor (BDNF) and HPA axis dysregulation may also be relevant (Makhija & Karunakaran, 2013).

The effect of acupuncture was investigated in Bacillus Calmette-Guerin (BCG) inoculated mice, an animal model of chronic systemic inflammation (Kwon, Lee, et al., 2012). Stimulation was at SP6 or HT7, acupuncture points on the limbs considered to have anxiety-reducing properties. Serotonin levels were unchanged by acupuncture but levels of dopamine and depression-like behavior were improved by stimulation of SP6. Effects were not seen at HT7 although this point had proved "active" in several previous studies as described earlier.

The inflammatory response has also been a focus of human studies, one of which reported that electroacupuncture modulated lymphocyte subpopulations and serum cytokine levels (Jong, Hwang, & Chen, 2006). However, the small size of the study suggests the findings were preliminary at best. A further study attempted to measure the effects of expectation on the proinflammatory immune response in healthy volunteers (Dennehy et al., 2009). Proinflammatory cytokine levels significantly increased in the acupuncture-expectation and acupuncture-only groups but not in a relaxation–expectation control group, indicating that expectation is less significant in an acupuncture response than has been suggested.

Thus, a range of potential neurobiological responses to acupuncture effects have been investigated, mainly in animal models, with varied results. In the following section, summaries are presented on research specific to each of the relevant psychiatric disorders, anxiety disorders, depression and bipolar disorder, schizophrenia, substance abuse, and eating and sleep disorders. These refer to studies from basic science (animal models) and evidence based on human studies and clinical trials.

3. SUMMARIES OF THE RESEARCH ON SPECIFIC PSYCHIATRIC DISORDERS

3.1. Anxiety disorders

Potential neurobiological targets for acupuncture in generalized or chronic anxiety include dysregulation of the HPA axis and reduced levels of GABA and serotonergic neurotransmission. Similar targets exist in PTSD: effects on neural pathways, the HPA axis, the autonomic nervous system, inflammation, and gene expression (Hollifield, 2011).

As described earlier, the effect of acupuncture on chronic stress-induced anxiety, the HPA axis, and neuropeptide Y expression has been investigated in animal models of anxiety (Eshkevari et al., 2013; Kim et al., 2009; Park et al., 2005, 2011). Electroacupuncture was also reported to affect the balance between GABA and monoamine neurotransmitters in anxiety induced by chronic stress, increasing GABA and decreasing monoamines (Zhou et al., 2008).

Investigations in humans provide some support for an effect of acupuncture in anxiety: a single acupuncture session reportedly improved levels of the cytokines, interleukin-2 (IL-2) and tumor necrosis factor-alpha (TNF-alpha), and plasma cortisol in anxiety (Arranz, Siboni, & De la Fuente, 2006). Imaging studies also provide some supportive evidence indicating that acupuncture may mediate anti-anxiety effects through modulation of the limbic–paralimbic–neocortical network (Fang et al., 2009). Clinical trials on generalized anxiety disorder and posttraumatic stress disorder have not, however, proved conclusive (Kim et al., 2013). Beneficial effects with changes in physiological measures have been reported in short-term anxiety, but the relevance to chronic anxiety problems is unclear (Pilkington, 2010).

3.2. Depression

Genetic and environmental factors are viewed as contributory factors, and structural changes in the brain, particularly in hippocampal volume, have been observed in imaging studies (Maletic et al., 2007; Nestler et al., 2002). However, as described earlier, the potentially relevant aspects of the neurobiology of depression for acupuncture include the monoamine hypothesis of depression, specifically deficits in serotonergic and noradrenergic neurotransmission. Chronic stress and the dysregulation of the hippocampus and HPA axis are also implicated (Nestler et al., 2002). Other potentially relevant factors include impairment in BDNF modulation of neurocircuits and the possible involvement of dopamine, neuropeptide Y and the amino acids, GABA, and glutamine.

The effects of acupuncture on each of these factors have been described earlier. The serotonergic system has been a particular focus with reported effects of electroacupuncture on serotonin (5-HT1 and 5-HT2) receptors in the cerebral cortex in a chronic stress depression rat model (Sun, Zhang, & Han, 2005) and successful use of a serotonin depletor to abolish the effect of acupuncture (Dos Santos et al., 2008). Downregulation of ACTH and cortisol has also been reported in response to electroacupuncture applied to the ear, providing further support for an effect on the HPA axis (Liu et al., 2013).

Changes in cortisol levels in depressed patients were reported in one of the earlier trials in depression comparing electroacupuncture and antidepressant (Han, Li, Luo, & Zhao, 2004) and in a more recent study (Vazquez, Gonzalez-Macias, Berlanga, & Aedo, 2011). A large number of randomized controlled trials (RCTs) have been conducted on acupuncture in depression but methodological limitations preclude firm conclusions on efficacy (Smith, Hay, & Macpherson, 2010). The majority of studies rely on clinical assessment of response but several have assessed biological markers. Levels of Galphai and Galphaq receptors were significantly higher in depressed compared with healthy individuals but were not changed by the treatment with drugs, electroacupuncture, or sham electroacupuncture (Song, Zhou, Fan, Luo, & Halbreich, 2007). Neuroimaging studies have, however, detected responses: changes in abnormal levels of metabolites and volume of the hippocampus due to electroacupuncture treatment (Duan, Tu, Jiao, & Qin, 2011) and modulation of parietal–temporal–limbic cortices in depressed individuals in response to laser acupuncture (Quah-Smith, Suo, Williams, & Sachdev, 2013).

3.3. Bipolar disorder

No studies in animals were located, possibly due to a paucity of animal models of bipolar disorder (Gould & Einat, 2007). Few clinical studies of acupuncture in bipolar disorder are available although bipolar patients have occasionally been included in trials in depression. Adjunctive treatment with acupuncture was, however, tested in two preliminary RCTs in 46 bipolar patients suffering with hypomania or depressive symptoms, respectively (Dennehy et al., 2009). Effects were measured by clinical response, which was similar after acupuncture at targeted points and at nonspecific points.

3.4. Schizophrenia

The dopamine and glutamate hypotheses have been proposed as the main neurobiological bases of schizophrenia; however, schizophrenia is still not well understood (Falkai & Moller, 2012). Studies of acupuncture's effects on both dopamine and glutamate have been discussed earlier. Few studies of acupuncture have specifically focused on schizophrenia. Neuroimaging studies have detected a modulating effect of acupuncture on GABA that may be relevant in schizophrenia due to the possible link between decreased GABAergic tone and interneuron dysfunction in the corticolimbic system (Hui, Marina, Liu, Rosen, & Kwong, 2010). Decreased activity in the dorsomedial PFC due to acupuncture stimulation may also be of relevance in schizophrenia (Hori, Takamoto, Urakawa, Ono, & Nishijo, 2010). The number and quality of clinical trials of acupuncture for schizophrenia is too low to draw conclusions (Lee, Shin, Ronan, & Ernst, 2009).

3.5. Substance abuse, alcohol, and nicotine dependence

3.5.1 Cocaine, heroin, and amphetamines

A previous review of the research on neurochemical and behavioral responses to acupuncture concluded that modulation of mesolimbic dopamine neurons was the primary mechanism for the suppression of reinforcing effects of abused drugs (Yang et al., 2008). Moreover, that serotonin, opioids, and GABA are implicated in modulating acupuncture-related dopamine release.

Evidence of potential modulation of the central dopaminergic system has been provided by an animal study in which inhibitory effects of acupuncture at HT7 on cocaine-induced locomotor activity were accompanied by a reduction in dopamine biosynthesis (Lee, Han, & Shim, 2009). Also supportive are the results of a study in which a dopamine receptor antagonist appeared to block the inhibitory effects of electroacupuncture on cocaine-induced seizures (Chen, Ivanic, Chuang, Lu, & Lin, 2013). Stimulation at the point HT7 but not at L15 reduced morphine self-administration via regulation of GABA receptors (Yoon et al., 2010). It also reduced stress-induced cocaine-seeking behavior via c-Fos expression and pCREB activation, again with an apparent lack of response at the control point (LI5) (Yoon et al., 2012). Two further studies of electroacupuncture reported suppression of heroin-seeking behavior and FosB expression in the nucleus accumbens (Hu et al., 2013; Liu et al., 2012).

Early studies of acupuncture in drug addiction in humans reported measurable biochemical changes: reduced ACTH and cortisol levels (Wen et al., 1979) and reduced met-enkephalin with unchanged beta-endorphin (Clement-Jones et al., 1979). Subsequently, much effort has been expended in evaluating clinical effects, but the evidence to date is inconclusive or of low quality (Gates, Smith, & Foxcroft, 2006; Lin, Chan, & Chen, 2012).

3.5.2 Alcohol dependence

A series of studies have investigated effects of acupuncture on alcohol (ethanol) withdrawal in animal models. Behavioral and biological effects have been reported. Biological changes include an influence on the GABA system with modulation of catecholamines in the amygdala, including inhibition or normalization of ethanol-induced dopamine release (Lee et al., 2008; Yoon et al., 2004; Zhao et al., 2006, 2011). Eleven clinical trials of acupuncture involving 1110 individual cases of alcohol dependence were identified but the results were equivocal (Cho & Whang, 2009).

3.5.3 Nicotine dependence

In vivo studies have focused on behavioral changes and biological effects. An expected increase in nicotine-induced locomotor activity and Fos-like-immunoreactivity in the nucleus accumbens and striatum was attenuated by acupuncture at ST36 but not at a control point suggesting modification of post-synaptic neuronal activity (Chae et al., 2004). Acupuncture also appeared to modulate both the anxiety-like behavior and CRF in the amygdala during nicotine withdrawal (Chae et al., 2008). Differing patterns of changes in the levels of various neurotransmitters in response to acupuncture and nicotine patches for smoking cessation were observed in a clinical trial, but these were not accompanied by differences in the smoking cessation rate (Cabioğlu, Kutlu, Albayrak, Marakoğlu, & Ergene, 2012). The results correlate with the conclusion of a review of clinical trials, that to date, there is no conclusive evidence on efficacy (White, Rampes, Liu, Stead, & Campbell, 2011).

3.6. Eating disorders and sleep disorders

Neuropeptide dysregulation and possible effects of serotonin have been discussed in connection with eating disorders (Kaye, 2008). The effect of acupuncture on these has been investigated, but few studies specifically on eating disorders were located, with the exception of several studies published in Chinese for which further details were not available and a pilot study that reported promising findings (Fogarty, Harris, Zaslawski, McAinch, & Stojanovska, 2010).

Insomnia is a result of autonomic and/or central nervous system arousal, and a role for acupuncture has been proposed based on its potential to influence these systems (Huang, Kutner, & Bliwise, 2011). As with eating disorders, many of the animal studies are published in Chinese. Large numbers of clinical trials have been completed but are not sufficiently rigorous to provide reliable evidence on effectiveness (Cheuk, Yeung, Chung, & Wong, 2012).

4. CONCLUSIONS

Investigation of the neurobiology of acupuncture has proved to be a research area in which significant effort has been invested. Knowledge of the neurobiology of acupuncture analgesia is more advanced than that of acupuncture in psychiatric illness. There are a number of reasons for this. Firstly, the neurobiology of conditions such as depression and schizophrenia is still to be fully elucidated. Drugs such as the SSRIs have proven benefit in depression, but simply increasing levels of neurotransmitters does not explain the delayed response to drug treatment and suggests that the response is far more complex involving longer-term adaptive changes. Similarly, for schizophrenia, different forms of this condition exist, and the involvement of the dopaminergic system, indicated by the effectiveness of agents such as the phenothiazines, does not fully explain the neurobiology.

While animal models have proved very valuable in revealing potential effects of acupuncture particularly in the area of pain, there are limitations in the models that are available for psychiatric illness. For example, fear rather than chronic anxiety may be the focus in animal studies of anxiety. Depression is modeled on such strategies as the forced swim test that has limitations in truly representing the human condition of depression; however, animal models for several disorders are lacking. The lack of models may help explain the limited numbers of studies of acupuncture in bipolar disorder, schizophrenia, and eating disorders. Specific considerations on research on acupuncture include the extent to which needling in animals reflects that in humans, for example, the precise location of acupuncture points. Neurobiological differences between species are also relevant, for example, a difference in the corticosteroid released in rodents compared with that released in humans.

The question of acupuncture points is particularly relevant for conditions such as anxiety and depression as these are often chosen based on traditional theories involving the "meridians" and "qi," concepts not well understood or accepted within current physiological and anatomical frameworks. Thus, reported differences in response to different points are difficult to explain in contemporary scientific terms, and some trials have not found significant differences between stimulation at acupuncture and nonspecific points. The studies described earlier illustrate some of the challenges involved in interpreting the results of research into the effect of acupuncture on the central nervous system particularly when the focus is on psychiatric disorders: specific doses, frequencies, and duration of acupuncture treatment vary as do the animal models utilized, while the rationale for selecting "active" and "sham" points is based on traditional theory and practice. Interpretation of results is further complicated by the use of electrical and manual stimulation of acupuncture points, each of which may generate a different range of responses.

A range of potentially relevant neurobiological responses have, however, been measured in animal studies. These have included reported changes in levels of various neurotransmitters including serotonin, norepinephrine, dopamine, GABA, and glutamate and effects on HPA dysregulation and inflammatory responses. What is unclear is the how these responses translate into clinical application. Rarely do clinical studies in humans use measures other than those based on self-report or clinical assessment. In both cases, due to the nature of acupuncture treatment, it is unclear as to what extent positive responses are due to expectation because of the lack of blinding of patients and acupuncturists to treatment. Thus, when analysis of these trials is conducted, even when a significant number of trials involving large numbers of participants are involved, the trials are judged to be poor in methodological terms precluding firm conclusions.

Overall, a range of neurobiological responses that have been reported in response to acupuncture appear to require further investigation and substantiation. This will, however, provide further corresponding challenges in interpretation for the reasons outlined earlier.

REFERENCES

Ahn, A. C., Colbert, A. P., Anderson, B. J., Martinsen, O. G., Hammerschlag, R., Cina, S., et al. (2008). Electrical properties of acupuncture points and meridians: A systematic review. *Bioelectromagnetics*, *29*(4), 245–256.

Arranz, L., Siboni, L., & De la Fuente, M. (2006). Improvement of the interleukin 2 and tumour necrosis factor alpha release by blood leukocytes as well as of plasma cortisol and antioxidant levels after acupuncture treatment in women suffering anxiety. *Journal of Applied Biomedicine*, *4*(3), 115–122.

Bai, L., Tian, J., Zhong, C., Xue, T., You, Y., Liu, Z., et al. (2010). Acupuncture modulates temporal neural responses in wide brain networks: Evidence from fMRI study. *Molecular Pain*, *6*, 73.

Cabioğlu, M. T., Kutlu, R., Albayrak, C., Marakoğlu, K., & Ergene, N. (2012). Changes in the levels of serum beta endorphin, serotonin, adrenaline, noradrenaline and dopamine during smoking cessation by electroacupuncture and nicotine patch. *Turkiye Klinikleri Journal of Cardiovascular Science*, *24*(2), 95–101.

Chae, Y., Yang, C. H., Kwon, Y. K., Kim, M. R., Pyun, K. H., Hahm, D. H., et al. (2004). Acupuncture attenuates repeated nicotine-induced behavioral sensitization and c-Fos expression in the nucleus accumbens and striatum of the rat. *Neuroscience Letters*, *358*(2), 87–90.

Chae, Y., Yeom, M., Han, J. H., Park, H. J., Hahm, D. H., Shim, I., et al. (2008). Effect of acupuncture on anxiety-like behavior during nicotine withdrawal and relevant mechanisms. *Neuroscience Letters*, *430*(2), 98–102.

Chen, Y. H., Ivanic, B., Chuang, C. M., Lu, D. Y., & Lin, J. G. (2013). Electroacupuncture reduces cocaine-induced seizures and mortality in mice. *Evidence-Based Complementary and Alternative Medicine*, *2013*, 134610.

Cheuk, D. K., Yeung, W. F., Chung, K. F., & Wong, V. (2012). Acupuncture for insomnia. *Cochrane Database of Systematic Reviews, 9.* http://dx.doi.org/10.1002/14651858.CD005472.pub3, CD005472.

Cho, S. H., & Whang, W. W. (2009). Acupuncture for alcohol dependence: A systematic review. *Alcoholism, Clinical and Experimental Research, 33*(8), 1305–1313.

Clement-Jones, V., McLoughlin, L., Lowry, P. J., Besser, G. M., Rees, L. H., & Wen, H. L. (1979). Acupuncture in heroin addicts; changes in Met-enkephalin and beta-endorphin in blood and cerebrospinal fluid. *Lancet, 2*(8139), 380–383.

Dennehy, E. B., Schnyer, R., Bernstein, I. H., Gonzalez, R., Shivakumar, G., Kelly, D. I., et al. (2009). The safety, acceptability, and effectiveness of acupuncture as an adjunctive treatment for acute symptoms in bipolar disorder. *The Journal of Clinical Psychiatry, 70*(6), 897–905.

Dos Santos, J. G., Jr., Kawano, F., Nishida, M. M., Yamamura, Y., Mello, L. E., & Tabosa, A. (2008). Antidepressive-like effects of electroacupuncture in rats. *Physiology & Behavior, 93*(1–2), 155–159.

Druss, B. G., & Rosenheck, R. A. (2000). Use of practitioner-based complementary therapies by persons reporting mental conditions in the United States. *Archives of General Psychiatry, 57*(7), 708–714.

Duan, D. M., Tu, Y., Jiao, S., & Qin, W. (2011). The relevance between symptoms and magnetic resonance imaging analysis of the hippocampus of depressed patients given electro-acupuncture combined with Fluoxetine intervention—A randomized, controlled trial. *Chinese Journal of Integrative Medicine, 17*(3), 190–199.

Eshkevari, L., Permaul, E., & Mulroney, S. E. (2013). Acupuncture blocks cold stress-induced increases in the hypothalamus-pituitary-adrenal axis in the rat. *The Journal of Endocrinology, 217*(1), 95–104.

Falkai, P., & Moller, H. J. (2012). Biomarkers and neurobiology of schizophrenia. *European Archives of Psychiatry and Clinical Neuroscience, 262*(5), 363–364.

Fang, J., Jin, Z., Wang, Y., Li, K., Kong, J., Nixon, E. E., et al. (2009). The salient characteristics of the central effects of acupuncture needling: Limbic-paralimbic-neocortical network modulation. *Human Brain Mapping, 30*(4), 1196–1206.

Farde, L. (1997). Brain imaging of schizophrenia–the dopamine hypothesis. *Schizophrenia Research, 28*(2–3), 157–162.

Feighner, J. P. (1999). Mechanism of action of antidepressant medications. *The Journal of Clinical Psychiatry, 60*(Suppl 4), 4–11, discussion 12–13.

Felger, J. C., & Lotrich, F. E. (2013). Inflammatory cytokines in depression: Neurobiological mechanisms and therapeutic implications. *Neuroscience, 246*, 199–229.

Filshie, J., & Cummings, M. (1999). Western medical acupuncture. In E. Ernst & A. White (Eds.), *Acupuncture: A scientific appraisal* (pp. 31–59). Oxford: Butterworth-Heinemann.

Flament, M. F., Bissada, H., & Spettigue, W. (2012). Evidence-based pharmacotherapy of eating disorders. *The International Journal of Neuropsychopharmacology, 15*(2), 189–207.

Fogarty, S., Harris, D., Zaslawski, C., McAinch, A. J., & Stojanovska, L. (2010). Acupuncture as an adjunct therapy in the treatment of eating disorders: A randomised cross-over pilot study. *Complementary Therapies in Medicine, 18*(6), 233–240.

Franco-Santana, L. E., Torres-Castillo, S., González-Trujano, M. E., & González-Ramírez, M. (2013). Stimulation of the Po-shen and Shen-hun scalp-acupuncture bands modifies levels of inhibitory and excitatory amino acids in the immature rat brain. *Neurochemistry International, 63*(4), 275–282.

Gates, S., Smith, L. A., & Foxcroft, D. R. (2006). Auricular acupuncture for cocaine dependence. *Cochrane Database of Systematic Reviews, 1*, CD005192.

Gilpin, N. W., & Koob, G. F. (2008). Neurobiology of alcohol dependence: Focus on motivational mechanisms. *Alcohol Research & Health, 31*(3), 185–195.

Gould, T. D., & Einat, H. (2007). Animal models of bipolar disorder and mood stabilizer efficacy: A critical need for improvement. *Neuroscience and Biobehavioral Reviews, 31*(6), 825–831.
Haenisch, B., & Bonisch, H. (2011). Depression and antidepressants: Insights from knockout of dopamine, serotonin or noradrenaline re-uptake transporters. *Pharmacology & Therapeutics, 129*(3), 352–368.
Han, J. S. (1986). Electroacupuncture: An alternative to antidepressants for treating affective diseases? *The International Journal of Neuroscience, 29*(1–2), 79–92.
Han, C., Li, X., Luo, H., & Zhao, X. (2004). Clinical study on electro-acupuncture treatment for 30 cases of mental depression. *Journal of Traditional Chinese Medicine, 24*(3), 172–176.
Hollifield, M. (2011). Acupuncture for posttraumatic stress disorder: Conceptual, clinical, and biological data support further research. *CNS Neuroscience & Therapeutics, 17*(6), 769–779.
Hori, E., Takamoto, K., Urakawa, S., Ono, T., & Nishijo, H. (2010). Effects of acupuncture on the brain hemodynamics. *Autonomic Neuroscience, 157*(1–2), 74–80.
Hu, A., Lai, M., Wei, J., Wang, L., Mao, H., Zhou, W., et al. (2013). The effect of electroacupuncture on extinction responding of heroin-seeking behavior and FosB expression in the nucleus accumbens core. *Neuroscience Letters, 534*, 252–257.
Huang, W., Kutner, N., & Bliwise, D. L. (2011). Autonomic activation in insomnia: The case for acupuncture. *Journal of Clinical Sleep Medicine*, 7(1), 95–102.
Hui, K. K., Marina, O., Liu, J., Rosen, B. R., & Kwong, K. K. (2010). Acupuncture, the limbic system, and the anticorrelated networks of the brain. *Autonomic Neuroscience, 157*(1–2), 81–90.
Ionescu, D. F., Niciu, M. J., Mathews, D. C., Richards, E. M., & Zarate, C. A., Jr. (2013). Neurobiology of anxious depression: A review. *Depression and Anxiety, 30*(4), 374–385.
Johnson, S. W., & North, R. A. (1992). Opioids excite dopamine neurons by hyperpolarization of local interneurons. *The Journal of Neuroscience, 12*(2), 483–488.
Jong, M. S., Hwang, S. J., & Chen, F. P. (2006). Effects of electro-acupuncture on serum cytokine level and peripheral blood lymphocyte subpopulation at immune-related and non-immune-related points. *Acupuncture & Electro-Therapeutics Research, 31*(1–2), 45–59.
Jorm, A. F., Griffiths, K. M., Christensen, H., Parslow, R. A., & Rogers, B. (2004). Actions taken to cope with depression at different levels of severity: A community survey. *Psychological Medicine, 34*(2), 293–299.
Kaye, W. (2008). Neurobiology of anorexia and bulimia nervosa. *Physiology & Behavior, 94*(1), 121–135.
Kim, Y. D., Heo, I., Shin, B. C., Crawford, C., Kang, H. W., & Lim, J. H. (2013). Acupuncture for posttraumatic stress disorder: A systematic review of randomized controlled trials and prospective clinical trials. *Evidence-Based Complementary and Alternative Medicine, 2013*, 615857.
Kim, H., Park, H. J., Han, S. M., Hahm, D. H., Lee, H. J., Kim, K. S., et al. (2009). The effects of acupuncture stimulation at PC6 (Neiguan) on chronic mild stress-induced biochemical and behavioral responses. *Neuroscience Letters, 460*(1), 56–60.
Kreek, M. J., Levran, O., Reed, B., Schlussman, S. D., Zhou, Y., & Butelman, E. R. (2012). Opiate addiction and cocaine addiction: Underlying molecular neurobiology and genetics. *The Journal of Clinical Investigation, 122*(10), 3387–3393.
Kwon, S., Kim, D., Park, H., Yoo, D., Park, H. J., Hahm, D. H., et al. (2012). Prefrontal-limbic change in dopamine turnover by acupuncture in maternally separated rat pups. *Neurochemical Research, 37*(10), 2092–2098.
Kwon, S., Lee, B., Yeom, M., Sur, B. J., Kim, M., Kim, S. T., et al. (2012). Modulatory effects of acupuncture on murine depression-like behavior following chronic systemic inflammation. *Brain Research, 1472*, 149–160.

Lee, B., Han, S. M., & Shim, I. (2009). Acupuncture attenuates cocaine-induced expression of behavioral sensitization in rats: Possible involvement of the dopaminergic system in the ventral tegmental area. *Neuroscience Letters, 449*(2), 128–132.

Lee, J. H., Kim, H. Y., Jang, E. Y., Choi, S. H., Han, C. H., Lee, B. H., et al. (2011). Effect of acupuncture on naloxone-precipitated withdrawal syndrome in morphine-experienced rats: The mediation of GABA receptors. *Neuroscience Letters, 504*(3), 301–305.

Lee, B., Shim, I., Lee, H. J., Yang, Y., & Hahm, D. H. (2009). Effects of acupuncture on chronic corticosterone-induced depression-like behavior and expression of neuropeptide Y in the rats. *Neuroscience Letters, 453*(3), 151–156.

Lee, M. S., Shin, B. C., Ronan, P., & Ernst, E. (2009). Acupuncture for schizophrenia: A systematic review and meta-analysis. *International Journal of Clinical Practice, 63*(11), 1622–1633.

Lee, B. H., Zhao, R. J., Moon, J. Y., Yoon, S. S., Kim, J. A., An, H., et al. (2008). Differential involvement of GABA system in mediating behavioral and neurochemical effect of acupuncture in ethanol-withdrawn rats. *Neuroscience Letters, 443*(3), 213–217.

Lim, S., Ryu, Y. H., Kim, S. T., Hong, M. S., & Park, H. J. (2003). Acupuncture increases neuropeptide Y expression in hippocampus of maternally-separated rats. *Neuroscience Letters, 343*(1), 49–52.

Lin, J. G., Chan, Y. Y., & Chen, Y. H. (2012). Acupuncture for the treatment of opiate addiction. *Evidence-Based Complementary and Alternative Medicine, 2012*, 739045.

Lin, J. G., & Chen, W. L. (2008). Acupuncture analgesia: A review of its mechanisms of actions. *The American Journal of Chinese Medicine, 36*(4), 635–645.

Liu, R. P., Fang, J. L., Rong, P. J., Zhao, Y., Meng, H., Ben, H., et al. (2013). Effects of electroacupuncture at auricular concha region on the depressive status of unpredictable chronic mild stress rat models. *Evidence-Based Complementary and Alternative Medicine, 2013*, 789674.

Liu, S., Zhu, F., Lai, M., Sun, L., Liu, Y., & Zhou, W. (2012). Electroacupuncture suppresses discrete cue-evoked heroin-seeking and Fos protein expression in the nucleus accumbens core in rats. *Evidence-Based Complementary and Alternative Medicine, 2012*, 286404.

Ma, S. X. (2004). Neurobiology of acupuncture: Toward CAM. *Evidence-Based Complementary and Alternative Medicine, 1*(1), 41–47.

MacPherson, H., & Schroer, S. (2007). Acupuncture as a complex intervention for depression: A consensus method to develop a standardised treatment protocol for a randomised controlled trial. *Complementary Therapies in Medicine, 15*(2), 92–100.

MacPherson, H., Sinclair-Lian, N., & Thomas, K. (2006). Patients seeking care from acupuncture practitioners in the UK: A national survey. *Complementary Therapies in Medicine, 14*(1), 20–30.

Makhija, K., & Karunakaran, S. (2013). The role of inflammatory cytokines on the aetiopathogenesis of depression. *The Australian and New Zealand journal of psychiatry, 47*(9), 828–839.

Maletic, V., Robinson, M., Oakes, T., Iyengar, S., Ball, S. G., & Russell, J. (2007). Neurobiology of depression: An integrated view of key findings. *International Journal of Clinical Practice, 61*(12), 2030–2040.

Mochcovitch, M. D., & Nardi, A. E. (2010). Selective serotonin-reuptake inhibitors in the treatment of panic disorder: A systematic review of placebo-controlled studies. *Expert Review of Neurotherapeutics, 10*(8), 1285–1293.

Mukaino, Y., Park, J., White, A., & Ernst, E. (2005). The effectiveness of acupuncture for depression—A systematic review of randomised controlled trials. *Acupuncture in Medicine, 23*(2), 70–76.

Nestler, E. J., Barrot, M., DiLeone, R. J., Eisch, A. J., Gold, S. J., & Monteggia, L. M. (2002). Neurobiology of depression. *Neuron, 34*(1), 13–25.

Park, H. J., Chae, Y., Jang, J., Shim, I., Lee, H., & Lim, S. (2005). The effect of acupuncture on anxiety and neuropeptide Y expression in the basolateral amygdala of maternally separated rats. *Neuroscience Letters*, *377*(3), 179–184.

Park, H. J., Chae, Y., Kim, J. W., Lee, H., & Chung, J. H. (2011). Effect of acupuncture on hypothalamic-pituitary-adrenal system in maternal separation rats. *Cellular and Molecular Neurobiology*, *31*(8), 1123–1127.

Park, H., Yoo, D., Kwon, S., Yoo, T. W., Park, H. J., Hahm, D. H., et al. (2012). Acupuncture stimulation at HT7 alleviates depression-induced behavioral changes via regulation of the serotonin system in the prefrontal cortex of maternally-separated rat pups. *The Journal of Physiological Sciences*, *62*(4), 351–357.

Pilkington, K. (2010). Anxiety, depression and acupuncture: A review of the clinical research. *Autonomic Neuroscience*, *157*(1–2), 91–95.

Quah-Smith, I., Suo, C., Williams, M. A., & Sachdev, P. S. (2013). The antidepressant effect of laser acupuncture: A comparison of the resting brain's default mode network in healthy and depressed subjects during functional magnetic resonance imaging. *Medical Acupuncture*, *25*(2), 124–133.

Ruhe, H. G., Mason, N. S., & Schene, A. H. (2007). Mood is indirectly related to serotonin, norepinephrine and dopamine levels in humans: A meta-analysis of monoamine depletion studies. *Molecular Psychiatry*, *12*(4), 331–359.

Shen, E. Y., & Lai, Y. J. (2007). The efficacy of frequency-specific acupuncture stimulation on extracellular dopamine concentration in striatum—A rat model study. *Neuroscience Letters*, *415*(2), 179–184.

Simon, G. E., Cherkin, D. C., Sherman, K. J., Eisenberg, D. M., Deyo, R. A., & Davis, R. B. (2004). Mental health visits to complementary and alternative medicine providers. *General Hospital Psychiatry*, *26*(3), 171–177.

Smith, C. A., Hay, P. P., & Macpherson, H. (2010). Acupuncture for depression. *Cochrane Database of Systematic Reviews*, (1), CD004046. http://dx.doi.org/10.1002/14651858.CD004046.pub3.

Song, Y., Zhou, D., Fan, J., Luo, H., & Halbreich, U. (2007). Effects of electroacupuncture and fluoxetine on the density of GTP-binding-proteins in platelet membrane in patients with major depressive disorder. *Journal of Affective Disorders*, *98*(3), 253–257.

Soomro, G. M., Altman, D., Rajagopal, S., & Oakley-Browne, M. (2008). Selective serotonin re-uptake inhibitors (SSRIs) versus placebo for obsessive compulsive disorder (OCD). *Cochrane Database of Systematic Reviews*(1). http://dx.doi.org/10.1002/14651858.CD001765.pub3, CD001765.

Sun, H., Zhang, Y., & Han, C. H. (2005). Effects of electro-acupuncture on the number and binding activity of 5-HT1 and 5-HT2 receptors in the cerebral cortex of chronic stress depression model rats. *International Journal of Clinical Acupuncture*, *14*(2), 113–118.

Unutzer, J., Klap, R., Sturm, R., Young, A. S., Marmon, T., Shatkin, J., et al. (2000). Mental disorders and the use of alternative medicine: Results from a national survey. *The American Journal of Psychiatry*, *157*(11), 1851–1857.

Vazquez, R. D., Gonzalez-Macias, L., Berlanga, C., & Aedo, F. J. (2011). Effect of acupuncture treatment on depression: Correlation between psychological outcomes and salivary cortisol levels. *Salud Mental*, *34*(1), 21–26.

Wang, S. M., Kain, Z. N., & White, P. (2008). Acupuncture analgesia: I. The scientific basis. *Anesthesia & Analgesia*, *106*(2), 602–610.

Wang, P. S., Lane, M., Olfson, M., Pincus, H. A., Wells, K. B., & Kessler, R. C. (2005). Twelve-month use of mental health services in the United States: Results from the National Comorbidity Survey Replication. *Archives of General Psychiatry*, *62*(6), 629–640.

Wen, H. L., Ho, W. K., Wong, H. K., Mehal, Z. D., Ng, Y. H., & Ma, L. (1979). Changes in adrenocorticotropic hormone (ACTH) and cortisol levels in drug addicts treated by a new and rapid detoxification procedure using acupuncture and naloxone. *Comparative Medicine East and West, 6*(3), 241–245.

White, A. R., Rampes, H., Liu, J. P., Stead, L. F., & Campbell, J. (2011). Acupuncture and related interventions for smoking cessation. *Cochrane Database of Systematic Reviews*(1). http://dx.doi.org/10.1002/14651858.CD000009.pub3, CD000009.

Wollweber, B. T., & Wetter, T. C. (2011). A brief review of neurobiological principles of insomnia. *Schweizer Archive Fur Neurologie und Psychiatrie, 162*(4), 139–141.

Wu, G., Feder, A., Wegener, G., Bailey, C., Saxena, S., Charney, D., et al. (2011). Central functions of neuropeptide Y in mood and anxiety disorders. *Expert Opinion on Therapeutic Targets, 15*(11), 1317–1331.

Yang, C. H., Lee, B. H., & Sohn, S. H. (2008). A possible mechanism underlying the effectiveness of acupuncture in the treatment of drug addiction. *Evidence-Based Complementary and Alternative Medicine, 5*(3), 257–266.

Yoon, S. S., Kim, H., Choi, K. H., Lee, B. H., Lee, Y. K., Lim, S. C., et al. (2010). Acupuncture suppresses morphine self-administration through the GABA receptors. *Brain Research Bulletin, 81*(6), 625–630.

Yoon, S. S., Kwon, Y. K., Kim, M. R., Shim, I., Kim, K. J., Lee, M. H., et al. (2004). Acupuncture-mediated inhibition of ethanol-induced dopamine release in the rat nucleus accumbens through the GABAB receptor. *Neuroscience Letters, 369*(3), 234–238.

Yoon, S. S., Yang, E. J., Lee, B. H., Jang, E. Y., Kim, H. Y., Choi, S. M., et al. (2012). Effects of acupuncture on stress-induced relapse to cocaine-seeking in rats. *Psychopharmacology (Berl), 222*(2), 303–311.

Yoshimoto, K., Fukuda, F., Hori, M., Kato, B., Kato, H., Hattori, H., et al. (2006). Acupuncture stimulates the release of serotonin, but not dopamine, in the rat nucleus accumbens. *The Tohoku Journal of Experimental Medicine, 208*(4), 321–326.

Zhao, Z. Q. (2008). Neural mechanism underlying acupuncture analgesia. *Progress in Neurobiology, 85*(4), 355–375.

Zhao, R. J., Yoon, S. S., Lee, B. H., Kwon, Y. K., Kim, K. J., Shim, I., et al. (2006). Acupuncture normalizes the release of accumbal dopamine during the withdrawal period and after the ethanol challenge in chronic ethanol-treated rats. *Neuroscience Letters, 395*(1), 28–32.

Zhao, Z. L., Zhao, G. W., Li, H. Z., Yang, X. D., Wu, Y. Y., Lin, F., et al. (2011). Acupuncture attenuates anxiety-like behavior by normalizing amygdaloid catecholamines during ethanol withdrawal in rats. *Evidence-Based Complementary and Alternative Medicine, 2011*, 429843.

Zhou, Q., Zhao, J., Cai, D., Jiang, Q., Yu, S. G., Wei, J., et al. (2008). The effect of electroacupuncture on the imbalance between monoamine neurotransmitters and GABA in the CNS of rats with chronic emotional stress-induced anxiety. *International Journal of Clinical Acupuncture, 17*(2), 79–84.

CHAPTER ELEVEN

Acupuncture for the Treatment of Insomnia

Kaicun Zhao[1]
Department of Natural Sciences, Middlesex University London, The Burroughs, Hendon, London, United Kingdom
[1]Corresponding author: e-mail address: k.zhao@mdx.ac.uk

Contents

Abstract

Insomnia appears to be a fast-spreading problem in the modern days, which not only affects people's living quality but also impairs people's working efficiency even causing disability. Pharmacological treatment is effective but frequently with significant side effects. Acupuncture is traditionally used for the treatment of insomnia in China and now is widely accepted in the Western countries. Many research works on clinical applications of acupuncture in the treatment of insomnia and the potential mechanisms underlying the acupuncture treatment have been reported. This chapter will try to provide a systematic review on the research findings.

A number of clinical studies, mainly randomized controlled clinical trials, have shown positive effects in acupuncture treatment of insomnia. Some of the studies demonstrated that acupuncture treatment appeared to be better than conventional pharmacological drugs in the improvement of insomnia. These encouraging findings are limited by the qualities problems of the methodology used in these clinical studies.

The clinical efficacy of acupuncture appeared to be supported by evidence obtained from basic neuroendocrinological studies. A number of studies have demonstrated that acupuncture may modulate a wide range of neuroendocrinological factors

International Review of Neurobiology, Volume 111
ISSN 0074-7742
http://dx.doi.org/10.1016/B978-0-12-411545-3.00011-0

following stimulation of acupoints. Evidence has suggested that the clinical efficacy of acupuncture in treatment of insomnia is potentially mediated by a variety of neurotransmitters including norepinephrine, melatonin, gamma-aminobutyric acid, and β-endorphin. However, due to the complexity, these findings are far from conclusive. More research is necessary. More rigors methodology and integrated approach to evaluate both clinical and basic research evidence are required for future studies.

1. INTRODUCTION

Acupuncture is a major part of traditional Chinese medicine (TCM) in addition to Chinese herbal medicines. This treatment modality is based on the meridian-collateral theories of TCM and involves stimulation of certain selected acupoints on the body by insertion and manipulations of sterilized solid needles on the points. Acupuncture has been widely used to treat a variety of clinical conditions in particular those involving pathological changes in neuroendocrinology, such as menopause, depression, and insomnia.

Insomnia is a common sleep disorder. According to the Diagnostic and Statistical Manual of Mental Disorders IV (DSM-IV), primary insomnia is a clinical condition with predominant complaint as difficulty in initiating or maintaining sleep, or nonrestorative sleep, for at least 1 month. The sleep disturbance (or associated daytime fatigue) causes clinically significant distress or impairment in social, occupational, or other important areas of functioning (American Psychiatric Association, 2000). The Great British Sleep Survey has shown that out of 11,129 surveyed participants with mean age of 39 years old, 5083 reported as having possible insomnia disorder (Espie, Kyle, Hames, Cyhlarova, & Benzeval, 2012). Similarly, according to NIH Consensus and State-of-the-Science Statements, there are multiple millions of individuals affected by insomnia each year in the United States (NIH, 2005). Another investigation crossing multiple countries of Asia and Africa has shown that the prevalence of insomnia in these countries is from 3.9% to 40% (Stranges, Tigbe, Gómez-Olivé, Thorogood, & Kandala, 2012).

Insomnia not only affects people's sleep at night but also impair people's daytime functions. It has found negative impact on energy and mood as the most common consequence of insomnia. Insomnia may cause difficulties in concentration, fatigue, and low work efficiency (Espie et al., 2012; Leger, Massuel, & Metlaine, 2006; Walsh, 2004). Studies also revealed significant associations between sleep disorders and depression and anxiety (Alfano, Zakem, Costa, Taylor, & Weems, 2009). The economic impact of insomnia is high.

For treatment of insomnia, cognitive and behavioral treatments are helpful. For severe insomnia, pharmacological treatment will be needed. In the United Kingdom, short-acting benzodiazepines or Z medicines, such as zopiclone, or melatonin may be prescribed. However, the medicines may have significant side effects such as hungover feeling and drowsiness during the day. As long-term use of these medicines will lose the clinical effectiveness, these medicines may not be suitable for chronic insomnia. Due to limitations of clinical efficacy and concerns of side effects of the medicines, more and more insomnia patients are seeking alternative treatments. Acupuncture is one of the popular choices for these patients (Gould & MacPherson, 2001; Vickers & Zollman, 1999).

With the widespread use of acupuncture in the treatment of insomnia, many basic and clinical researches have been done attempting to examine the clinical efficacy and the potential action mechanisms underlying the acupuncture treatment. Systematic reviews on acupuncture treatment of insomnia have also been published with majority are based on randomized controlled trials (RCTs). This chapter tries to discuss the clinical efficacy of acupuncture treatment of insomnia in combination with the potential neuroendocrinology mechanisms based on the available research findings in both clinical observations and related biochemical studies.

2. ACUPUNCTURE TREATMENT OF INSOMNIA AND CLINICAL EFFICACY

Chinese medicine is a medical system with its own theories about physiology and pathology of human body. The theories are mainly based on long-term holistic clinical observations rather than on the body anatomy and biochemical principles. Therefore, interpretations to clinical conditions and its diagnosis as well as clinical treatments appear very different between TCM and the conventional medicine. The differences bring a lot more complexities to evaluations of the clinical efficacy of TCM treatment including acupuncture treatment. For instance, TCM treat patients with individualized treatment plan, but proper clinical trial require standardized intervention. Another common difficulty for clinical trial of acupuncture treatment is how to find out a suitable placebo control.

In Chinese medicine, insomnia is believed to be a pathological hyperarousal or restless condition related to the functions of heart and brain. There are different causative reasons that can lead to insomnia. Acupuncture is able to regulate the functioning of the heart and brain through stimulations of certain acupoints on the body. In the past years, many clinical studies including RCTs have been reported on acupuncture treatment of insomnia. Most of the reports have demonstrated positive clinical effects in the treatment of insomnia by acupuncture.

In clinical studies of acupuncture treatment of insomnia, most frequently used outcome measurements include Pittsburgh Sleep Quality Index (PSQI) and Insomnia Severity Index (ISI). These are subjective measurement. Actigraphy and polysomnography as objective measurements are also frequently used in clinical studies for the treatment of insomnia. Many clinical studies have reported that in comparison with the baseline value of pretreatment, ISI and PSQI scores in total as well as the scores for all items were significantly reduced following acupuncture treatment of insomnia patients (Huang, Kutner, & Bliwise, 2009; Huo, Guo, & Li, 2013; Lee et al., 2009; Luo, Zhang, & Lai, 2010; Tu, Chung, Yang, & Tzeng, 2012; Xiao, Luo, & Shen, 2013; Yao, Zhang, & Chen, 2012; Yeung et al., 2011). Significant differences were also found between acupuncture treatment and no-treatment control groups. The reduction was accompanied with a higher therapeutic effective rate in improvement of insomnia (Xiao et al., 2013). Acupuncture treatment also reduced the falling asleep latency and increased the sleep duration and sleep efficiency (Yao et al., 2012).

Studies also suggested that acupuncture treatment can be as effective as or superior to conventional medicines such as with oral administration of trazodone, zolpidem, or estazolam for sleep quality and daytime functioning (Luo et al., 2010; Tu et al., 2012; Xuan, Guo, Wang, & Wu, 2007).

In consistency with the subjective outcome measurements, results from polysomnography (PSG) showed that in comparison with control of sham acupuncture, the real acupuncture treatment significantly increased the percentage of the N3 + 4 stage sleep and effectively improved sleep quality in postmenopausal women with insomnia (Hachul et al., 2013).

In the past years, a number of systematic reviews and meta-analysis of the clinical studies have also been published (Cao, Pan, Li, & Liu, 2009; Cheuk, Yeung, Chung, & Wong, 2012; Huang et al., 2009; Yeung, Chung, Leung, Zhang, & Law, 2009; Yeung et al., 2012; Yeung, Chung, Zhang, Yap, & Law, 2009). Although it has been noted that the majority of clinical trials showed significant effectiveness (Huang et al., 2009), conclusions from most of these systemic reviews are inconclusive about the clinical effectiveness in acupuncture treatment of insomnia. This is mainly due to poor quality of the methodology used in the clinical researches.

2.1. Problem of placebo control

While majority of the clinical studies reported positive results in acupuncture treatment of insomnia, there are also conflict research findings reported following acupuncture treatment of insomnia patients. Some studies showed that while acupuncture clearly resulted in significant improvement in subjective sleep measures such as PQSI, it failed to show any difference in objective actigraphy measures in comparison with the control of noninvasive placebo acupuncture (Yeung et al., 2011; Zollman, Larson, Wasek-Throm, Cyborski, & Bode, 2012).

In an RCT study, Yeung et al. (2011) have found that there were no differences between the electroacupuncture and minimal acupuncture (superficial needling at nonacupuncture points) groups, although both of the groups showed significant improvement in treatment of insomnia and also showed significant difference from the noninvasive placebo acupuncture group. This phenomenon is a common problem in clinical trials of acupuncture treatment. Similar findings were also found in RCT studies on acupuncture treatment of headache (Melchart et al., 2005). These research findings have led some researchers to suggest that the effects shown in the clinical trials could be purely caused by nonspecific stimulation of needling, while TCM theory in acupuncture may not play any role in it (Yeung et al., 2011).

The arguments here are whether or not the sham or minimum acupuncture is appropriate. First, regardless inserting needles superficially or deeper, the needles will produce stimulations that may still be effective to trigger the body's responses. Secondly, in acupuncture practice, it is common to use a combination of multiple acupoints for the treatment of various clinical conditions. It is known that different acupoints may have same or similar activities, particularly, acupoints located on the same meridians. Even with nonacupoints, they may still have certain kinds of activities depending on the locations. The overlapping in activities between different points of the body is far from clear. In addition, according to Chinese medicine theories, acupuncture meridians are not simply isolated lines in the body but are functional pathways including related skin area, tissues of muscles and tendons as well as specifically pertained organs. All these have made it very difficult to have an appropriate selection of points for placebo control in clinical studies of acupuncture treatment.

2.2. Methods for stimulation of acupoints

Here are two aspects in acupuncture treatment, one is the stimulation and another one is the acupoints. Various methods are used as stimulation means in clinical practice. The most commonly used stimulation method is insertion of a needle on relevant acupoints (Hachul et al., 2013; Tu et al., 2012; Zollman et al., 2012). Other stimulation methods include electrical acupuncture (Yeung et al., 2011), acupressure, catgut embedding (Wei, 2010; Yao et al., 2012), acupoints penetrating (Dong, Wang, Sun, & Liu, 2008; Zhou, Shi, Li, & Guan, 2010), auricular acupuncture (Jiang, Ma, & Zuo, 2010; Lee, Shin, Suen, Park, & Ernst, 2008; Sjöling, Rolleri, & Englund, 2008), and moxibustion (Ju, Chi, & Liu, 2009). The clinical trials or observations have shown significant effectiveness in the treatment of insomnia by using these different stimulation methods. There may be some differences in extent of therapeutic effectiveness among the different stimulation methods, for instance, studies have shown that penetration needling appears to have better result in improvement of insomnia in comparison with conventional acupuncture (Zhou et al., 2010). However, from the available evidence, it seems the stimulation mode is not a key issue for achieving clinical positive results in acupuncture treatment of insomnia.

2.3. Selection of acupoints

Another important aspect in acupuncture treatment is the selection of acupoints. Literature studies have revealed a wide range of acupoints that have been selected for the treatment of insomnia. According to Chinese medicine and acupuncture theories, patients are treated individually depending on the individual conditions of the particular patient. Every individual may have their own specific conditions. This may lead to different selection of acupoints. The variation in acupoint selection brings problems to comparison of different clinical trials. As shown in Table 11.1, a total of 30 points were used in 20 clinical trials during the period of 2007–2013. Acupoints used for insomnia treatment with a frequency higher than 20% are Baihui (GV-20), Sishencong (EX-HN1), Shenmen (He-7), Shenting (GV-24), and Sanyinjiao (SP-6) among the 30 acupoints used in the clinical trials. Acupoints used in these clinical studies ranged from single point to combination of more than eight points. All these studies claimed the acupoints used were effective in improvement of insomnia. The variances in acupoint selections brought more complexities to evaluation of the effectiveness of acupuncture treatment of insomnia.

Table 11.1 Frequency of acupoints used in acupuncture treatment of insomnia

Acupoints	Frequency of clinical use	References
Baihui (GV-20)	50.0	Zollman et al. (2012), Yeung, Chung, Leung, et al. (2009), Yeung, Chung, Zhang, et al. (2009), Yeung et al. (2011), Yao et al. (2012), Luo et al. (2010), Wei (2010), Ju et al. (2009), Ruan (2009), Xuan et al. (2007), Gao et al. (2007), Gao, Ren, and Wang (2010)
Sishencong (EX-HN1)	41.7	Yeung, Chung, Leung, et al. (2009), Yeung, Chung, Zhang, et al. (2009), Yeung et al. (2011), Ruan et al. (2009), Xiao et al. (2013), Yan et al. (2010), Wei (2010), Dong et al. (2008), Tang, Liu, and Liu (2007), Gao et al. (2010, 2007)
Shenmen (HT-7)	37.5	Tu et al. (2012), Ruan et al. (2009), Xiao et al. (2013), Yao et al. (2012), Luo et al. (2010), Yan et al. (2010), Lee et al. (2009), Ruan (2009), Xuan et al. (2007), Gao et al. (2010, 2007)
Shenting (GV-24)	25.0	Wei (2010), Ruan (2009), Dong et al. (2008), Xuan et al. (2007)
Sanyinjiao (SP-6)	20.8	Ruan et al. (2009), Xiao et al. (2013), Yan et al. (2010), Gao et al. (2010, 2007)
Neiguan (PC-6)	12.5	Yao et al. (2012), Lee et al. (2009), Ruan (2009)
Shenmai (BL-62)	12.5	Ruan et al. (2009), Yan et al. (2010)
Yintang (EX-HN3)	12.5	Yeung, Chung, Leung, et al. (2009), Yeung, Chung, Zhang, et al. (2009), Yeung et al. (2011), Dong et al. (2008)
Zhaohai (KI-6)	12.5	Ruan et al. (2009), Yan et al. (2010), Gao et al. (2010, 2007)
Auricular Shenmen	12.5	Yeung, Chung, Leung, et al. (2009), Yeung, Chung, Zhang, et al. (2009), Yeung et al. (2011), Gao et al. (2007)
Anmian (EX)	8.3	Yeung, Chung, Leung, et al. (2009), Yeung, Chung, Zhang, et al. (2009), Yeung et al. (2011)
Fengchi (GB-20)	8.3	Yao et al. (2012)
Guanyuan (CV-4)	8.3	Wang et al. (2008), Zhong, Cai, Li, and Lü (2008)

Continued

Table 11.1 Frequency of acupoints used in acupuncture treatment of insomnia—cont'd

Acupoints	Frequency of clinical use	References
Hegu (LI-4)	8.3	Zollman et al. (2012), Luo et al. (2010)
Qihai (CV-6)	8.3	Wang et al. (2008), Zhong et al. (2008)
Shaohai (HT-3)	8.3	Zollman et al. (2012), Ruan (2009)
Taichong (LV-3)	8.3	Zollman et al. (2012), Luo et al. (2010)
Taixi (KI-3)	8.3	Zollman et al. (2012), Ruan et al. (2009)
Zhongwan (CV-12)	8.3	Wang et al. (2008), Zhong et al. (2008)
Dazhui (GV-14)	4.2	Gao et al. (2010, 2007)
Huaroumen (ST-24)	4.2	Wang et al. (2008)
Jianshi (PC-7)	4.2	Zollman et al. (2012)
Kunlun (BL-60)	4.2	Zollman et al. (2012)
Qiangjian (GV-18)	4.2	Dong et al. (2008)
Qipang	4.2	Wang et al. (2008)
Shangqu (KI-17)	4.2	Wang et al. (2008)
Toulinqi (GB-15)	4.2	Dong et al. (2008)
Xiafengshidian	4.2	Wang et al. (2008)
Xiawan (CV-10)	4.2	Wang et al. (2008)
Xinshu (BL-15)	4.2	Ruan (2009)
Zusanli (St-36)	4.2	Tu et al. (2012)
Auricular Tranquilizer	4.2	Zollman et al. (2012)

3. THE POTENTIAL MECHANISMS UNDERLYING THE ACUPUNCTURE TREATMENT

Insomnia is associated with a number of neuroendocrinological disorders. It has been suggested that both the central nervous system (CNS) and the autonomic nervous system are involved in the pathology of insomnia. There is substantial evidence showing multiple neurotransmitters may be involved in the pathophysiology of insomnia including of sympathetic, GABAergic, opiate, and melatonin. Although it is far from clear about the underlying action mechanism of acupuncture treatment of insomnia, there are many studies suggesting acupuncture may improve insomnia by acting on these nervous systems and modulate the activities of neurotransmitters.

3.1. Sympathetic system and hypothalamic–pituitary–adrenal axis

Evidence has shown that insomnia is accompanied by sympathetic hyperactivity. Studies have found that insomnia patients have increased metabolic activation with elevated heart rate and sympathetic nervous system activation during sleep (Bonnet & Arand, 2010; Roth, 2007). Studies have also demonstrated that norepinephrine in CNS induces wakefulness via β- and α_1-receptors but suppression of CNS norepinephrine release results in profound sedation (Berridge, Schmeichel, & España, 2012).

Kung, Yang, Chiu, and Kuo (2011) reported a clinical trial on treatment of insomnia using auricular acupuncture. The study found that the improvement of insomnia (as indicated by PQSI) was related to the increased cardiac parasympathetic activity and decreased the sympathetic activity. A double-blind RCT has shown that intradermal acupuncture on Shenmen (HT-7) and Neiguan (PC-6) stabilized the sympathetic hyperactivities of insomnia patients as indicated by a decreased ratio of low-frequency power/high-frequency power in heart rate variability analysis. This change was closely related to the significant improvement of insomnia as indicated by ISI (Lee et al., 2009). Other acupoints that have been suggested to have similar activities are Sishencong (EX-HN1) (Wang, Kuo, & Yang, 2002) and Xinshu (BL-15) (Hsu, Weng, Liu, Tsai, & Chang, 2006). Cabrini, Gioia, Gemma, Cedrati, and Crivellari (2006) investigated the sedative effect of acupuncture in healthy volunteers by monitoring the bispectral index and heart rate. However, no differences were observed between the true acupuncture and the sham control. This result may suggest that the sedative action of acupuncture only can be seen in anxiety and insomnia patients.

Experimental study demonstrated that electroacupuncture at Neiguan (PC-6) and Xinshu (BL-15), or at Shenmen (HT-7), or at Quchi (L1-11) and Zusanli (ST-36) increased the contents of hypothalamic norepinephrine (NE), dopamine (DA), and serotonin (5-HT) (Li, Hu, Cai, Wu, & Wang, 2012; Wang et al., 2011; Zhou et al., 1995). These results are controversial with the sedative effects of these acupoints in the treatment of insomnia. More research works need to be done in the future.

Hypothalamic–pituitary–adrenal (HPA) axis is an important neuroendocrine system involved in the regulation of sleep. It has been suggested that activation of HPA system may lead to insomnia. Stress stimulates HPA system and very commonly induces insomnia (Han, Kim, & Shim, 2012). Animal studies have found electroacupuncture on acupoints of Baihui (GV-20) and Yintang (EX-HN3) can effectively reduce adrenal cortisol content and downregulate hippocampal expression of glucocorticoid receptor mRNA (You, Shi, Han, Jia, & Tu, 2010). Another study in rat has shown that acupuncture at Shenmen (HT-7) reduced plasma corticosterone and adrenocorticotropin hormone (ACTH) levels (Park et al., 2011). These findings provide further supports to the clinical use of these acupoints in the treatment of insomnia as described earlier.

3.2. Gamma-aminobutyric acid

Gamma-aminobutyric acid (GABA) is a neurotransmitter usually responsible for subduing physiological activities in the brain. It exerts its activities through binding to GABA(A) receptor. Research has found that average brain GABA levels can be as much as 30% lower in primary insomnia patients than that in the normal control. It was also shown that the brain GABA levels were negatively correlated with wake after sleep onset recorded on PSG of the insomnia patients (Winkelman et al., 2008).

GABA is an important neuromediator in regulating sleep. Some medications currently used in clinical practice for the treatment of insomnia, such as benzodiazepine, zolpidem, and eszopiclone, are based on their positive modulating activities on GABA(A) receptor. These medicines relieve insomnia by potentiating brain GABA activity, which in turn exerts an inhibitory effect on sympathetic NE transmission, therefore reducing hyperarousal (Mitchell & Weinshenker, 2010). Direct evidence on NE release has been shown in an animal studies with midazolam (Kubota et al., 1999).

Recent evidence has shown that acupuncture significantly increased GABA level and upregulated the expression of GABA(A) receptor in hypothalamic neurons of rat. The positive acupoints tested for this activity include Sanyinjiao (SP-6), Neiguan (PC-6), Zusanli (ST-36), Shenmen (HT-7), and the combination of Shenmai (BL-62) and Zhaohai (KI-6). This finding may provide explanation to the insomnia relieving effect of these acupoints in clinical acupuncture practice. Among these acupoints, Shenmen (HT-7) and the combination of Shenmai–Zhaohai (BL-62–KI-6) were found to be the most effective acupoints with significantly superior upregulating results (Zhou, Gao, Wang, & Ren, 2012).

There are more experimental studies that provide further evidence suggesting acupuncture on Shenmen (HT-7) may cause stimulation of GABA neurotransmissions (Lee, Zhao, et al., 2008; Yoon et al., 2010). Acupuncture on Baihui (GV-20) showed a preventive effect on the impairment of cortical GABAergic neurons in an ischemic stroke model (Zhang et al., 2011). Another experiment with Parkinsonian rat model demonstrated that electroacupuncture on Baihui (GV-20) induced an increase of brain GABA levels (Du, Sun, Jia, Wang, & Wang, 2011). Increased GABA activities in the brain are expected to produce sedative effects, contributing to insomnia releasing.

3.3. Melatonin

Melatonin is a hormone present naturally in the human body. This hormone is secreted from the pineal gland in the brain. Melatonin is important in helping to regulate the body's circadian rhythm of sleep–wake cycle, therefore playing an important role in maintaining normal sleep. Normally, melatonin is secreted during the night and certain level of melatonin in the blood is essential for normal sleep. It has been found that nocturnal melatonin production was reduced in elderly insomnia patients, suggesting a possible association between this hormone and primary insomnia (Abbasi et al., 2012; Riemann et al., 2002).

Melatonin has been used for the treatment of insomnia. Results from a clinical trial have demonstrated the efficacy and safety of a prolonged release of melatonin in the treatment of primary insomnia (Cummings, 2012; Lemoine, Garfinkel, Laudon, Nir, & Zisapel, 2011). A meta-analysis has shown that melatonin decreases sleep onset latency, increases total sleep time, and improves overall sleep quality (Ferracioli-Oda, Qawasmi, & Bloch, 2013).

A clinical trial has shown that the endogenous nocturnal melatonin secretion of anxious patients was significantly increased following 5 weeks of acupuncture treatment. The increase was associated with significant improvements in sleep onset latency, total sleep time, and sleep efficiency as measured in polysomnography. At the same time, patients' anxiety conditions were also significantly reduced (Spence et al., 2004). It was suggested that acupressure on Shenmen (HT-7) may help to normalize the nocturnal secretion of melatonin in insomniacs and this may be involved in the improvement of sleep quality (Nordio & Romanelli, 2008).

3.4. Other neuroendocrinological factors

Studies have also suggested some other neuroendocrinological factors might mediate the effects of acupuncture in the treatment of insomnia. Opioidergic neurotransmissions in the brain have been found to be involved in the regulation of sleep–wake rhythm, and endogenous opioids may play a role in maintaining normal sleep (Greco et al., 2008).

Cheng, Yi, Lin, and Chang (2011) have reported that in a rat model, electroacupuncture on Anmian (EX-17) increased nonrapid eye movement (NREM) sleep, and this effect was suggested to be mediated by a pathway involving cholinergic activation, stimulation of opioidergic neurons to secret β-endorphin in both the brainstem and hippocampus, and regulation of sleep via μ-opioid receptors (Cheng et al., 2011). Another studies with a rat model also showed that electroacupuncture on Zusanli (ST-36) and Sanyinjiao (SP-6) significantly increased NREM sleep, rapid eye movement (REM) sleep, and total sleep time during acute morphine withdrawal. This action was presumed to be mediated by the electroacupuncture facilitated release of endogenous opioids (Li et al., 2011).

4. DISCUSSIONS

Insomnia is a common disease affecting large numbers of the population. Although there are effective pharmacological medications available, significant side effects have limited their clinical applications and long-term use (Jacob et al., 2012). Acupuncture is a Chinese medicine modality and has been used in the treatment of various diseases including insomnia for thousands of years. To find out why and how this old modality works, many clinical and experimental studies have been conducted in the past years. As described earlier, however, there are still no clear conclusions about the effectiveness of acupuncture treatment on insomnia. This is largely due to the quality of methodologies used in the previous clinical studies. There are several weaknesses that exist in the methodology of previous research studies including small sample size, questionable placebo control or sham control, arguable use of standardized intervention, and variety of acupoint selection. Some of these difficulties are caused by the conflict between the scientific "golden standards" required for RCT and the traditional clinical practice of acupuncture, for example, individualized treatment principle in Chinese medicine is against the RCT requirement for a standardized intervention. For future studies, it is desperately needed to find a solution to overcome these difficulties.

Placebo control is important for proper clinical trial and is frequently problematic for clinical research of Chinese medicine and acupuncture. Acupuncture is a physical intervention. It has been known that physical intervention is more likely to produce placebo effect in clinical trials (Hróbjartsson & Gøtzsche, 2010). As mentioned earlier in this chapter, sham needling may not be an appropriate placebo control for acupuncture clinical trial. It is believed that a noninvasive placebo intervention would be more suitable for acupuncture clinical studies.

In view of the previous clinical trials in acupuncture treatment of insomnia, subjective outcome measurements were used in most of the studies. It appears that subjective measurements are more susceptible to placebo effects (Hróbjartsson & Gøtzsche, 2010). Induction of more objective outcome measurements would be helpful in reducing placebo effects. In the case of insomnia, actigraphy and polysomnography are useful objective measurements for clinical studies.

As reviewed in this chapter, it can be seen that a very wide range of acupoint selections have been used in the clinical trials in addition to the variety of stimulation or manipulation methods. This has brought a lot of difficulties to the interpretation and comparison of the published research studies. At present, evidence from the available studies is not able to provide a clear and conclusive picture about the specificities of the acupoints and meridians used in the studies. Similar questions have been raised in a recently published literature review on acupuncture effect and central autonomic regulation (Li et al., 2013). In most of the research studies, the acupoint selections were based on Chinese medicine theories. The variations in acupoint selections may be due to different acupuncture styles such as body acupuncture, abdominal acupuncture, or scalp acupuncture. In addition, Chinese medicine treats patients individually according to specific Chinese medicine syndrome differentiations. It is always a debatable issue for RCT using standardized intervention as this may not represent the real acupuncture practice. To solve this problem, it would be helpful to recruit clinical trial participants by using combined diagnosis standards from both Chinese medicine and conventional medicine, for instance, insomnia patients with a further differentiation using Chinese medicine diagnostic principles. In this case, acupoint selection can be standardized according to the diagnosis. Indeed, a clinical study was done with participants selected as insomnia due to deficiency of both the heart and spleen (Lu & Liu, 2008).

In this article, the neuroendocrinological activities of the acupoints commonly used for treatment of insomnia have been reviewed. In comparison with the results of clinical trials, it can be seen that most frequently used acupoints also demonstrated certain kinds of neuroendocrinological activities. Acupuncture can cause responses from a wide range of neuroendocrinological pathways, even with a single acupoint such as Shenmen (HT-7). With a combination of different acupoints, the induced neuroendocrinological reactions will be more complicated. At present, the full picture of the specificities of the involved acupoints is still far from completely clear. There are also some confusing findings that need to be solved in the future.

5. SUMMARY AND CONCLUSIONS

This chapter is trying to review the research with a consideration of integrating the available research results to obtain more comprehensive understanding of the effectiveness of acupuncture treatment of insomnia. Although clear conclusion is still hard to be made based on the available studies due to low quality in methodology, the efficacy of acupuncture at the most commonly used acupoints for insomnia can be further supported by evidence from neuroendocrinological activities of these acupoints. Although many difficulties exist, further investigations on efficacy of acupuncture treatment of insomnia are warranted. It would be beneficial to the future studies if integrated approaches are applied to explore both clinical effectiveness and the underlying neurophysiological mechanism for acupuncture treatment of insomnia.

REFERENCES

Abbasi, B., Kimiagar, M., Sadeghniiat, K., Shirazi, M. M., Hedayati, M., & Rashidkhani, B. (2012). The effect of magnesium supplementation on primary insomnia in elderly:

A double-blind placebo-controlled clinical trial. *Journal of Research in Medical Sciences, 17*(12), 1161–1169.

Alfano, C. A., Zakem, A. H., Costa, N. M., Taylor, L. K., & Weems, C. F. (2009). Sleep problems and their relation to cognitive factors, anxiety, and depressive symptoms in children and adolescents. *Depression and Anxiety, 26*(6), 503–512.

American Psychiatric Association, (2000). *Diagnostic and Statistical Manual of Mental Disorders IV Text Revision (DSM-IV-TR)* (4th-sub ed.). Washington: American Psychiatric Association.

Berridge, C. W., Schmeichel, B. E., & España, R. A. (2012). Noradrenergic modulation of wakefulness/arousal. *Sleep Medicine Reviews, 16*(2), 187–197.

Bonnet, M. H., & Arand, D. L. (2010). Hyperarousal and insomnia: state of the science. *Sleep Med Rev, 14*(1), 9–15.

Cabrini, L., Gioia, L., Gemma, M., Cedrati, V., & Crivellari, M. (2006). Bispectral index evaluation of the sedative effect of acupuncture in healthy volunteers. *Journal of Clinical Monitoring and Computing, 20*(5), 311–315.

Cao, H., Pan, X., Li, H., & Liu, J. (2009). Acupuncture for treatment of insomnia: A systematic review of randomized controlled trials. *Journal of Alternative and Complementary Medicine, 15*(11), 1171–1186.

Cheng, C. H., Yi, P. L., Lin, J. G., & Chang, F. C. (2011). Endogenous opiates in the nucleus tractus solitarius mediate electroacupuncture-induced sleep activities in rats. *Evidence-Based Complementary and Alternative Medicine, 2011*, 159209.

Cheuk, D. K., Yeung, W. F., Chung, K. F., & Wong, V. (2012). Acupuncture for insomnia. *Cochrane Database of Systematic Reviews,* (9).

Cummings, C. (2012). Melatonin for the management of sleep disorders in children and adolescents. *Paediatrics and Child Health, 17*, 331–333.

Dong, J. P., Wang, S., Sun, W. Y., & Liu, F. (2008). Randomized controlled observation on head point-through-point therapy for treatment of insomnia. *Zhongguo Zhen Jiu, 28*(3), 159–162.

Du, J., Sun, Z. L., Jia, J., Wang, X., & Wang, X. M. (2011). High-frequency electro-acupuncture stimulation modulates intracerebral γ-aminobutyric acid content in rat model of Parkinson's disease. *Sheng Li Xue Bao, 63*(4), 305–310.

Espie, C. A., Kyle, S. D., Hames, P., Cyhlarova, E., & Benzeval, M. (2012). The daytime impact of DSM-5 insomnia disorder: Comparative analysis of insomnia subtypes from the Great British Sleep Survey. *The Journal of Clinical Psychiatry, 73*(12), e1478–e1484.

Ferracioli-Oda, E., Qawasmi, A., & Bloch, M. H. (2013). Meta-analysis: Melatonin for the treatment of primary sleep disorders. *PLoS One, 8*(5), e63773.

Gao, X. Y., Ren, S., & Wang, P. Y. (2010). Acupuncture treatment of insomnia by regulating the defensive-qi and strengthening the brain and the spinal cord. *Journal of Traditional Chinese Medicine, 30*(3), 222–227.

Gao, X. Y., Wei, Y. L., Shao, S. J., Li, X. R., Zhang, H. J., Wang, J., et al. (2007). Multiple central clinical studies on the needling method for regulating wei and strengthening brain for treatment of insomnia. *Zhongguo Zhen Jiu, 27*(8), 623–625.

Gould, A., & MacPherson, H. (2001). Patient perspectives on outcomes after treatment with acupuncture. *Journal of Alternative and Complementary Medicine*, 7, 261–268.

Greco, M. A., Fuller, P. M., Jhou, T. C., Martin-Schild, S., Zadina, J. E., Hu, Z., et al. (2008). Opioidergic projections to sleep-active neurons in the ventrolateral preoptic nucleus. *Brain Research, 1245*, 96–107.

Hachul, H., Garcia, T. K., Maciel, A. L., Yagihara, F., Tufik, S., & Bittencourt, L. (2013). Acupuncture improves sleep in postmenopause in a randomized, double-blind, placebo-controlled study. *Climacteric, 16*(1), 36–40.

Han, K. S., Kim, L., & Shim, I. (2012). Stress and sleep disorder. *Experimental Neurobiology, 21*(4), 141–150.

Hróbjartsson, A., & Gøtzsche, P. C. (2010). Placebo interventions for all clinical conditions. *Cochrane Database of Systematic Reviews,* (1), CD003974.
Hsu, C. C., Weng, C. S., Liu, T. S., Tsai, Y. S., & Chang, Y. H. (2006). Effects of electrical acupuncture on acupoint BL15 evaluated in terms of heart rate variability, pulse rate variability and skin conductance response. *The American Journal of Chinese Medicine, 34,* 23–36.
Huang, W., Kutner, N., & Bliwise, D. L. (2009). A systematic review of the effects of acupuncture in treating insomnia. *Sleep Medicine Reviews, 13*(1), 73–104.
Huo, Z. J., Guo, J., & Li, D. (2013). Effects of acupuncture with meridian acupoints and three Anmian acupoints on insomnia and related depression and anxiety state. *Chinese Journal of Integrative Medicine, 19*(3), 187–191.
Jacob, T. C., Michels, G., Silayeva, L., Haydon, J., Succol, F., & Moss, S. J. (2012). Benzodiazepine treatment induces subtype-specific changes in GABA(A) receptor trafficking and decreases synaptic inhibition. *Proceedings of the National Academy of Sciences of the United States of America, 109*(45), 18595–18600.
Jiang, B., Ma, Z. H., & Zuo, F. (2010). Auricular acupuncture for insomnia: A randomized controlled trial. *Zhonghua Liu Xing Bing Xue Za Zhi, 31*(12), 1400–1402.
Ju, Y. L., Chi, X., & Liu, J. X. (2009). Forty cases of insomnia treated by suspended moxibustion at Baihui (GV 20). *Journal of Traditional Chinese Medicine, 29*(2), 95–96.
Kubota, T., Hirota, K., Yoshida, H., Takahashi, S., Anzawa, N., Ohkawa, H., et al. (1999). Effects of sedatives on noradrenaline release from the medial prefrontal cortex in rats. *Psychopharmacology, 146,* 335–338.
Kung, Y. Y., Yang, C. C., Chiu, J. H., & Kuo, T. B. (2011). The relationship of subjective sleep quality and cardiac autonomic nervous system in postmenopausal women with insomnia under auricular acupressure. *Menopause, 18*(6), 638–645.
Lee, S. Y., Baek, Y. H., Park, S. U., Moon, S. K., Park, J. M., Kim, Y. S., et al. (2009). Intradermal acupuncture on shen-men and nei-kuan acupoints improves insomnia in stroke patients by reducing the sympathetic nervous activity: A randomized clinical trial. *The American Journal of Chinese Medicine, 37*(6), 1013–1021.
Lee, M. S., Shin, B. C., Suen, L. K., Park, T. Y., & Ernst, E. (2008). Auricular acupuncture for insomnia: A systematic review. *International Journal of Clinical Practice, 62*(11), 1744–1752.
Lee, B. H., Zhao, R. J., Moon, J. Y., Yoon, S. S., Kim, J. A., An, H., et al. (2008). Differential involvement of GABA system in mediating behavioral and neurochemical effect of acupuncture in ethanol-withdrawn rats. *Neuroscience Letters, 443*(3), 213–217.
Leger, D., Massuel, M. A., & Metlaine, A. (2006). Professional correlates of insomnia. *Sleep, 29,* 171–178.
Lemoine, P., Garfinkel, D., Laudon, M., Nir, T., & Zisapel, N. (2011). This article has been cited by other articles in PMC. Prolonged-release melatonin for insomnia—An open-label long-term study of efficacy, safety, and withdrawal. *Therapeutics and Clinical Risk Management, 7,* 301–311.
Li, M., Hu, L., Cai, R. L., Wu, Z. J., & Wang, K. M. (2012). Effects of electroacupuncture at PC6 and BL15 on nerve electrical activity in spinal dorsal root and norepinephrine and dopamine contents in paraventricular nucleus of hypothalamus in rats with acute myocardial ischemia. *Zhong Xi Yi Jie He Xue Bao, 10*(8), 874–879.
Li, Q. Q., Shi, G. X., Xu, Q., Wang, J., Liu, C. Z., & Wang, L. P. (2013). Acupuncture effect and central autonomic regulation. *Evidence-Based Complementary and Alternative Medicine, 2013,* 267959.
Li, Y. J., Zhong, F., Yu, P., Han, J. S., Cui, C. L., & Wu, L. Z. (2011). Electroacupuncture treatment normalized sleep disturbance in morphine withdrawal rats. *Evidence-Based Complementary and Alternative Medicine, 2011,* 361054.
Lu, M., & Liu, X. (2008). Insomnia due to deficiency of both the heart and spleen treated by acupuncture-moxibustion and Chinese tuina. *Journal of Traditional Chinese Medicine, 28*(1), 10–12.

Luo, W. Z., Zhang, Q. Z., & Lai, X. S. (2010). Effect of acupuncture treatment of relieving depression and regulating mind on insomnia accompanied with depressive disorders. *Zhongguo Zhen Jiu*, *30*(11), 899–903.

Melchart, D., Streng, A., Hoppe, A., Brinkhaus, B., Witt, C., Wagenpfeil, S., et al. (2005). Acupuncture in patients with tension-type headache: Randomised controlled trial. *BMJ*, *331*, 376–382.

Mitchell, H. A., & Weinshenker, D. (2010). Good night and good luck: Norepinephrine in sleep pharmacology. *Biochemical Pharmacology*, *79*(6), 801–809.

NIH, (2005). NIH state-of-the-science conference statement on manifestations and management of chronic insomnia in adults. *NIH Consensus and State-of-the-Science Statements*, *22*(2), 1–30.

Nordio, M., & Romanelli, F. (2008). Efficacy of wrists overnight compression (HT 7 point) on insomniacs: Possible role of melatonin? *Minerva Medica*, *99*(6), 539–547.

Park, H. J., Park, H. J., Chae, Y., Kim, J. W., Lee, H., & Chung, J. H. (2011). Effect of acupuncture on hypothalamic-pituitary-adrenal system in maternal separation rats. *Cellular and Molecular Neurobiology*, *31*(8), 1123–1127.

Riemann, D., Klein, T., Rodenbeck, A., Feige, B., Horny, A., Hummel, R., et al. (2002). Nocturnal cortisol and melatonin secretion in primary insomnia. *Psychiatry Research*, *113*(1–2), 17–27.

Roth, T. (2007). Insomnia: Definition, prevalence, etiology, and consequences. *Journal of Clinical Sleep Medicine*, *3*(5), s7–s10.

Ruan, J. W. (2009). Clinical observation on acupuncture regulating "double vitality" for treatment of insomnia. *Zhongguo Zhen Jiu*, *29*(5), 371–373.

Ruan, J. W., Wang, C. H., Liao, X. X., Yan, Y. S., Hu, Y. H., Rao, Z. D., et al. (2009). Electroacupuncture treatment of chronic insomniacs. *Chinese Medical Journal*, *122*(23), 2869–2873.

Sjöling, M., Rolleri, M., & Englund, E. (2008). Auricular acupuncture versus sham acupuncture in the treatment of women who have insomnia. *Journal of Alternative and Complementary Medicine*, *14*(1), 39–46.

Spence, D. W., Kayumov, L., Chen, A., Lowe, A., Jain, U., Katzman, M. A., et al. (2004). Acupuncture increases nocturnal melatonin secretion and reduces insomnia and anxiety: A preliminary report. *The Journal of Neuropsychiatry and Clinical Neurosciences*, *16*(1), 19–28.

Stranges, S., Tigbe, W., Gómez-Olivé, F. X., Thorogood, M., & Kandala, N. B. (2012). Sleep problems: An emerging global epidemic? Findings from the in-depth WHO-SAGE study among more than 40,000 older adults from 8 countries across Africa and Asia. *Sleep*, *35*(8), 1173–1181.

Tang, S. C., Liu, J. M., & Liu, G. L. (2007). Clinical observation on effect of electric acupuncture at Sishencong in treating insomnia. *Zhongguo Zhong Xi Yi Jie He Za Zhi*. *27*(11), 1030–1032.

Tu, J. H., Chung, W. C., Yang, C. Y., & Tzeng, D. S. (2012). A comparison between acupuncture versus zolpidem in the treatment of primary insomnia. *Asian Journal of Psychiatry*, *5*(3), 231–235.

Vickers, A., & Zollman, C. (1999). ABC of complementary medicine. Acupuncture. *BMJ*, *319*, 973–976.

Walsh, J. K. (2004). Clinical and socioeconomic correlates of insomnia. *The Journal of Clinical Psychiatry*, *65*(Suppl. 8), 13–19.

Wang, J. D., Kuo, T. B. J., & Yang, C. C. H. (2002). An alternative method to enhance vagal activities and suppress sympathetic activities in humans. *Autonomic Neuroscience*, *100*, 90–95.

Wang, X. Y., Yuan, S. H., Yang, H. Y., Sun, Y. M., Cheng, F. P., Zhang, C. L., et al. (2008). Abdominal acupuncture for insomnia in women: A randomized controlled clinical trial. *Acupuncture & Electro-Therapeutics Research*, *33*(1–2), 33–41.

Wang, K. M., Liu, J., Wu, Z. J., Wang, Y. L., Chen, Y. N., He, L., et al. (2011). Relatively specific effect of electroacupuncture of different acupoints on hypothalamic monoamine neurotransmitters in myocardial ischemia rats. *Zhen Ci Yan Jiu*, *36*(3), 205–208.

Wei, Y. (2010). Clinical observation on acupoint catgut embedding at head-acupoint combined with massage of sole for treatment of refractory insomnia. *Zhongguo Zhen Jiu*, *30*(2), 117–120.

Winkelman, J. W., Buxton, O. M., Jensen, J. E., Benson, K. L., O'Connor, S. P., Wang, W., et al. (2008). Reduced brain GABA in primary insomnia: Preliminary data from 4T proton magnetic resonance spectroscopy (^{1}H-MRS). *Sleep*, *31*(11), 1499–1506.

Xiao, B. B., Luo, X. J., & Shen, Y. T. (2013). Efficacy observation on refractory insomnia treated with the balance needling therapy. *Zhongguo Zhen Jiu*, *33*(2), 101–104.

Xuan, Y. B., Guo, J., Wang, L. P., & Wu, X. (2007). Randomized and controlled study on effect of acupuncture on sleep quality in the patient of primary insomnia. *Zhongguo Zhen Jiu*, *27*(12), 886–888.

Yan, X. K., Zhang, Y., Yu, L., Yang, B., Chen, C., & Wang, F. C. (2010). Effect of "tranquilization needling" on the sleep quality in patients with insomnia of heart-spleen deficiency type. *Zhen Ci Yan Jiu*, *35*(3), 222–225.

Yao, H. F., Zhang, H. F., & Chen, X. L. (2012). Observation on therapeutic effect of scalp-acupoint catgut embedding for 33 cases of insomnia patients. *Zhen Ci Yan Jiu*, *37*(5), 394–397.

Yeung, W. F., Chung, K. F., Leung, Y. K., Zhang, S. P., & Law, A. C. (2009). Traditional needle acupuncture treatment for insomnia: A systematic review of randomized controlled trials. *Sleep Medicine*, *10*(7), 694–704.

Yeung, W. F., Chung, K. F., Poon, M. M., Ho, F. Y., Zhang, S. P., Zhang, Z. J., et al. (2012). Acupressure, reflexology, and auricular acupressure for insomnia: A systematic review of randomized controlled trials. *Sleep Medicine*, *13*(8), 971–984.

Yeung, W. F., Chung, K. F., Tso, K. C., Zhang, S. P., Zhang, Z. J., & Ho, L. M. (2011). Electroacupuncture for residual insomnia associated with major depressive disorder: A randomized controlled trial. *Sleep*, *34*(6), 807–815.

Yeung, W. F., Chung, K. F., Zhang, S. P., Yap, T. G., & Law, A. C. (2009). Electroacupuncture for primary insomnia: A randomized controlled trial. *Sleep*, *32*(8), 1039–1047.

Yoon, S. S., Kim, H., Choi, K. H., Lee, B. H., Lee, Y. K., Lim, S. C., et al. (2010). Acupuncture suppresses morphine self-administration through the GABA receptors. *Brain Research Bulletin*, *81*(6), 625–630.

You, W., Shi, Y. J., Han, Y. J., Jia, B. H., & Tu, Y. (2010). Effect of electroacupuncture of "Baihui" (GV 20)-"Yintang" (EX-HN 3) on the expression of glucocorticoid and glucocorticoid receptor mRNA of the chronic stress model rats. *Zhen Ci Yan Jiu*, *35*(4), 261–266.

Zhang, S., Li, G., Xu, X., Chang, M., Zhang, C., & Sun, F. (2011). Acupuncture to point Baihui prevents ischemia-induced functional impairment of cortical GABAergic neurons. *Journal of the Neurological Sciences*, *307*(1–2), 139–143.

Zhong, Z. G., Cai, H., Li, X. L., & Lü, D. (2008). Effect of acupuncture combined with massage of sole on sleeping quality of the patient with insomnia. *Zhongguo Zhen Jiu*, *28*(6), 411–413.

Zhou, Y. L., Gao, X. Y., Wang, P. Y., & Ren, S. (2012). Effect of acupuncture at different acupoints on expression of hypothalamic GABA and GABA(A) receptor proteins in insomnia rats. *Zhen Ci Yan Jiu*, *37*(4), 302–307.

Zhou, Z. L., Shi, X., Li, S. D., & Guan, L. (2010). Scalp penetration acupuncture for insomnia: A randomized controlled trial. *Zhong Xi Yi Jie He Xue Bao*, *8*(2), 126–130.

Zhou, Y., Wang, Y., Fang, Z., Xia, C., Liu, B., Chen, Q., et al. (1995). Influence of acupuncture on blood pressure, contents of NE, DA and 5-HT of SHR and the interrelation between blood pressure and whole blood viscosity. *Zhen Ci Yan Jiu*, *20*(3), 55–61.

Zollman, F. S., Larson, E. B., Wasek-Throm, L. K., Cyborski, C. M., & Bode, R. K. (2012). Acupuncture for treatment of insomnia in patients with traumatic brain injury: A pilot intervention study. *The Journal of Head Trauma Rehabilitation*, *27*(2), 135–142.

CHAPTER TWELVE

Acupuncture for the Treatment of Drug Addiction

Cai-Lian Cui[*,†,1], **Liu-Zhen Wu**[*,†], **Yi-jing Li**[*,†]
[*]Neuroscience Research Institute & Department of Neurobiology, School of Basic Medical Sciences, Peking University, Beijing, PR China
[†]Key Laboratory for Neuroscience, Ministry of Education /National Health and Family Planning Commission, Peking University, Beijing, PR China
[1]Corresponding author: e-mail address: clcui@bjmu.edu.cn

Contents

Abstract

Over the last four decades, there has been an increasing interest in acupuncture treatment of substance abuse around the world. Three important steps can be identified in this field. Dr. Wen of Hong Kong was the first (1972) to report that acupuncture at four body points and two ear points combined with electric stimulation can relieve opioid withdrawal signs in the addicts. The second major step was made by Dr. M. Smith in New York, the head of the National Acupuncture Detoxification Association (NADA) of the United States, who finalized a protocol (1985), using only ear points without electric stimulation for the treatment of cocaine dependence. The recent advance in this

International Review of Neurobiology, Volume 111
ISSN 0074-7742
http://dx.doi.org/10.1016/B978-0-12-411545-3.00012-2

field was made by Dr. Han of the Peking University, Beijing, who characterized a protocol (2005), using electric stimulation of identified frequencies on body points to ameliorate heroin withdrawal signs and reduce relapse of heroin use. In this chapter, the efficacy of acupuncture and related techniques for the treatment of drug dependence in experimental settings and clinical practice will be reviewed, and the possible mechanisms underlying this effect be discussed.

1. INTRODUCTION

Acupuncture is an essential part of traditional Chinese medicine, performed by inserting thin solid needles to specific documented points of the body. The needle can be manipulated manually (manual acupuncture) or by connecting it to an electric stimulator for "electroacupuncture" (EA). Acupuncture or EA stimulation typically elicits a composite of sensations termed *deqi*, which manifests as numbness, heaviness, distention, and soreness, believed to be a trait characteristic in achieving a therapeutic effect (Kong et al., 2005). The practice of manual acupuncture has accumulated affluent clinical experience for thousands of years in China. It spread to Japan in the sixth century, then to Europe in the seventeenth century. It is now practiced in 120 countries and regions and has become an important modality for treatment throughout the world (Shi et al., 2006). As we have known that acupuncture is most commonly used for its analgesic effects, however, its medical applications are by no means limited to pain treatment. Some 35 years ago, it was serendipitously observed that acupuncture could relieve opiate withdrawal signs in the addicts (Wen & Cheung, 1973). This finding has been given much attention by clinicians and research scientists around the world (D'Alberto, 2004; Margolin, 2003; McLellan, Grossman, Blaine, & Haverkos, 1993). Acupuncture is different from the other pharmacological treatments of drug addiction. For example, while methadone maintenance treatment is aiming at long-term replacement for heroin, acupuncture attempts to strengthen the endogenous opioid system and eventually get rid of the drug. Different types of protocols have been developed since then. In this chapter, we will briefly analyze the history of development of the technique, the site of stimulation (body or ear points), the mode of stimulation (needle insertion only, needle insertion plus manipulation, and electric stimulation via needles or via skin), the parameter of electric stimulation (frequency, pulse width, and intensity), the target substance of abuse (heroin and cocaine), and the stage of drug dependence (physical or psychic dependence) (Cui, Wu, & Luo, 2008). In order to compare the results with pharmacological treatments, the intervention was also evaluated including the dependence liability of the acupuncture per se.

2. DEVELOPMENT OF ACUPUNCTURE TREATMENT OF DRUG ADDICTION

Although acupuncture has been used in China for thousands of years, acupuncture for the treatment of addiction in the modern era began with an incidental discovery by Wen, a neurosurgeon in Hong Kong (Wen, 1975). In early November 1972, a 50-year-old man with brain concussion was admitted to the neurosurgical unit of the Kwong Wah Hospital. Knowing that the patient is an opium addict of 5 years' duration, he was given tincture of opium to relieve his withdrawal syndrome in the ward. After the cerebral concussion had improved, the patient was asked whether he would agree to do cingulotomy to relieve his drug abuse problem. He agreed. During the operation for surgery, instead of local anesthesia being injected under the scalp, acupuncture anesthesia (analgesia) was used. Four needles were inserted into the right hand (IL-4 and SI-3) and in the arm (EH-4 and TB-9), and another two needles into the right ear (brainstem and shenmen). Stimulation with an electric stimulator was carried out for half an hour. During stimulation, the patient voluntarily stated that his withdrawal signs are completely gone. The operation was canceled and the patient returned to the ward. In the evening, the patient had another withdrawal syndrome. Acupuncture and electric stimulation were carried out again, and the withdrawal symptoms disappeared. Wen and Cheung pursued this unexpected finding and conducted a series of studies examining the effect of acupuncture for heroin addiction and concluded that acupuncture did relieve heroin withdrawal syndrome (Wen, 1977, 1979).

In 1974, Wen and Cheung's acupuncture protocol described earlier was introduced into the Lincoln Recovery Center in Bronx, New York. Guided by clinical observation, an acupuncture treatment protocol evolved with two modifications: (1) Acupuncture was applied only at the ear, not at the body. Five needles were inserted bilaterally into the outer ear or auricle at points termed *kidney*, *liver*, *lung*, *shenmen*, and *sympathetic*. (2) Since they found that electric stimulation of the needles did not seem to increase treatment effectiveness, it was given up. The protocol was apparently effective across different substances of abuse. In 1985, under the guidance of Dr. Michael Smith, the Director of the Detoxification Center, the National

Acupuncture Detoxification Association (NADA) was formed and the five-point auricular protocol—now known widely as the "NADA protocol"—was codified. An extensive training procedure was established. Individuals who receive the 70 h NADA training are known as "acupuncture detoxification specialists," to distinguish them from acupuncturists who have undergone multiyear training in order to perform full-body acupuncture. The NADA protocol was later adopted in many clinical settings in Western countries including residential programs, acute detoxification facilities, and outpatient programs (Margolin, 2003).

As mentioned earlier, acupuncture was originally developed in the 1970s as a treatment for opiate withdrawal syndrome; there have been no recent studies of acupuncture for opiate addiction in the United States and in Europe, perhaps due to the fact that cocaine addiction has excessively surpassed that of heroin and no effective pharmacological treatment is available for it. The following are some results and evaluations for the effectiveness of auricular acupuncture treatment for cocaine/crack.

While the NADA protocol was primarily aimed at detoxification of drug addicts by acupuncture that seemed to facilitate compliance with a drug-free treatment program (Konefal, Duncan, & Clemence, 1994), Shwartz, Saitz, Mulvey, and Brannigan (1999) observed the withdrawal syndrome in the cocaine or heroin addicts within the outpatient detox programs that include acupuncture or short-term residential programs. The result from the programs showed that acupuncture detoxification programs are a useful component of a substance abuse treatment system, and the risk of relapse was lower in patients with access to acupuncture.

Subsequently, in a controlled study conducted at Yale Medical School (Avants, Margolin, Holford, & Kosten, 2000), 82 patients addicted to cocaine and maintained on methadone were randomized to one of the following treatment conditions: auricular acupuncture (a four-point subset of the NADA protocol), needle insertion control (four needles inserted into the rim or helix of the auricle), and relaxation control ("nature" videos with relaxing music). The treatments were provided five times per week for 8 weeks. During the study, patients attended a weekly coping skills group and received standard care in the methadone program (methadone plus weekly individual counseling). The main outcome, cocaine use, was assessed by urine samples three times per week. Treatment credibility and therapeutic alliance were also assessed. Overall, 52 of 82 patients completed the study: 13 in the acupuncture condition, 17 in the needle insertion control condition, and 22 in the relaxation control. Analysis of urine screens showed that patients assigned to the NADA acupuncture condition used significantly less cocaine over the course of the study compared with the two control conditions. In the last week of treatment, patients receiving the NADA protocol were significantly more likely to provide three cocaine-free urine screens (54%) compared with the needle insertion control group (24%) and the relaxation group (9%). However, one interpretation of these findings was that less motivated patients dropped out of "true" acupuncture condition but were retained in the other conditions.

The Cocaine Alternative Treatment Study (CATS) (Margolin et al., 2002) was a large-scale, multisite study based on a design similar to the Yale study. In this study, 620 patients addicted to cocaine were enrolled from 6 treatment sites across the country: 412 of the patients were "primary" cocaine-dependent, and 208 were also opiate-dependent and maintained on methadone. Patients were randomized to the same three treatment conditions used in the previous Yale study: true acupuncture, helix needle insertion control (sham acupuncture), and relaxation control for a treatment period lasting for 8 weeks. Individual counseling was offered to patients at the primary cocaine sites, and the patients maintained on methadone received standard care as offered in their methadone program. The primary outcome measure was cocaine use as assessed by urine screens three times per week. Results showed that, unlike the previous study (Avants et al., 2000), there were no differences in outcome among any of the three treatment conditions, overall or at any of the six sites during the study or at a 3- or 6-month follow-up. In the last week of treatment, 24%, 31%, and 29% of patients in the true acupuncture, sham acupuncture, and relaxation control conditions, respectively, were abstinent from cocaine. This large study seemed to provide negative evidence regarding acupuncture's efficacy for the treatment of cocaine addiction. Margolin, Avants, and Holfordnts, and Holford (2002) reported after doing further investigation that treatment context may influence the outcome. As for the mechanism of action of the auricular acupuncture on treating cocaine addiction, there were just a few reports (Kosten, Rounsaville, Babor, Spitzer, & Williams, 1987; Oleson, 2001), showing that stimulation of the auricular branch of the vagus nerve would cause release of endogenous opioids and other neuropeptides (Wen, Ho, Ling, Mehal, & Ng, 1980).

Although a review summarized by D'Alberto in 2004 stated that the previous trials based on the NADA protocol could not confirm the effectiveness of acupuncture on treatment of cocaine abuse (D'Alberto, 2004), drug abuse was still kept in the record of World Health Organization (WHO) as one of the 42 medical problems suitable for acupuncture treatment. The major putative advantages are inexpensive, simple, and without side effects (Brumbaugh, 1993).

3. ACUPUNCTURE TREATMENT OF DRUG ADDICTION IN PEKING UNIVERSITY

Large-scale research on acupuncture analgesia in China was triggered by the clinical application of "acupuncture anesthesia," when a group of medical professionals started to use acupuncture for the prevention of surgically induced pain in the late 1950s. In 1965, Dr. Han of the Department of Physiology, Beijing Medical University, and his colleagues were involved in the study of the mechanisms underlying the acupuncture anesthesia. After about 30 years of research, it was made clear that acupuncture (both manual and electroacupuncture) can accelerate the production and release of opioid peptides in the central nervous system (CNS) to induce an antinociceptive effect. This effect was frequency-dependent, that is, low-frequency (2 Hz) stimulation accelerated the production of enkephalin and endorphin, whereas high-frequency (100 Hz) upregulated the level of dynorphin (Han, 2003, 2004). It was in the year of 1990 that China started to report cases of drug addiction. If acupuncture can accelerate the release of endorphins, it is natural to consider why not make use of acupuncture to relieve withdrawal syndrome of opiate addicts during drug abstinence. This thought was put into action in rats made dependent on morphine and in humans.

3.1. Acupuncture for the relief of withdrawal syndrome: Rat experiment

3.1.1 Immediate effect induced by single EA treatment

Rats were made dependent to morphine by repeated injections of morphine for 5 days. An injection of the opioid receptor antagonist naloxone would precipitate a serious episode of withdrawal syndrome. The original perspective was that 2 Hz would be more effective than 100 Hz in suppressing withdrawal syndrome, if the effect of EA was to accelerate the release of morphine-like opioid peptides (enkephalin and endorphin) to replace morphine, thus ameliorating abstinence syndrome. Surprisingly, the results showed that 2 Hz EA was only marginally effective in reducing withdrawal syndrome in two of five signs (Han & Zhang, 1993), whereas 100 Hz EA produced a dramatic suppression of all five withdrawal signs. In other words, 100 Hz EA was far more effective than 2 Hz EA in suppressing withdrawal syndrome (Han & Zhang, 1993; Wu, Cui, Tian, Ji, & Han, 1999). Literature survey revealed that dynorphin can suppress the withdrawal syndrome in heroin-dependent humans (Wen & Ho, 1982) and in morphine-dependent animals via the κ-opioid receptors and the site of action is in the spinal cord (Green & Lee, 1988). Indeed, the naloxone-precipitated withdrawal syndrome can be suppressed by spinal intrathecal administration of a κ-opioid receptor agonist U-50488, whereas the κ-opioid antagonist nor-BNI elicited a dose-dependent augmentation of naloxone-precipitated withdrawal (Cui, Wu, & Han, 2000). The latter result implies that an endogenous κ-agonist, most probably dynorphin, exerts a tonic suppressive effect on morphine withdrawal syndrome at the spinal level.

3.1.2 Cumulative effect of multiple treatments with EA

Multiple sessions of EA within a day are more effective in relieving withdrawal syndrome compared with single session of EA. Wang et al. (2011) further investigated the role of dynorphin and κ-opioid receptor in cumulative effect of 100 Hz EA. Results showed that (1) 100 Hz EA administered 12–24 h after the last morphine injection suppressed the withdrawal syndrome in rats, multiple sessions of EA were more effective than single session. This effect still could be observed even 7 days after abstinence. (2) A downregulation of preprodynorphin (PPD) mRNA level was observed in spinal cord, periaqueductal gray (PAG), and hypothalamus 60 h after the last morphine injection, which could be reversed by multiple sessions, but not a single session of EA. (3) Accompanied with the decrease of PPD mRNA level, there was an upregulation of p-CREB in the three CNS regions, which was abolished by 100 Hz EA treatment. The findings suggest that downregulation of p-CREB and acceleration of dynorphin synthesis in the spinal cord, PAG, and hypothalamus may be implicated in the cumulative effect of multiple 100 Hz EA treatment for opioid detoxification.

Dopamine (DA) neurons in ventral tegmental area (VTA) have been broadly implicated in the initiation and development of drug addiction. A hypofunction of the VTA DA neurons is thought to contribute to the negative consequences of acute and protracted opiate withdrawal (Nestler, 1997, 2004). Therefore, we also investigated the effect of EA on the neuronal morphology and function of VTA DA neurons. In rats that received chronic morphine treatment followed by spontaneous withdrawal for 14 days, the size of DA neurons in the VTA was decreased and the ultrastructure of the neurons here was altered (Chu et al., 2008, 2007).

Meanwhile, intravenous injection of 1 mg/kg morphine to the rats showed no significant changes on firing rate and burst firing of the VTA DA neurons (Hu et al., 2009). 100 Hz EA, while suppressing morphine withdrawal, could facilitate the recovery of the VTA dopaminergic cells, normalize the reactivity of the VTA DA neurons to morphine, and upregulate the BDNF protein level in the VTA. These results implied that the detoxification effect of 100 Hz EA may be associated with the activation of endogenous BDNF (Chu et al., 2008, 2007). Nevertheless, 0.5 mg/kg of morphine failed to induce CPP in chronic morphine-withdrawn rats, suggesting a diminished sensitivity in response to morphine-induced reinforcing behavior. Repeated 100 Hz EA could recover the sensitivity of morphine-induced CPP in morphine-withdrawn rats (Hu et al., 2009).

In addition to drug-induced neural impairment, chronic opioid administration could result in immune suppression (Odunayo, Dodam, Kerl, & DeClue, 2010; Wang, Barke, Ma, Charboneau, & Roy, 2008) and increased susceptibility to virus (Macias et al., 2008; Rivera-Amill, Silverstein, Noel, Kumar, & Kumar, 2010) and bacteria (Wang et al., 2008) in heroin addicts. If acupuncture could normalize the immune dysfunction after chronic drug exposure, that would help the addicts recover from the detoxication. Li et al. found that chronic morphine-induced decrease of splenic T lymphocyte proliferation and IL-2 production can be significantly raised by 5 days of 2 Hz EA treatment (once per day) and the fluctuation of CD4+/CD8+ ratio is also run to the baseline level by the EA (Li, Zhang, Cui, Han, & Wu, 2011).

Additionally, sleep disturbance is a significant symptom of acute and protracted withdrawal syndrome. It is positively correlated with increased potential of relapse to heroin abuse. Individuals with shorter sleep time seem more likely to leave treatment earlier (Beswick et al., 2003). In fact, transcutaneous electric acupoint stimulation (TEAS) was reported to show an immediate hypnotic effect during the treatment in one-third of the heroin addicts (Wu, Cui, & Han, 1995). The animal studies showed that nonrapid eye movement (NREM) sleep, rapid eye movement (REM) sleep, and total sleep time decreased dramatically, while the sleep latency prolonged significantly during acute morphine withdrawal. Both 2 and 100 Hz EA produced a significant increase in NREM sleep, REM sleep, and total sleep time (Li, Zhong, et al., 2011). If the problem of sleep disturbance is resolved, the risk of relapse may be reduced. Therefore, the improvement of sleep profile during morphine withdrawal by 2 and 100 Hz EA may provide useful information for the design of appropriate care for opiate addicts in drug abstinence.

3.2. Acupuncture for the relief of withdrawal syndrome: Human observation

Encouraged by the results obtained in rats, acupuncture was used in heroin addicts for detoxification. A technical problem was that the drug addicts subjected to withdrawal were very irritable and it was inconvenient for precise needle insertion. Moreover, the needle inserted into the point was sometimes broken by excessive agitation. The metallic needle was thus replaced by self-sticking electrodes placed at the skin over the acupoint. Experimental data showed that this TEAS was at least as effective as, if not more effective than, EA (Wang, Mao, & Han, 1992). The device used for this purpose was named Han's acupoint nerve stimulator.

3.2.1 Immediate effect induced by single TEAS treatment

For acupoint stimulation, one pair of skin electrodes (29 × 29 mm) was placed on the Hegu point (LI-4, at the dorsum of the hand) and the other at the palmar side on Laogong point (P-8, opposite to LI-4), to complete an electric circuit. Another pair of electrodes were placed on Neiguan (P-6, located at the palmar side of the forearm, 2 in. proximal to the palmar groove, between the tendons of the palmaris longus and flexor carpi radialis) and Waiguan (TE-5, on the dorsal surface of the forearm opposite the P-6) point to complete a circuit (Figure 12.1). These points are known to calm down the heart rate according to the theories of traditional Chinese medicine. A "dense-and-disperse, DD" mode of stimulation was administered, in which 2 Hz stimulation alternated automatically with 100 Hz stimulation, each lasting for 3 s. This mode of stimulation could release all four kinds

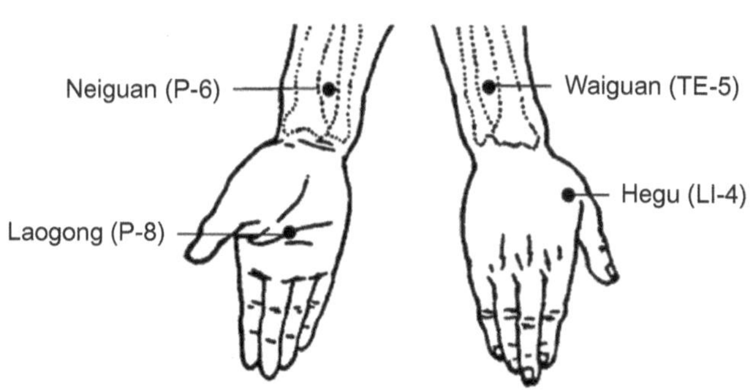

Figure 12.1 Sketch map indicating the location of acupoints: LI-4 (at the dorsum of the hand on the thenar eminence) and P-6 (at the palmar side of the forearm, 2 in. proximal to the palmar groove, between the tendons of the palmaris longus and flexor carpi radialis) for the application of TEAS stimulation.

of opioid peptides in the CNS (Han & Wang, 1992), producing a maximal opioidergic effect. The control group received the same treatment of placing the skin electrodes on site, except that the electrodes were disconnected from the electric circuitry. One of the most prominent signs of withdrawal was tachycardia. The average heart rate of the patients during opioid withdrawal was 109 beats per minute (BPM) before treatment. The DD mode stimulation produced an immediate suppression of the tachycardia within 5–10 min, reaching 90 BPM at the end of 30 min. This effect is statistically very significant, even though it is short-lasting, and returns to its original level in another 30 min (Wu, Cui, & Han, 1996).

3.2.2 *Cumulative effect of multiple treatments with TEAS (Wu et al., 1996)*

In another study, the same treatment was given 30 min a day for 10 consecutive days. Heart rate was measured with an electrocardiogram before (PRE) and immediately after (POST) the TEAS stimulation. The difference between the PRE and the POST of the same day indicated the immediate effect, while the decay of the PRE over the whole period of observation showed the long-lasting effect. It took 4 days for the heart rate of the patients in TEAS group to return to the normal level. In the control group receiving mock TEAS, heart rate did not come down to a level of 100 BPM until 8 days after the starting of the "treatment" (Han & Cui, 2011). The results suggest that repeated daily TEAS treatment is effective in reducing the tachycardia of heroin withdrawal syndrome, with an effective order of DD > 100 Hz > 2 Hz.

Another objective parameter for measuring the severity of heroin withdrawal is the body weight. The heroin addicts recruited in this study were 17–35 years old; their average body weight was 49–51 kg. In the control group receiving mock TEAS, body weight remained unchanged in the 10-day observation period, whereas a net increase of 5 kg was recorded in the TEAS-treated groups. This increase of body weight (~10%) is apparently due to the reduction of the withdrawal syndrome, which leads to an increase of food and water intake.

To obtain a quantitative estimate of the effect of TEAS, the following protocol was established in another clinical setup, with a total observation period of 14 days (Wu, Cui, & Han, 1999): (a) TEAS of 2/100 Hz, 30 min per session, was used three to four times a day on the first 2 days, then two to three times a day on days 3–7, and twice a day on days 8–14. (b) A four-channel TEAS device was used instead of two-channel device. The acupoints used were Hegu/Laogong (LI-4/P-8) for left (or right) hand, Neiguan/Waiguan (P-6/TE-5) for right (or left) arm, and Xingjian/Sanyinjiao (LV-2/SP-6) for both legs. (c) The intensity of stimulation was twice the threshold (4–6 mA being the threshold value [*T*]) for the first 2 days and 2.5–3.0 *T* thereafter. (d) Buprenorphine (Buprenex) i. m. was ordered *p.r.n.* Twenty-eight heroin addicts were randomly divided into two groups, receiving buprenorphine only or TEAS plus buprenorphine. In the buprenorphine-only group, the total dose of buprenorphine requested in 14 days averaged to 12.91 ± 1.34 mg, whereas that of the TEAS plus buprenorphine group averaged only to 1.01 ± 0.09 mg, which was consumed only in the first 5 days. In other words, the total amount of buprenorphine used in the TEAS group was only 7.8% of that needed in the pure buprenorphine group, showing a reduction of more than 90%. This can be taken as a quantitative estimate of the effect of TEAS on opioid detoxification. Similar observations were made in another group of heroin addicts using methadone (Wu, Cui, & Han, 2001) instead of buprenorphine. The total dose of methadone used in the control group averaged to 202 ± 15 mg, whereas that in the TEAS plus methadone group was only 50.5 ± 8.2 mg, with a 75% reduction ($p < 0.001$).

3.3. Acupuncture for suppression of psychic dependence: Animal studies

It is well known that drug addiction is a chronic, recurrent condition with a high rate of relapse even after prolonged drug-free periods. According to statistics from different sources, the relapse rate can be as high as 95–99%. Therefore, the final goal for treating drug addiction is not just detoxification (relieving or curing the withdrawal syndrome) but to relieve craving (psychic dependence) and to remove the compulsive behavior of drug seeking and taking after the detoxification. There are several animal models (Markou et al., 1993) that can be used to study the problem of craving and relapse to drugs of abuse. Conditioned place preference (CPP) is one of the frequently used models. In a two-chamber or three-chamber experimental apparatus, the drug (unconditioned stimulus) is injected to the animal in one of the chambers and normal saline to the other chamber. After repeated training, the rat will choose to stay longer on the drug-associated side than in a chamber associated with normal saline. The ratio between the time spent in the drug-associated and the saline-associated side can be taken as an index for the degree of "craving." Using this model, experiments were conducted to test whether acupuncture can suppress the expression of morphine-induced

CPP. Wang, Luo, Xia, and Han, Xia, and Han (2000) first observed the effect of EA at 2, 100 or 2/100 Hz (DD) for 30 min, with intensity increasing stepwise from 1 mA to 2 mA and to 3 mA within 30 min, on morphine CPP expression in rats. The result showed that the CPP was significantly suppressed by EA of 2 and 2/100 Hz, but not of 100 Hz. Three control groups of rats received one of the following: (a) restraining in the holder for 30 min, (b) holder restraining plus needle insertion without electric stimulation, or (c) intermittent electric stimulation on the feet (foot shock). None of the three control groups showed any suppression of the CPP. The results suggest that it is the low-frequency component of the EA that suppressed the CPP expression. Since the effect of 2 Hz EA could be completely reversed by the systemic injection of opioid receptor antagonist naloxone at a small dose of 1 mg/kg, which is sufficient to block the μ- and δ-receptors, but not the κ-receptors, it seems evident that the effect of EA is mediated by endogenously released μ- and δ-opioid agonists, most likely endorphins and enkephalins, to ease "craving" for exogenous opioids (in this case, the morphine). Further studies showed that (1) a single session of 2 Hz EA produced a significant increase of the content of enkephalin in the nucleus accumbens (NAc) of morphine-induced CPP rats and this effect was stronger in three consecutive sessions of EA; (2) intracerebroventricular injection of the μ-opioid receptor antagonist CTAP or δ-opioid receptor antagonist NTI, but not κ-opioid receptor antagonist nor-BNI, dose-dependently reversed the inhibitory effects of 2 Hz EA on the expression of morphine-induced CPP; (3) three consecutive sessions of 2 Hz EA upregulated the mRNA level of preproenkephalin in the NAc of morphine-induced CPP rats. The results suggest that the inhibitory effects of 2 Hz EA on the expression of the morphine CPP are mediated by μ- and δ-opioid receptors, but not κ-opioid receptor, possibly via accelerating both the release and synthesis of enkephalin in the NAc (Liang et al., 2010). Considering the close interaction between opioids and DA systems in the central nervous system (Di Chiara & North, 1992; Shalev, Grimm, & Shaham, 2002), it would be interesting to examine whether the DA system participates in the inhibitory effect of EA on morphine-induced CPP. While the content of DA in the NAc was significantly increased in rats showing a profound expression of CPP (Ma et al., 2009; Ma, Shi, Han, & Cui, 2008), it dropped to the baseline level when the CPP was blocked by 2 Hz EA. Accompanying the decrease in the content of DA, there was a simultaneous decrease of its metabolites DOPAC and HVA down to the baseline level, suggesting a total quenching of activities of the mesolimbic dopaminergic system (MLDS) (Ma et al., 2008). These results suggest that MLDS seems to play important roles in the mechanisms underlying EA's suppression of the psychic dependence in rat. Another issue deserving attention is that the effect of EA could still be revealed 12 h after the episode of EA, suggesting that this effect lasts longer than acupuncture-induced analgesia (Han, 2003), which usually disappears within 60 min after the end of stimulation.

In practical life, craving and relapse can be easily induced by stress or by a very small dose of opioids. This phenomenon can be reproduced in animals using the CPP model. Wang et al. (2000) reported that morphine-induced CPP disappeared after a 9-day extinction period. The extinguished CPP could be easily reinstated by foot shock stress or by a small dose of morphine or amphetamine (Wang, Luo, Zhang, & Han, 2000). It is worth mentioning that while both drug priming and foot shock stress can reactivate morphine CPP, the underlying mechanisms may be different. The drug priming-induced reactivation could be totally blocked by the destruction of the mesolimbic DA system, including the VTA and the shell part of the NAc, while the foot shock-induced reactivation of CPP depended on the integrity of the central nucleus of amygdala (Wang, Luo, Ge, Fu, & Han, 2002).

It was interesting to find that while a single session of 100 Hz EA was ineffective to block morphine CPP (Wang et al., 2000), multiple treatments of 100 Hz EA (once a day for 3 days) became effective (Shi et al., 2003). This effect can be blocked by the δ- and κ-opioid antagonists, but not the μ-opioid antagonists. Both 2 and 100 Hz EA (once a day for 3 days) could also inhibit the CPP reinstated by drug priming (Shi et al., 2004).

In a further study, Chen, Liang, Wang, Han, and Cui (2005) confirmed that the efficacy of EA to suppress morphine-induced CPP depends not only on the frequency of EA (2 Hz better than 100 Hz) but also on the total session of EA being administered (5 times > 3 times > single session). This may be related with the degree of activation of the genes encoding opioid peptides. By measuring the mRNA level of preproenkephalin (PPE) and PPD in the NAc of morphine CCP rats, it was found that 2 and 100 Hz EA can selectively elevate the level of PPE- and PPD-mNRA, respectively (Shi et al., 2004).

3.4. Acupuncture for suppression of psychic dependence: Human studies (Wang, Zhang, Ge, & Luo, 2003)

In a clinical trial using TEAS for the suppression of opiate craving in humans, a total of 117 heroin addicts who had completed the process of detoxification for more than 1 month were recruited. They were randomly and evenly assigned into four groups. Three groups received TEAS treatment of different frequencies (2, 100, and 2/100 Hz). Self-sticking skin electrodes were placed on four acupoints: Hegu and Laogong (palmar side of the Hegu point) in the left (or right) hand to complete a circuit and Neiguan and Weiguan in the opposite arm to complete a circuit. The control group was processed as in the previous groups except that the intensity was minimal (15 Hz, threshold stimulation for 3 min, and then switched to 1 mA thereafter) to serve as a mock TEAS control. A visual analogue scale (VAS) was used to assess the degree of craving. There was a very slow decline of the VAS in the mock TEAS control group in a period of 1 month. A dramatic decline of the degree of craving was observed in the groups receiving 2 and 2/100 Hz electric stimulation, but not in the group receiving 100 Hz stimulation. In summary, the results observed in humans were in line with the findings obtained in the rat: low-frequency TEAS is more effective than high-frequency TEAS in suppressing the morphine-induced CPP (Wang et al., 2003).

Meanwhile, a local rehabilitation center under the auspices of the Peking University Neuroscience Research Institute was established in Hainan province. The staffs kept in close contact with all the previous drug addicts discharged from the detoxification center. They could get TEAS treatment from the rehabilitation center for free and *ad libitum*. As an alternative, they could buy a unit of the device at an affordable cost and apply TEAS at home under the staff's continuing supervision. A follow-up study was conducted on a group of 56 subjects who used TEAS at home. Using monthly urine test as an outcome criterion, the relapse rates at 3rd, 6th, 9th, and 12th months were 50.0%, 71.4%, 80.4%, and 83.9%, respectively. Those showing negative urine tests for 12 or 24 consecutive months were given a naloxone test (0.4 mg subcutaneously × 2 at 15 min interval) to further confirm their heroin-free status. Compared with the 94% relapse rate at 6 months and more than 98% relapse rate at 1 year in the majority of reports on heroin addiction (without methadone maintenance) (Sun, Ye, & Qin, 2001), an 83.9% relapse rate (16.1% success rate) at 1 year in the present investigation is encouraging (Han, Trachtenberg, & Lowinson, 2005).

Similar studies (Han, Wu, & Cui, 2003) were performed in treatment centers located in Zhanjiang county, Guangdong province, and in Shanghai, the 1-year success rate was between 20% and 30%, respectively, depending on the degree of follow-up psychological and medical care offered to the patients. Obviously, EA or TEAS has emerged as a highly hopeful foreground on the treatment for opioid addiction. A well-designed, multicenter clinical trial is warranted in the future. Whether TEAS is effective in reducing the craving for other drugs of abuse, such as cocaine, and other substance of abuse, like alcohol and cigarette smoking, deserves further investigation.

4. SOME TECHNICAL COMMENTS

4.1. Ear acupuncture versus body acupuncture

Although the ear is not included in the area of distribution of the classical 14 meridian system of the ancient Chinese medicine, it is certainly worthwhile to try ear acupuncture in medical practice. However, there is at least no evidence to limit the site of acupuncture only at the ear, and not at the body sites. In fact, the original protocol of Wen and Cheung (1973) used four body points and two ear points for treatment. In the United States, a person who received 70 h of NADA ear protocol training obtained a mid-level title of "acu-detox specialist" allowed only to administer the ear protocol (Margolin, 2003). It would be difficult for them to treat other complaints with conventional body acupuncture. In fact, the vast possibility of using body acupuncture should not be neglected.

4.2. Needle staying versus mechanical or electric stimulation

The original protocol of Wen used acupuncture combined with electric stimulation (Wen & Cheung, 1973), now called EA, whereas the NADA protocol preferred to leave the needle in place without further stimulation. It is true that the ear concha, composed of cartilage and skin, is very sensitive to pain even when the needle is remained untouched. However, an early study conducted in 1973 in Shanghai had revealed that in order to produce a significant analgesic effect, manual manipulation or electric stimulation applied on the needle is indispensable (The effect of ear acupuncture on the pain threshold of the skin at thoracic and abdominal region. Theoretical study on acupuncture anesthesia, 1973).

4.3. Acupuncture and EA versus TEAS

A series of study showed that the manipulation of the needle triggers a train of nerve impulses transmitted along the afferent nerve to the CNS. Injection of local anesthetics deep into the acupoint or along the nerve trunk would block the effect of acupuncture (Effect of acupuncture on pain threshold of human skin, 1973). If nerve activation accounts for the transmission of the acupuncture signals, then similar effects should be produced whether the nerve impulses are generated (1) by manipulation of the needle, (2) by electric stimulation via the needle, or (3) by electrodes placed on the skin over the point that forces the current to pass through the same underlying tissue and producing a sense of *deqi*. In a rat experiment, the analgesic effect of EA and TEAS was compared. There was no significant difference between them, whether on the efficacy of antinociception or the underlying mechanisms (Wang et al., 1992).

Compared to manual needling, EA or TEAS has the advantage of being easy to master with high reproducibility. To take an example, it is now evident that the frequency is a very important and precise parameter of the electric stimulation: high-frequency (100 Hz) stimulation is best for reducing the physical dependence, low-frequency (2 Hz) for the psychic dependence, and DD for both in opioid addicts. Another example is that while 2 Hz EA is more effective than 100 Hz EA to suppress morphine-induced CPP in rats, the reverse is true when EA is used to suppress cocaine-induced CPP (Ren et al., 2002). Besides, it may take years to master the sophisticated technique of manual needling, according to traditional Chinese medicine.

Compared to EA, the advantages of using the transcutaneous stimulation for the treatment of drug addiction are at least twofold: (1) Treatments can be performed at home by the patient under the auspice of the physician, thus saving time and effort to be treated in the hospital; (2) the possibility of cross infection through needles is reduced from minimal to zero.

4.4. How frequently should acupuncture be used for the treatment of drug dependence?

Although the use of single session of EA or TEAS is effective in reducing opiate withdrawal, it is recommended to use it three times a day in the first 5 days, two times a day in the next 5 days, and at least once a day for the rest of the time for a total of 2 weeks. The session applied immediately before sleep is critical since this would facilitate a good sleep (Zhong, Wu, & Han, 2006).

4.5. Treatment for physical versus psychic dependence

The effect of acupuncture for reducing withdrawal syndrome is easy to observe and straightforward as it was described by Wen in his first incidental experience (Wen & Cheung, 1973). The most important advantage of TEAS resides on its ability to suppress craving and hence to reduce the chance of relapse, since there is so far no effective means to prevent relapse without introducing another drug of addition, such as methadone. The drug addicts always confess that they have no confidence on themselves. What is important to them is not the past, nor the future, but the present. They often asked the question of "Can you solve my problem without taking the drug?" Once they feel that acupuncture or TEAS can resolve, or partially resolve their problem in 30 min without taking an addictive drug, they will start to develop confidence. After several episodes of effective protection, the confidence and loyalties to this nonpharmacological treatment become increasingly evident, and the compliance to acupuncture is built up quickly. In other words, they start to seek help not from outside, but from within their own brain (increased release of endorphins). This is one of the most important psychological perspectives for the treatment of drug abuse with acupuncture.

4.6. Unexpected physiologic changes that help to build up psychological confidence

Two unexpected changes in the body function were found. One is the disappearance of injection marks within 2 weeks of TEAS treatment (Wu et al., 1999), and the other is the recovery of the depressed sexual function (Wu, Cui, & Han, 2000). These changes provided with positive psychological impact and confidence and would certainly be helpful in the establishment of their normal life, thus delaying the occurrence of relapse to drug. The underlying mechanisms for the previously mentioned efficacies of TEAS have not yet been clarified.

4.7. Does EA or TEAS per se induce dependence?

If EA can accelerate the release of endogenous endorphins to relief craving, a natural question would be "Does EA itself produce dependence?" Chen et al. (2005) used CPP model to clarify this question. Twenty-eight rats were randomly divided into two groups. The EA group was confined in the holder for EA stimulation for 30 min and then immediately put into one of the chambers for 45 min. The control group was treated in the same way except with no electric stimulation administered to the needles and then put into another chamber. The treatment was executed in 4 consecutive days. In the fifth day, the rats were put into the midway chamber to test their preference between the two chambers. Most of the rats stayed longer in the chamber paired with EA ($p < 0.05$), indicating that they seemed to like the experience of EA. However, to the best of our knowledge, there is no single report showing dependence to EA or TEAS. Therefore, the risk of producing dependence to EA or TEAS appears to be minority.

We had known that CPP involves the mechanisms of learning and memory. Theoretically, a blockade of the expression of CPP may be caused by the deterioration of the memory. To assess whether EA could affect learning and memory, Chen et al. (2005) performed the Morris water maze (MWM) experiment to observe the effect of EA on the consolidation of spatial memory. EA was administered after the completion of MWM training for 7 days. On days 8–12, the rats were given EA or restraint once a day for 5 days, with the blank control group leaving in the home cage in the same time period. Results showed that in the control group and restraint group, there was a marked prolongation of the escape latency, implying a natural fading of memory with time. In the EA group, no significant change was observed as compared to the pre-EA value, suggesting that EA did not produce dysfunction or worsening of memory but, instead, caused a strengthening of memory maintenance or consolidation.

5. CONCLUSIONS

Science is rapidly progressing, so should be the idea and technology of acupuncture. The identification of "specific" points on the ear is a discovery of the twentieth century. The use of needles made of stainless steel rather than silver, and the application of electric stimulation in lieu of manual needling, is natural to modern physicians. The principle of acupuncture treatment of drug abuse is based more on the evidence obtained from clinical observation as well as basic research in animals and humans. Meanwhile, the ancient rules of the traditional acupuncture should certainly be considered seriously.

In order to best serve the drug addicts seeking for painless drug abstinence, the following should be considered in setting up criteria for the protocols of acupuncture treatment: effectiveness, standardization and reproducibility, low-cost, time-saving, least side effects, and complete drug abstinence achievement. The TEAS protocol introduced by the Neuroscience Research Institute at Peking University could be considered as one of the choices when a large-scale, multicenter research program is planned in the future.

REFERENCES

Avants, S. K., Margolin, A., Holford, T. R., & Kosten, T. R. (2000). A randomized controlled trial of auricular acupuncture for cocaine dependence. *Archives of Internal Medicine, 160*(15), 2305–2312.

Beswick, T., Best, D., Rees, S., Bearn, J., Gossop, M., & Strang, J. (2003). Major disruptions of sleep during treatment of the opiate withdrawal syndrome: Differences between methadone and lofexidine detoxification treatments. *Addiction Biology, 8*(1), 49–57.

Brumbaugh, A. G. (1993). Acupuncture: New perspectives in chemical dependency treatment. *Journal of Substance Abuse Treatment, 10*(1), 35–43.

Chen, J. H., Liang, J., Wang, G. B., Han, J. S., & Cui, C. L. (2005). Repeated 2 Hz peripheral electrical stimulations suppress morphine-induced CPP and improve spatial memory ability in rats. *Experimental Neurology, 194*(2), 550–556.

Chu, N. N., Xia, W., Yu, P., Hu, L., Zhang, R., & Cui, C. L. (2008). Chronic morphine-induced neuronal morphological changes in the ventral tegmental area in rats are reversed by electroacupuncture treatment. *Addiction Biology, 13*(1), 47–51.

Chu, N. N., Zuo, Y. F., Meng, L., Lee, D. Y., Han, J. S., & Cui, C. L. (2007). Peripheral electrical stimulation reversed the cell size reduction and increased BDNF level in the ventral tegmental area in chronic morphine-treated rats. *Brain Research, 1182*, 90–98.

Cui, C. L., Wu, L. Z., & Han, J. S. (2000). Spinal kappa-opioid system plays an important role in suppressing morphine withdrawal syndrome in the rat. *Neuroscience Letters, 295*, 42–48.

Cui, C. L., Wu, L. Z., & Luo, F. (2008). Acupuncture for the treatment of drug addiction. *Neurochemical Research, 33*(10), 2013–2022.

D'Alberto, A. (2004). Auricular acupuncture in the treatment of cocaine/crack abuse: A review of the efficacy, the use of the National Acupuncture Detoxification Association protocol, and the selection of sham points. *Journal of Alternative and Complementary Medicine, 10*(6), 985–1000.

Di Chiara, G., & North, R. A. (1992). Neurobiology of opiate abuse. *Trends in Pharmacological Sciences, 13*(5), 185–193.

Green, P. G., & Lee, N. M. (1988). Dynorphin A-(1-13) attenuates withdrawal in morphine-dependent rats: Effect of route of administration. *European Journal of Pharmacology, 145*(3), 267–272.

Han, J. S. (2003). Acupuncture: Neuropeptide release produced by electrical stimulation of different frequencies. *Trends in Neurosciences, 26*(1), 17–22.

Han, J. S. (2004). Acupuncture and endorphins. *Neuroscience Letters, 361*(1–3), 258–261.

Han, J. S., & Cui, C. L. (2011). Acupuncture. In E. Strain & P. Ruiz (Eds.), *Lowinson and Ruiz's Substance abuse—A comprehensive text book* (5th ed., pp. 466–476). Philadelphia: Lippincott Williams & Wilkins.

Han, J. S., Trachtenberg, A. I., & Lowinson, J. H. (2005). Acupuncture. In J. H. Lowinson, P. Ruiz, R. B. Millman, & J. G. Langrod (Eds.), *Substance abuse—A comprehensive text book* (4th ed., pp. 743–783). Philadelphia: Lippincott Williams & Wilkins.

Han, J. S., & Wang, Q. (1992). Mobilization of specific neuropeptides by peripheral stimulation of different frequencies. *News in Physiological Sciences, 7*, 176–180.

Han, J. S., Wu, L. Z., & Cui, C. L. (2003). Prevention of relapse to heroin by Han's acupoint nerve stimulation. *Chinese Journal of Integrative Medicine, 2*(1), 56.

Han, J. S., & Zhang, R. L. (1993). Suppression of morphine abstinence syndrome by body electroacupuncture of different frequencies in rats. *Drug and Alcohol Dependence, 31*(2), 169–175.

Hu, L., Chu, N. N., Sun, L. L., Zhang, R., Han, J. S., & Cui, C. L. (2009). Electroacupuncture treatment reverses morphine-induced physiological changes in dopaminergic neurons within the ventral tegmental area. *Addiction Biology, 14*(4), 431–437.

Konefal, J., Duncan, R., & Clemence, C. (1994). The impact of the addition of an acupuncture treatment program to an existing metro-Dade County outpatient substance abuse treatment facility. *Journal of Addictive Diseases, 13*(3), 71–99.

Kong, J., Fufa, D. T., Gerber, A. J., Rosman, I. S., Vangel, M. G., Gracely, R. H., et al. (2005). Psychophysical outcomes from a randomized pilot study of manual, electro, and sham acupuncture treatment on experimentally induced thermal pain. *The Journal of Pain, 6*(1), 55–64.

Kosten, T. R., Rounsaville, B. J., Babor, T. F., Spitzer, R. L., & Williams, J. B. (1987). Substance-use disorders in DSM-III-R. Evidence for the dependence syndrome across different psychoactive substances. *The British Journal of Psychiatry, 151*, 834–843.

Li, H. Y., Zhang, R., Cui, C. L., Han, J. S., & Wu, L. Z. (2011). Damage of splenic T lymphocyte proliferation and differentiation and its normalization by electroacupuncture in morphine-dependent mice mode. *Evidence-Based Complementary and Alternative Medicine, 2011*, 424092.

Li, Y. J., Zhong, F., Yu, P., Han, J. S., Cui, C. L., & Wu, L. Z. (2011). Electroacupuncture treatment normalized sleep disturbance in morphine withdrawal rats. *Evidence-Based Complementary and Alternative Medicine, 2011*, 361054.

Liang, J., Ping, X. J., Li, Y. J., Ma, Y. Y., Wu, L. Z., Han, J. S., et al. (2010). Morphine-induced conditioned place preference in rats is inhibited by electroacupuncture at 2 Hz: Role of enkephalin in the nucleus accumbens. *Neuropharmacology, 58*(1), 233–240.

Ma, Y. Y., Meng, L., Guo, C. Y., Han, J. S., Lee, D. Y., & Cui, C. L. (2009). Dose- and time-dependent, context-induced elevation of dopamine and its metabolites in the nucleus accumbens of morphine-induced CPP rats. *Behavioural Brain Research, 204*(1), 192–199.

Ma, Y. Y., Shi, X. D., Han, J. S., & Cui, C. L. (2008). Peripheral electrical stimulation-induced suppression of morphine-induced CCP in rats: A role for dopamine in the nucleus accumbens. *Brain Research, 1212*, 63–70.

Macias, J., Palacios, R. B., Claro, E., Vargas, J., Vergara, S., Mira, J. A., et al. (2008). High prevalence of hepatitis C virus infection among noninjecting drug users: Association with sharing the inhalation implements of crack. *Liver International, 28*(6), 781–786.

Margolin, A. (2003). Acupuncture for substance abuse. *Current Psychiatry Reports, 5*(5), 333–339.

Margolin, A., Avants, S. K., & Holford, T. R. (2002). Interpreting conflicting findings from clinical trials of auricular acupuncture for cocaine addiction: Does treatment context influence outcome? *Journal of Alternative and Complementary Medicine, 8*(2), 111–121.

Margolin, A., Kleber, H. D., Avants, S. K., Konefal, J., Gawin, F., Stark, E., et al. (2002). Acupuncture for the treatment of cocaine addiction: A randomized controlled trial. *JAMA, 287*(1), 55–63.

Markou, A., Weiss, F., Gold, L. H., Caine, S. B., Schulteis, G., & Koob, G. F. (1993). Animal models of drug craving. *Psychopharmacology, 112*(2–3), 163–182.

McLellan, A. T., Grossman, D. S., Blaine, J. D., & Haverkos, H. W. (1993). Acupuncture treatment for drug abuse: A technical review. *Journal of Substance Abuse Treatment, 10*(6), 569–576.

Nestler, E. J. (1997). Molecular mechanisms of opiate and cocaine addiction. *Current Opinion in Neurobiology*, 7(5), 713–719.

Nestler, E. J. (2004). Historical review: Molecular and cellular mechanisms of opiate and cocaine addiction. *Trends in Pharmacological Sciences, 25*(4), 210–218.

Odunayo, A., Dodam, J. R., Kerl, M. E., & DeClue, A. E. (2010). Immunomodulatory effects of opioids. *Journal of Veterinary Emergency and Critical Care (San Antonio, Tex), 20*(4), 376–385.

Oleson, T. (2001). Neurophysiological basis of auricular acupuncture. In G. Stux & R. Hammerschlag (Eds.), *Clinical acupuncture: Scientific basis* (pp. 97–112). Berlin: Springer Verlag.

Ren, Y. H., Wang, B., Luo, F., Cui, C. L., Zheng, J. W., & Han, J. S. (2002). Peripheral electric stimulation attenuates the expression of cocaine-induced place preference in rats. *Brain Research, 957*(1), 129–135.

Research Group of Acupuncture Anesthesia, Peking Medical College (1973). Effect of acupuncture on pain threshold of human skin. *Chronobiology International, 3*, 151–157.

Research Group of Ear Acupuncture, Jiangsu College of New Medicine (1973). *The effect of ear acupuncture on the pain threshold of the skin at thoracic and abdominal region. Theoretical study on acupuncture anesthesia.* (pp.27–32). Shanghai, China: Shanghai People's Press.

Rivera-Amill, V., Silverstein, P. S., Noel, R. J., Kumar, S., & Kumar, A. (2010). Morphine and rapid disease progression in nonhuman primate model of AIDS: Inverse correlation between disease progression and virus evolution. *Journal of Neuroimmune Pharmacology, 5*(1), 122–132.

Shalev, U., Grimm, J. W., & Shaham, Y. (2002). Neurobiology of relapse to heroin and cocaine seeking: A review. *Pharmacological Reviews, 54*(1), 1–42.

Shi, J., Liu, Y. L., Fang, Y. X., Xu, G. Z., Zhai, H. F., & Lu, L. (2006). Traditional Chinese medicine in treatment of opiate addiction. *Acta Pharmacologica Sinica, 27*(10), 1303–1308.

Shi, X. D., Ren, W., Wang, G. B., Luo, F., Han, J. S., & Cui, C. L. (2003). Brain opioid-receptors are involved in mediating peripheral electric stimulation-induced inhibition of morphine conditioned place preference in rats. *Brain Research, 981*(1–2), 23–29.

Shi, X. D., Wang, G. B., Ma, Y. Y., Ren, W., Luo, F., Cui, C. L., et al. (2004). Repeated peripheral electrical stimulations suppress both morphine-induced CPP and reinstatement of extinguished CPP in rats: Accelerated expression of PPE and PPD mRNA in NAc implicated. *Brain Research Molecular Brain Research, 130*(1–2), 124–133.

Shwartz, M., Saitz, R., Mulvey, K., & Brannigan, P. (1999). The value of acupuncture detoxification programs in a substance abuse treatment system. *Journal of Substance Abuse Treatment, 17*(4), 305–312.

Sun, B., Ye, Y., & Qin, L. (2001). An analysis of relapse factors of 615 heroin addicts. *Chinese Journal of Drug Dependence, 10*(3), 214–216.

Wang, J., Barke, R. A., Ma, J., Charboneau, R., & Roy, S. (2008). Opiate abuse, innate immunity, and bacterial infectious diseases. *Archivum Immunologiae et Therapiae Experimentalis, 56*(5), 299–309.

Wang, B., Luo, F., Ge, X. C., Fu, A. H., & Han, J. S. (2002). Effects of lesions of various brain areas on drug priming or footshock-induced reactivation of extinguished conditioned place preference. *Brain Research, 950*(1–2), 1–9.

Wang, B., Luo, F., Xia, Y. Q., & Han, J. S. (2000). Peripheral electric stimulation inhibits morphine-induced place preference in rats. *Neuroreport, 11*(5), 1017–1020.

Wang, B., Luo, F., Zhang, W. T., & Han, J. S. (2000). Stress or drug priming induces reinstatement of extinguished conditioned place preference. *Neuroreport, 11*(12), 2781–2784.

Wang, J. Q., Mao, L., & Han, J. S. (1992). Comparison of the antinociceptive effects induced by electroacupuncture and transcutaneous electrical nerve stimulation in the rat. *The International Journal of Neuroscience, 65*(1–4), 117–129.

Wang, G. B., Wu, L. Z., Yu, P., Li, Y. J., Ping, X. J., & Cui, C. L. (2011). Multiple 100 Hz electroacupuncture treatments produced cumulative effect on the suppression of morphine withdrawal syndrome: Central preprodynorphin mRNA and p-CREB implicated. *Peptides, 32*(4), 713–721.

Wang, B., Zhang, B. G., Ge, X. C., & Luo, F. (2003). Inhibition by peripheral electric stimulation of the reinstatement of morphine induced place preference in rats and drug-craving in heroin addicts. *Journal of Peking University (Health Sciences), 85*(3), 241–247.

Wen, H. L. (1975). Role of acupuncture in narcotic withdrawal. *Medycyna Pracy, 2*, 15–16.

Wen, H. L. (1977). Fast detoxification of drug abuse by acupuncture and electrical stimulation (A.E.S.) in combination with naloxone. *Modern Medicine of Asia, 13*, 13–17.
Wen, H. L. (1979). Acupuncture and electrical stimulation (AES) outpatient detoxification. *Modern Medicine of Asia, 15*, 39–43.
Wen, H. L., & Cheung, S. (1973). Treatment of drug addiction by acupuncture and electrical stimulation. *Asian Journal of Medicine, 9*, 138–141.
Wen, H. L., & Ho, W. K. (1982). Suppression of withdrawal symptoms by dynorphin in heroin addicts. *European Journal of Pharmacology, 82*(3–4), 183–186.
Wen, H. L., Ho, W. K., Ling, N., Mehal, Z. D., & Ng, Y. H. (1980). Immunoassayable beta-endorphin level in the plasma and CSF of heroin addicted and normal subjects before and after electroacupuncture. *The American Journal of Chinese Medicine, 8*(1–2), 154–159.
Wu, L. Z., Cui, C. L., & Han, J. S. (1995). Han's acupoint nerve stimulator (HANS) for the treatment of opiate withdrawal syndrome. *Chinese Journal of Pain Medicine, 1*, 30–38.
Wu, L. Z., Cui, C. L., & Han, J. S. (1996). Effect of Han's acupoint nerve stimulator (HANS) on the heart rate of 75 inpatients during heroin withdrawal. *Chinese Journal of Pain Medicine, 2*, 98–102.
Wu, L. Z., Cui, C. L., & Han, J. S. (1999). Treatment on heroin addicts by 4-channel Han's Acupoint Nerve Stimulator (HANS). *Journal of Beijing Medical University, 31*, 239–242.
Wu, L. Z., Cui, C. L., & Han, J. S. (2000). Effect of 2/100 Hz transcutaneous electric nerve stimulation on sexual dysfunction and serum sex hormone of heroin addicts. *Chinese Journal of Integrated Traditional and Western Medicine, 20*, 15–18.
Wu, L. Z., Cui, C. L., & Han, J. S. (2001). Reduction of methadone dosage and relief of depression and anxiety by 2/100 Hz TENS for heroin detoxification. *Chinese Journal of Drug Dependence, 10*, 124–126.
Wu, L. Z., Cui, C. L., Tian, J. B., Ji, D., & Han, J. S. (1999). Suppression of morphine withdrawal by electroacupuncture in rats: Dynorphin and kappa-opioid receptor implicated. *Brain Research, 851*(1–2), 290–296.
Zhong, F., Wu, L. Z., & Han, J. S. (2006). Suppression of cue-induced heroin craving and cue-reactivity by single-trial transcutaneous electrical nerve stimulation at 2 Hz. *Addiction Biology, 11*(2), 184–189.

CHAPTER THIRTEEN

Acupuncture Regulation of Blood Pressure: Two Decades of Research

John C. Longhurst[*,1], **Stephanie Tjen-A-Looi**[†]
[*]Department of Medicine, Physiology and Biophysics, and Pharmacology, School of Medicine, University of California, Medical Sciences, Irvine, California, USA
[†]Department of Medicine, School of Medicine, University of California, Irvine, California, USA
[1]Corresponding author: e-mail address: jcl@uci.edu

Contents

Abstract

Although mechanisms underlying acupuncture regulation of pain have been studied by a number of laboratories in many countries, much less is known about its ability to modulate cardiovascular function. In the last two decades, our laboratory has systematically investigated the peripheral and central neural mechanisms underlying acupuncture regulation of blood pressure. These observations account for acupuncture's distant actions and, to some extent, its local actions, with respect to the site of needling. Four fundamental findings have advanced our knowledge. First, point-specific effects of acupuncture underlie its cardiovascular actions. Second, variable regions in the supraspinal and spinal central nervous system that receive input from somatic afferent stimulation account for acupuncture's ability to modulate blood pressure. Thus, depending on the underlying situation, for example, high or low blood pressure, acupuncture modifies autonomic outflow by reducing activity in brain stem

International Review of Neurobiology, Volume 111
ISSN 0074-7742
http://dx.doi.org/10.1016/B978-0-12-411545-3.00013-4

nuclei that participate in the primary response. Third, repetitive acupuncture through a molecular mechanism can cause prolonged cardiovascular effects that far outlast acupuncture stimulation. Fourth, there is a range of cardiovascular responsiveness to acupuncture that depends, at least in part, on interactions between neural modulators that synaptically regulate autonomic function in the brain stem. Thus, acupuncture has the capability of profoundly regulating cardiovascular function in patients with disease, for example, hypertension, and the experimental laboratory is directing best approaches to study its actions in humans.

1. INTRODUCTION

If I (senior author) had been asked in the early 1990s to participate in studies concerned with acupuncture, I would have refused. My early education in acupuncture in medical school in the United States was mainly in my class called "quackery in medicine." I had been taught that acupuncture was mainly a placebo therapy sought by desperate patients who were trying any therapy for diseases like chronic pain and cancer that often were not well treated by traditional Western medical methods. All that began to change in 1992 when I traveled to China and Shanghai for the first time. I began to talk with traditional Oriental doctors (OMDs) and see acupuncture practiced in clinics of traditional Chinese medicine (TCM) hospitals. When I asked about the evidence base behind these Eastern practices, I was told by OMDs that they did not need scientific or clinical studies to know that acupuncture worked since it had been practiced for well over 2000 years. Empirical practice was acceptable to them, but for a Western physician and scientist, the absence of an evidence-based approach was inconsistent with my principles of medical practice. Thus, my first response to Dr. Peng Li's request that I consider research collaboration was thank you, but I do not wish to study a therapy that has no science behind it. Dr. Li, however, was persistent. He gave me his curriculum vitae and encouraged me to read the literature more carefully, thus began my journey in acupuncture research that now has lasted approximately two decades. I learned that Dr. Li and a small cadre of Chinese scientists, like Dr. J S Han in Beijing, over the previous 30–50 years had conducted a number of rigorous studies suggesting that acupuncture has a neural basis. Dr. Li eventually participated in two sabbaticals in the United States and in 1999 joined my laboratory on a full-time basis to conduct research on the neural mechanisms underlying acupuncture's cardiovascular actions.

2. EARLY OBSERVATIONS

Small clinical studies in the early 1990s suggested that acupuncture could reduce myocardial ischemia in patients with coronary artery disease (Richter, Herlitz, & Hjalmarson, 1991). Our first experimental study in a model of demand-induced ischemia demonstrated that myocardial ischemia could be completely reversed by a single application of median nerve stimulation or electroacupuncture (EA) at the pericardial meridian, points 5 and 6 or PC5 and PC6 (Chao et al., 1999; Li, Pitsillides, Rendig, Pan, & Longhurst, 1998). The mechanism underlying the EA anti-ischemia response was a decrease in oxygen demand (reduced blood pressure response), rather than an increase in coronary blood flow. In a follow-up investigation, we found that intravenous naloxone, a nonspecific opioid antagonist, reversed the beneficial action of EA in myocardial ischemia (Chao et al., 1999). Thus, the opioid system appears to play an important role in acupuncture's cardiovascular action.

3. ACUPUNCTURE MODULATES ABNORMALITIES OF BLOOD PRESSURE

Since these initial investigations, we subsequently have evaluated acupuncture's effect on blood pressure because this appears to be one of its main cardiovascular actions. Multiple observations indicate that acupuncture reduces elevations in blood pressure in acutely anesthetized animals (Crisostomo, Li, Tjen-A-Looi, & Longhurst, 2005; Li, Tjen-A-Looi, Guo, Fu, & Longhurst, 2009; Tjen-A-Looi, Guo, Li, & Longhurst, 2013; Tjen-A-Looi, Li, & Longhurst, 2004, 2009). EA also lowers blood pressure in chronically hypertensive patients (Flachskampf et al., 2007; Li & Longhurst, 2007) and likely in animal models of sustained hypertension (Li, Chi, Tjen-A-Looi, & Longhurst, 2013), when it is applied at acupuncture points that provide strong input to cardiovascular regions of the brain that regulate sympathetic outflow (Tjen-A-Looi et al., 2004; Zhou, Tjen-A-Looi, & Longhurst, 2005). Thus, in a point-specific manner, acupuncture is able to lower reflex elevations in blood pressure by 40–50% (7–15 mmHg) when it stimulates underlying neural pathways of select acupoints that project to brain stem regions associated with regulation of sympathetic outflow (Tjen-A-Looi et al., 2004). Conversely, when acupoints are stimulated that have little input to premotor sympathetic neurons (Tjen-A-Looi et al., 2004) or when sympathetic outflow is not elevated but rather is at steady-state levels associated with normotension, acupuncture has little or no influence on blood pressure (Li, Ayannusi, Reed, & Longhurst, 2004; Tjen-A-Looi et al., 2004). Point-specific systemic actions of acupuncture are mediated in large part by mixed (motor and sensory) nerve bundles, like the median nerve underlying PC5–PC6 acupoints or the deep peroneal nerve underlying ST36–ST37 acupoints (Li et al., 1998; Tjen-A-Looi, Fu, Zhou, & Longhurst, 2005; Zhou, Fu, Tjen-A-Looi, Li, & Longhurst, 2005). The sensory neural component of deep somatic nerves, involving group III finely myelinated and group IV unmyelinated afferents, is the critical pathway in these nerves since local anesthesia but not motor paralysis eliminates acupuncture's effects on pain and cardiovascular function (Han & Terenius, 1982; Li et al., 1998; Tjen-A-Looi et al., 2005). More superficial nerves, like cutaneous branches, do not project to cardiovascular brain stem regions, such as the rostral ventrolateral medulla (rVLM), and therefore exert little influence over blood pressure (Tjen-A-Looi et al., 2004).

Like hypertension, acupuncture modulates low blood pressure. Two experimental models show that acupuncture has predominately a parasympatholytic or a combined sympatho- and parasympatholytic action depending on the underlying condition (Tjen-A-Looi et al., 2013; Tjen-A-Looi, Li, Li, & Longhurst, 2012). For example, gastric distension in hypercapnic–acidotic rats (Tjen-A-Looi et al., 2013; Tjen-A-Looi, Hsiao, & Longhurst, 2011) or stimulation of a cardiopulmonary reflex with the 5-hydroxytrypatamine agonist, phenylbiguanide (to simulate vasovagal syncope; Tjen-A-Looi et al., 2012), leads to depressor–bradycardia reflexes due to increased parasympathetic tone and withdrawal of sympathetic outflow. Both cardiovascular inhibitory reflexes are reversed by approximately 40% following application of acupuncture, interestingly at the same acupoints used to control elevations in blood pressure. These studies reveal several novel pieces of information. First, EA regulates abnormalities in blood pressure through its actions on both sympathetic and parasympathetic centers in the medulla. Second, in addition to modifying elevated blood pressure, EA can modulate low blood pressure, if the elevations are related to changes in sympathetic and/or parasympathetic outflow. Third, acupuncture seems to have actions in medullary cardiovascular centers that regulate sympathetic and parasympathetic outflow when there is a change above or below steady-state levels of basal discharge activity. Thus, much like the ancient TCM principles, acupuncture seems to normalize autonomic discharge activity and in the process leads to homeostasis of cardiovascular function, thereby reducing cardiovascular dysfunction.

4. NEUROTRANSMITTER SYSTEMS IN CNS PROCESSING OF SOMATIC INPUT DURING ACUPUNCTURE CONTRIBUTE TO ITS UNIQUE ACTIONS

A number of regions in the hypothalamus, midbrain, and medulla receive somatic input during EA. These include the arcuate and paraventricular nuclei in the hypothalamus (Li, Tjen-A-Looi, Guo, & Longhurst, 2010; Li, Tjen-A-Looi, & Longhurst, 2006; Tjen-A-Looi, Guo, & Longhurst, 2013), ventrolateral periaqueductal gray (vlPAG) in the midbrain (Li et al., 2009; Li, Tjen-A-Looi, Guo, et al., 2010), nucleus tractus solitarii (Tjen-A-Looi et al., 2012), rVLM (Moazzami, Tjen-A-Looi, Guo, & Longhurst, 2010; Tjen-A-Looi, Li, & Longhurst, 2003; Tjen-A-Looi et al., 2004), cVLM (Tjen-A-Looi et al., 2013), medullary raphe, particularly the raphe pallidus (Li, Tjen-A-Looi, & Longhurst, 2010b; Moazzami et al., 2010), and nucleus ambiguus (Guo, Li, & Longhurst, 2012; Tjen-A-Looi et al., 2012) in the medulla as well as the dorsal horn and intermediolateral columns of the spinal cord (Zhou, Mahajan, & Longhurst, 2009). Evoked somatic input is largely excitatory in these regions and relies most commonly on glutamate (Zhou, Fu, Guo, & Longhurst, 2007; Zhou, Fu, Tjen-A-Looi, Guo, & Longhurst, 2006), other excitatory neurotransmitters like acetylcholine (Li, Tjen-A-Looi, Guo, et al., 2010), or presynaptic disinhibition of inhibitory neurotransmitters (endocannabinoids acting to reduce release of γ-aminobutyric acid or GABA) (Fu & Longhurst, 2009; Tjen-A-Looi et al., 2009). Acupuncture activates neurons in these regions, through a network of projections extending from the hypothalamus to more caudal regions like the vlPAG and rVLM and through inhibitory neurotransmitter mechanisms, including opioids (Li et al., 2009; Li, Tjen-A-Looi, & Longhurst, 2001; Tjen-A-Looi et al., 2003; Tjen-A-Looi, Li, & Longhurst, 2007), GABA (Fu & Longhurst, 2009; Tjen-A-Looi et al., 2007, 2009), nociceptin (Crisostomo et al., 2005; Tjen-A-Looi et al., 2007), or serotonin (Guo, Moazzami, Tjen-A-Looi, & Longhurst, 2008; Moazzami et al., 2010), acting postsynaptically to either directly or indirectly modulate autonomic outflow. Thus, somatic afferent-evoked excitatory input ultimately modulates autonomic outflow. Acupuncture's actions are manifested when basal autonomic tone, and hence arterial blood pressure, has been increased or decreased. This unique feature of acupuncture separates its actions from any pharmacological therapy used to treat hyper- or hypotension since drugs either lower or raise blood pressure, but the same drug does not do both.

5. PHYSIOLOGICAL AND MOLECULAR MECHANISMS CONTRIBUTING TO ACUPUNCTURE'S CARDIOVASCULAR EFFECTS

Application of EA for 30 min typically impacts cardiovascular function for 1–1.5 h (Li, Tjen-A-Looi, & Longhurst, 2010a; Tjen-A-Looi, Li, & Longhurst, 2006; Tjen-A-Looi et al., 2004; Zhou, Tjen-A-Looi, et al., 2005) and in this manner can be differentiated from brief somatosensory stimulation whose blood pressure effects last no longer than the period of stimulation (Johansson, 1962; Michikami et al., 2006; Sato & Schmidt, 1973). At least two mechanisms account for this prolonged action. First, there is a reciprocating reinforcing circuit between the ventral hypothalamic arcuate nucleus and the midbrain vlPAG (Guo & Longhurst, 2010; Li, Tjen-A-Looi, Guo, et al., 2010). Second, acupuncture leads to prolonged release of constitutive or preformed inhibitory neurotransmitters like opioids and GABA in the rVLM and possibly other regions where these peptides are synthesized (Li et al., 2010a; Tjen-A-Looi et al., 2007). Other modulatory messengers like nociceptin function only during the immediate period during the 30 min application of EA but not in the 30–90 min period following stimulation when there is a residual EA-cardiovascular response (Tjen-A-Looi et al., 2007). However, clinical acupuncture typically involves repetitive needling, once or twice a week over a period of several weeks or months. Acupuncture application once weekly for 30 min over a 2-month period appears to reduce blood pressure of subjects with mild to moderate hypertension, a response that is slow in onset (occurring after 2–4 weeks) and prolonged in duration, with decreases that can last days to weeks depending on how many times it is applied (Li et al., 2013; Li & Longhurst, 2007). Once again, opioids seem to be at least one explanation for the lengthy acupuncture hypotensive response. In this regard, recurrent somatic afferent nerve-evoked activation of rVLM neurons, consequent to each weekly application of EA, stimulates transcriptional regulation to heighten production of preproenkephalin (PPE), the mRNA of met- and leu-enkephalin. There is a transient increase in tissue PPE after one 30 min application of EA at PC5–PC6 in anesthetized animals 90 min after terminating the stimulus, indicating that transcription in neurons of the rVLM is activated, although at a time after the cardiovascular response has dissipated (Li et al., 2010a). Conscious rats receiving acupuncture at ST36–ST37 twice over a 3-day period demonstrate increases in PPE 24 h after the final stimulation and elevated tissue met-enkephalin 24 and 48 h following EA (Li, Tjen-A-Looi, Guo, & Longhurst, 2012). Preliminary studies of cold-induced hypertensive rats designed to more closely approximate clinical acupuncture suggest that EA applied twice weekly over a 5-week period to the same acupoints activates transcription to increase PPE for at least 72 h following termination of stimulation (Li et al., 2013). A course of acupuncture therapy thus has the potential to induce prolonged salutatory hemodynamic responses through molecular mechanisms or neural plasticity/adaptation in brain stem regions concerned with regulation of presympathetic neural activity.

6. CHEMICAL MECHANISMS OF ACUPUNCTURE RESPONSIVENESS

Not everyone responds to acupuncture. In fact, despite optimal technique, no more than ~70% of subjects treated with acupuncture demonstrate analgesia or meaningful blood pressure reductions (Andersson, Hansson, Homgren, & Reberg, 1976; Han & Terenius, 1982; Li et al., 2004; Ng, Katims, & Lee, 1992; Thomas & Lundeberg, 1994; Ulett, 1989; Ulett, Han, & Han, 1998). Hypo- or nonresponsiveness to acupuncture could be related to subpar actions of several of the modulatory neurotransmitter systems, including opioids and GABA, each of which contributes to prolonged actions of this stimulus. The gut hormone, cholecystokinin (CCK), occurs as an octapeptide in the brain, where it antagonizes the actions of opioids (Heinricher, McGaraughty, & Tortorici, 2001; Noble, Derrien, & Roques, 1993). CCK-8 in brain stem regions may regulate sympathetic outflow (Sartor & Verberne, 2002; Sartor & Verberne, 2006). Although the location of its action has not been identified, this peptide appears to contribute to hyporesponsiveness during acupuncture analgesia (Han, 1995; Bian, Sun, & Han, 1993; Han, Ding, & Fan, 1986). Furthermore, recent studies show that antagonism of CCK-8 in the rVLM converts EA cardiovascular nonresponders to responders through an opioid mechanism (Li et al., 2013). The complex interplay between pro- and antiopioid stimuli and other chemical mediators in brain stem autonomic centers thus determines the extent of acupuncture's influence on blood pressure.

7. CONCLUSIONS AND FUTURE DIRECTIONS OF ACUPUNCTURE MECHANISTIC RESEARCH

A fair question to pose is the relevance of basic science research in acupuncture since it has been around for literally thousands of years and therefore must be of benefit. One answer might be that elucidation of mechanisms underlying acupuncture's actions helps skeptics like our group understand and possibly accept this ancient therapeutic technique. But an equally compelling response could be that, by better understanding how acupuncture relieves pain or reduces elevated blood pressure, some of the controversies can be settled and its usefulness maximized. In the succeeding text are examples of each.

Acupuncture is based on a theory of meridians, along which acupoints are positioned that, when stimulated, alter the energy, called Qi, flowing through the meridians and principal Chinese organs. TCM theory states that imbalances in the body change the free flow of Qi and acupuncture can restore the flow, to improve harmony and health. However, for this aspect of TCM theory to be accepted in Western countries, meridians and acupoints have to be identified and Qi measured. This has never been possible, at least to the satisfaction of Western (and many Eastern) scientists and physicians. Despite some controversy, meridians, acupoints, and Qi have not been reliably identified by any technique (Longhurst, 2010). While the absence of proof of existence of these critical components dismays many practitioners and champions of TCM theory, other students have examined alternative explanations using modern investigative methods. The majority of these investigations indicate that sensory neural pathways underlying meridians and associated acupoints are stimulated during needle application (Li et al., 1998; Tjen-A-Looi et al., 2005). Sensory neural information transmitted centrally accounts for acupuncture's distant actions on pain and cardiovascular dysfunction (Han & Terenius, 1982; Li et al., 2009) since blockade at any level, peripherally or centrally, abolishes its effect (Chiang, Chang, Cicero, & Yano, 1973; Han & Terenius, 1982; Tjen-A-Looi et al., 2006). But the nervous system does not provide the entire explanation of acupuncture's clinical actions.

Acupuncture needles are frequently placed outside traditional meridians and acupoints when clinicians are treating areas of tenderness, also called Ashi points (Travell & Simons, 1992). Certainly, if pain exists, then nerves must be present to transmit the sensation. However, large neural pathways generally do not exist beneath Ashi points of stimulation, so it is reasonable to consider other more local actions of acupuncture. Over the last decade, a series of studies have shown that, at least with needle manipulation during manual acupuncture, connective tissue elements wrap around the needle to increase tension as the needle is rotated and to cause tissue deformation (Davis, Churchill, Badger, Dunn, & Langevin, 2012; Langevin et al., 2011). This grabbing of collagen and other subcutaneous tissue components appears to stimulate fibrocytes and other local tissue elements to activate cytokines and release chemicals that could function in local responses to acupuncture (Bouffard et al., 2008; Goldman et al., 2010). One local chemical is adenosine, which is released during acupuncture to reduce experimental pain in rodents (Goldman et al., 2010). Acupuncture's local action therefore may be separate from neural stimulation, or is it? Is it possible that chemicals released during either mechanical or EA can depolarize sensory nerves to generate either local axonal reflexes or even distant actions? In other word, is it possible that chemicals released at the point of needling rely on neural elements for their distant and perhaps local mechanisms of action? Further research, some of which is in progress, is required to answer these questions. However, the potential of acupuncture-induced chemical-afferent neural interaction is favored by multiple studies showing that chemicals released in other conditions, like exercise and ischemia (Fletcher, Stahl, & Longhurst, 1993; Fu, O'Neill, & Longhurst, 1997; Gee, Tjen-A-Looi, Hill, Chahal, & Longhurst, 2002; Longhurst, Aung-Din, & Mitchell, 1981; O'Neil, Stebbins, Bonigut, Halliwell, & Longhurst, 1996; Rotto, Schultz, Longhurst, & Kaufman, 1990; Stebbins, Carretero, Mindroiu, & Longhurst, 1990; Stebbins & Longhurst, 1986; Stebbins, Maruoka, & Longhurst, 1986; Symons, Theodossy, Longhurst, & Stebbins, 1991), in many organ systems lead to sensory nerve depolarization and a wide range of local and distant autonomic adjustments.

How can acupuncture be optimized so that it exerts maximal effects on cardiovascular function? In this regard, we need to ask how it can be best applied. What are the optimal places of acupoint stimulation? How long should it be applied and how frequently? And is there an optimal frequency of stimulation to maximally lower or raise blood pressure? Studies of point specificity show that PC5, PC6, ST36, and ST37 provide the greatest input to regions of the brain like the rVLM that regulate sympathetic outflow (Tjen-A-Looi et al., 2004; Zhou, Fu et al., 2005). Thus, current clinical studies evaluating acupuncture's role in treating hypertension focus on these points (Li & Longhurst, 2007). This standardized approach is at odds with usual TCM diagnosis and therapy that lead to stimulation of variable combinations of acupoints following tongue and pulse examination. But when these diagnosis-based approaches are employed, the results appear to be suboptimal (Macklin et al., 2006). Ongoing studies are examining the optimal duration of acupuncture stimulation. Typically, changes in blood pressure, if they occur, begin after 10–15 min of stimulation (Li, Tjen-A-Looi, Guo, et al., 2010; Moazzami et al., 2010; Zhou et al., 2009), and 30–40 min seems to yield significant decreases or increases in blood pressure (Li, Tjen-A-Looi, Guo, et al., 2010; Tjen-A-Looi et al., 2004, 2006, 2009). However, it is not known if longer stimulations would be more effective. Thirty minutes of manual and EA, when matched for frequency of stimulation, yields comparable hypotensive responses, and low frequencies of stimulation in the 2–4 Hz range are most effective (Zhou, Tjen-A-Looi, et al., 2005). An experimental study comparing the value of stimulating two sets of acupoints bilaterally (PC5–PC6 and ST36–ST37) demonstrated no added value in the reduction of reflex-induced hypertension vs. stimulating one set of points (PC5–PC6) (Zhou, Fu, et al., 2005). However, because stimulation of these two sets of acupoints evokes large changes individually (Tjen-A-Looi et al., 2004), they have been stimulated in combination in an ongoing clinical study on the chance that clinically the combination would be more powerful than either one alone (Li & Longhurst, 2007). Clearly, more research needs to be conducted on the value of stimulating acupoint combinations, especially since this is the most common form of treatment by TCM practitioners. The other, harder question to answer is how should the optimal combination of points be chosen? Should they be selected based on experimental studies showing greatest input to regions of the brain that can evoke acupuncture-related modulation, or should they be determined by ancient practices of physical diagnosis used by TCM practitioners? We argue that specifically with clinical trials, a standardized evidence-based approach to acupoint selection should be used to maximize effect and to limit the number of patients that have to be studied. It is difficult, however, to sway TCM practitioners who operate based on guidance gained from millennia of experience.

Once blood pressure has been altered to the greatest extent possible, how often should reinforcement therapy occur? An ongoing study suggests that blood pressure lowered after 8 weeks of weekly application remains low one month later (Li & Longhurst, 2007). Thus, it might be possible to apply ongoing therapy once each month, a hypothesis that appears to be confirmed in early follow-up stimulation protocols (unpublished data).

8. FINAL THOUGHTS

Despite optimal acupuncture technique, not every patient will respond similarly. Some will respond a lot and some little or not at all. As discussed earlier, knowledge of the neural and chemical mechanisms of acupuncture's clinical actions will help optimize responses and potentially can even convert nonresponders into responders. However, we are just beginning to understand the complex neural and nonneural mechanisms underlying acupuncture's action. Much more research is necessary. What is certain, however, is that our attitude about acupuncture has shifted. Once we began to replicate studies in our laboratory, consensus developed that acupuncture does have a real therapeutic action. Furthermore, application of standard Western techniques using methods incorporating modern science has revealed many of the physical, chemical, and biological mechanisms that underlie acupuncture's cardiovascular actions. Much more work is necessary, however, to allow a full understanding of the complex scope of acupuncture's clinical application so that we can optimally use this integrative technique to treat the many patients who today are demanding therapy.

ACKNOWLEDGMENTS

Research described in this chapter was supported by NIH R01-HL63313 and R01-HL72125, the Adolph Coors Foundation, the Peterson Family Foundation, the Larry K. Dodge Chair of Integrative Biology, and the Susan Samueli Chair of Integrative Medicine.

REFERENCES

Andersson, S. A., Hansson, G., Homgren, E., & Reberg, O. (1976). Evaluation of the pain suppressive effect of different frequencies of peripheral electrical stimulation in chronic pain conditions. *ACTA Orthopaedica Scandinavica*, *47*, 149–157.

Bian, J. T., Sun, M. Z., & Han, J. S. (1993). Reversal of electroacupuncture tolerance by CCK-8 antiserum: An electrophysiological study on pain-related neurons in nucleus parafascicularis of the rat. *International Journal of Neuroscience*, *72*, 15–29.

Bouffard, N. A., Cutroneo, K. R., Badger, G. J., White, S. L., Buttolph, T. R., Ehrlich, H. P., et al. (2008). Tissue stretch decreases soluble TGF-beta1 and type-1 procollagen in mouse subcutaneous connective tissue: Evidence from ex vivo and in vivo models. *Journal of Cellular Physiology*, *214*, 389–395.

Chao, D. M., Shen, L. L., Tjen-A-Looi, S. C., Pitsillides, K. F., Li, P., & Longhurst, J. C. (1999). Naloxone reverses inhibitory effect of electroacupuncture on sympathetic cardiovascular reflex responses. *American Journal of Physiology*, *276*, H2127–H2134.

Chiang, C. Y., Chang, H. T., Cicero, T. J., & Yano, I. (1973). Peripheral afferent pathway for acupuncture analgesia. *Scientia Sinica*, *16*, 210–217.

Crisostomo, M., Li, P., Tjen-A-Looi, S. C., & Longhurst, J. C. (2005). Nociceptin in rVLM mediates electroacupuncture inhibition of cardiovascular reflex excitatory response in rats. *Journal of Applied Physiology*, *98*, 2056–2063.

Davis, R. T., Churchill, D. L., Badger, G. J., Dunn, J., & Langevin, H. M. (2012). A new method for quantifying the needling component of acupuncture treatments. *Acupuncture in Medicine, 30*, 113–119.

Flachskampf, F. A., Gallasch, J., Gefeller, O., Gan, J., Mao, J., Pfahlberg, A., et al. (2007). Randomized trial of acupuncture to lower blood pressure. *Circulation, 115*, 3121–3129.

Fletcher, M. P., Stahl, G. L., & Longhurst, J. C. (1993). C5a-induced myocardial ischemia: Role for CD18-dependent PMN localization and PMN-platelet interactions. *American Journal of Physiology, 265*, H1750–H1761.

Fu, L.-W., & Longhurst, J. C. (2009). Electroacupuncture modulates vlPAG release of GABA through presynaptic cannabinoid CB_1 receptor. *Journal of Applied Physiology, 106*, 1800–1809.

Fu, L.-W., O'Neill, C. A., & Longhurst, J. C. (1997). Increased histamine and 5-HT in portal vein plasma and mesenteric lymph during brief ischemia and reperfusion. *American Journal of Physiology, 273*, H1135–H1141.

Gee, B., Tjen-A-Looi, S. C., Hill, J., Chahal, P., & Longhurst, J. C. (2002). Role of spinal NMDA and non-NMDA receptors in the pressor reflex response to abdominal ischemia. *American Journal of Physiology, 282*, R850–R857.

Goldman, N., Chen, M., Fujita, T., Xu, Q., Peng, W., Liu, W., et al. (2010). Adenosine A1 receptors mediate local anti-nociceptive effects of acupuncture. *Natural Neuroscience, 13*, 883–888.

Guo, Z.-L., Li, M., & Longhurst, J. C. (2012). Nucleus ambiguus cholinergic neurons activated by acupuncture: Relation to enkephalin. *Brain Research, 1442*, 25–35.

Guo, Z.-L., & Longhurst, J. C. (2010). Activation of reciprocal pathways between arcuate nucleus and ventrolateral periaqueductal gray during electroacupuncture: Involvement of VGLUT3. *Brain Research, 1360*, 77–88.

Guo, Z.-L., Moazzami, A., Tjen-A-Looi, S. C., & Longhurst, J. C. (2008). Responses of opioid and serotonin containing medullary raphe neurons to electroacupuncture. *Brain Research, 1229*, 125–136.

Han, J. S. (1995). Cholecystokinin octapeptide (CCK-8): A negative feedback control mechanism for opioid analgesia. *Progress in Brain Research, 105*, 263–271.

Han, J.-S., Ding, X. Z., & Fan, S. G. (1986). Cholecystokinin octapeptide (CCK-8): Antagonism to electroacupuncture analgesia and a possible role in electroacupuncture tolerance. *Pain, 27*, 101–115.

Han, J.-S., & Terenius, L. (1982). Neurochemical basis of acupuncture analgesia. *Annual Review of Pharmacology and Toxicology, 22*, 193–220.

Heinricher, M. M., McGaraughty, S., & Tortorici, V. (2001). Circuitry underlying antiopioid actions of cholecystokinin within the rostral ventromedial medulla. *Journal of Neurophysiology, 85*, 280–286.

Johansson, B. (1962). Circulatory responses to stimulation of somatic afferents. *Acta Physiologica Scandinavia, 57*(Suppl. 198), 1–91.

Langevin, H. M., Wayne, P. M., MacPherson, H., Schnyer, R., Milley, R. M., Napadow, V., et al. (2011). Paradoxes in acupuncture research: Strategies for moving forward. *Evidence-Based Complementary and Alternative Medicine, 2011*, 180805.

Li, P., Ayannusi, O., Reed, C., & Longhurst, J. C. (2004). Inhibitory effect of electroacupuncture (EA) on the pressor response induced by exercise stress. *Clinical Autonomic Research, 14*, 182–188.

Li, M., Chi, S., Tjen-A-Looi, S. C., & Longhurst, J. C. (2013). Repetitive electroacupuncture attenuates cold-induced hypertension and simultaneously enhances rVLM preproenkephalin mRNA expression. *FASEB Journal, 27*, 926.14.

Li, P., & Longhurst, J. C. (2007). Long-lasting inhibitory effect of EA on blood pressure in patients with mild to moderate hypertension. *Society for Neuroscience, 37*.

Li, P., Pitsillides, K. F., Rendig, S. V., Pan, H.-L., & Longhurst, J. C. (1998). Reversal of reflex-induced myocardial ischemia by median nerve stimulation: A feline model of electroacupuncture. *Circulation*, *97*, 1186–1194.

Li, P., Tjen-A-Looi, S. C., Guo, Z. L., Fu, L.-W., & Longhurst, J. C. (2009). Long-loop pathways in cardiovascular electroacupuncture responses. *Journal of Applied Physiology*, *106*, 620–630.

Li, P., Tjen-A-Looi, S. C., Guo, Z. L., & Longhurst, J. C. (2010). An arcuate-ventrolateral periaqueductal gray reciprocal circuit participates in electroacupuncture cardiovascular inhibition. *Autonomic Neuroscience: Basic & Clinical*, *158*, 13–23.

Li, M., Tjen-A-Looi, S. C., Guo, Z. L., & Longhurst, J. C. (2012). Repetitive electroacupuncture causes prolonged increased met-enkephalin expression in the rVLM of conscious rats. *Autonomic Neuroscience*, *170*, 30–35.

Li, P., Tjen-A-Looi, S. C., & Longhurst, J. C. (2001). Rostral ventrolateral medullary opioid receptor subtypes in the inhibitory effect of electroacupuncture on reflex autonomic response in cats. *Autonomic Neuroscience*, *89*, 38–47.

Li, P., Tjen-A-Looi, S. C., & Longhurst, J. C. (2006). Excitatory projections from arcuate nucleus to ventrolateral periaqueductal gray in electroacupuncture inhibition of cardiovascular reflexes. *American Journal of Physiology*, *209*, H2535–H2542.

Li, M., Tjen-A-Looi, S. C., & Longhurst, J. C. (2010a). Electroacupuncture enhances preproenkephalin mRNA expression in rostral ventrolateral medulla of rats. *Neuroscience Letters*, *477*, 61–65.

Li, P., Tjen-A-Looi, S. C., & Longhurst, J. C. (2010b). Nucleus raphé pallidus participates in midbrain-medullary cardiovascular sympathoinhibition during electroacupuncture. *American Journal of Physiology*, *299*, R1369–R1376.

Longhurst, J. C. (2010). Defining meridians: A modern basis of understanding. *Journal of Acupuncture and Meridian Studies*, *3*, 67–74.

Longhurst, J. C., Aung-Din, R., & Mitchell, J. H. (1981). Static exercise in anesthetized dogs, a cause of reflex alpha-adrenergic coronary vasoconstriction. *Basic Research in Cardiology*, *76*, 762–766.

Macklin, E. A., Wayne, P. M., Kalish, L. A., Valaskatgis, P., Thompson, J., Pian-Smith, M., et al. (2006). Stop hypertension with the acupuncture research program (SHARP): results of a randomized controlled clinical trial. *Hypertension*, *48*, 838–845.

Michikami, D., Kamiya, A., Kawada, T., Inagaki, M., Shishido, T., Yamamoto, K., et al. (2006). Short-term electroacupuncture at Zusanli resets the arterial baroreflex neural arc toward lower sympathetic nerve activity. *American Journal of Physiology Heart and Circulatory Physiology*, *291*, H318–H326.

Moazzami, A., Tjen-A-Looi, S. C., Guo, Z.-L., & Longhurst, J. C. (2010). Serotonergic projection from nucleus raphe pallidus to rostral ventrolateral medulla modulates cardiovascular reflex responses during acupuncture. *Journal of Applied Physiology*, *108*, 1336–1346.

Ng, L., Katims, J., & Lee, M. (1992). Acupuncture. In G. Aronoff (Ed.), *Evaluation and treatment of chronic pain* (2nd ed., pp. 291–298). Baltimore: Williams and Wilkins.

Noble, F., Derrien, M., & Roques, B. P. (1993). Modulation of opioid antinociception by CCK at the supraspinal level: evidence of regulatory mechanisms between CCK and enkephalin systems in the control of pain. *British Journal of Pharmacology*, *109*, 1064–1070.

O'Neil, C. A., Stebbins, C. L., Bonigut, S., Halliwell, B., & Longhurst, J. C. (1996). Production of hydroxyl radicals in contracting skeletal muscle of cats. *Journal of Applied Physiology*, *81*, 1197–1206.

Richter, A., Herlitz, J., & Hjalmarson, A. (1991). Effect of acupuncture in patients with angina pectoris. *European Heart Journal*, *12*, 175–178.

Rotto, D., Schultz, H., Longhurst, J. C., & Kaufman, M. (1990). Sensitization of group III muscle afferents to static contraction by arachidonic acid metabolism. *Journal of Applied Physiology, 68*, 861–867.

Sartor, D. M., & Verberne, A. J. (2002). Cholecystokinin selectively affects presympathetic vasomotor neurons and sympathetic vasomotor outflow. *American Journal of Physiology, 282*, R1174–R1184.

Sartor, D. M., & Verberne, A. J. (2006). The sympathoinhibitory effects of systemic cholecystokinin are dependent on neurons in the caudal ventrolateral medulla in the rat. *American Journal of Physiology Regulatory, Integrative and Comparative Physiology, 291*, R1390–R1398.

Sato, A., & Schmidt, R. F. (1973). Somatosympathetic reflexes: Afferent fibers, central pathways, discharge characteristics. *Physiological Reviews, 53*, 916–947.

Stebbins, C., Carretero, O., Mindroiu, T., & Longhurst, J. C. (1990). Bradykinin release from contracting skeletal muscle of the cat. *Journal of Applied Physiology, 69*, 1225–1230.

Stebbins, C. L., & Longhurst, J. C. (1986). Bradykinin in the reflex cardiovascular responses to static muscular contraction. *Journal of Applied Physiology, 61*, 271–279.

Stebbins, C., Maruoka, Y., & Longhurst, J. C. (1986). Prostaglandins contribute to cardiovascular reflexes evoked by static muscular contraction. *Circulation Research, 59*, 645–654.

Symons, J., Theodossy, S., Longhurst, J. C., & Stebbins, C. (1991). Intramuscular accumulation of prostaglandins during static contraction of the cat triceps surae. *Journal of Applied Physiology, 71*, 1837–1842.

Thomas, M., & Lundeberg, T. (1994). Importance of modes of acupuncture in the treatment of chronic nociceptive low back pain. *Acta Anaesthesiologica Scandinavica, 38*, 63–69.

Tjen-A-Looi, S. C., Fu, L.-W., Zhou, W., & Longhurst, J. C. (2005). Role of unmyelinated fibers in electroacupuncture cardiovascular responses. *Autonomic Neuroscience, 118*, 43–50.

Tjen-A-Looi, S. C., Guo, Z. L., Li, M., & Longhurst, J. C. (2013). Medullary GABAergic mechanisms contribute to electroacupuncture modulation of cardiovascular depressor responses during gastric distension in rats. *American Journal of Physiology: Regulatory, Integrative, and Comparative Physiology, 304*, R321–R332.

Tjen-A-Looi, S. C., Guo, Z. L., & Longhurst, J. C. (2013). Paraventricular nucleus contributes to acupuncture modulation of sympathoexcitatory cardiovascular reflexes. *FASEB Journal, 27*, LB692.

Tjen-A-Looi, S. C., Hsiao, A. F., & Longhurst, J. C. (2011). Central and peripheral mechanisms underlying gastric distention inhibitory reflex responses in hypercapnic-acidotic rats. *American Journal of Physiology, 300*, H1003–H1012.

Tjen-A-Looi, S. C., Li, P., Li, M., & Longhurst, J. C. (2012). Modulation of cardiopulmonary depressor reflex in nucleus ambiguus by electroacupuncture: Roles of opioids and gamma aminobutyric acid. *American Journal of Physiology, 302*, R833–R844.

Tjen-A-Looi, S. C., Li, P., & Longhurst, J. C. (2003). Prolonged inhibition of rostral ventral lateral medullary premotor sympathetic neuron by electroacupuncture in cats. *Autonomic Neuroscience, 106*, 119–131.

Tjen-A-Looi, S. C., Li, P., & Longhurst, J. C. (2004). Medullary substrate and differential cardiovascular response during stimulation of specific acupoints. *American Journal of Physiology, 287*, R852–R862.

Tjen-A-Looi, S. C., Li, P., & Longhurst, J. C. (2006). Midbrain vlPAG inhibits rVLM cardiovascular sympathoexcitatory responses during acupuncture. *American Journal of Physiology, 290*, H2543–H2553.

Tjen-A-Looi, S. C., Li, P., & Longhurst, J. C. (2007). Role of medullary GABA, opioids, and nociceptin in prolonged inhibition of cardiovascular sympathoexcitatory reflexes during electroacupuncture in cats. *American Journal of Physiology, 293*, H3627–H3635.

Tjen-A-Looi, S. C., Li, P., & Longhurst, J. C. (2009). Processing cardiovascular information in the vlPAG during electroacupuncture in rats: Roles of endocannabinoids and GABA. *Journal of Applied Physiology, 106*, 1793–1799.
Travell, J., & Simons, D. (1992). *Myofascial pain and dysfunction: The trigger point manual*. Baltimore: Lippincott Williams & Wilkins.
Ulett, G. (1989). Acupuncture. In C. Tollison & M. Kriegel (Eds.), *Interdisciplinary rehabilitation of low back pain* (pp. 85–100). Baltimore: Williams and Wilkins.
Ulett, G., Han, S., & Han, J.-S. (1998). Electroacupuncture: Mechanisms and clinical application. *Biological Psychiatry, 44*, 129–138.
Zhou, W., Fu, L. W., Guo, Z. L., & Longhurst, J. C. (2007). Role of glutamate in the rostral ventrolateral medulla in acupuncture-related modulation of visceral reflex sympathoexcitation. *American Journal of Physiology Heart and Circulatory Physiology, 292*, H1868–H1875.
Zhou, W., Fu, L. W., Tjen-A-Looi, S. C., Guo, Z. L., & Longhurst, J. C. (2006). Role of glutamate in a visceral sympathoexcitatory reflex in rostral ventrolateral medulla of cats. *American Journal of Physiology, 291*, H1309–H1318.
Zhou, W., Fu, L.-W., Tjen-A-Looi, S. C., Li, P., & Longhurst, J. C. (2005). Afferent mechanisms underlying stimulation modality-related modulation of acupuncture-related cardiovascular responses. *Journal of Applied Physiology, 98*, 872–880.
Zhou, W., Mahajan, A., & Longhurst, J. C. (2009). Spinal nociceptin mediates electroacupuncture-related modulation of visceral sympathoexcitatory reflex responses in rats. *American Journal of Physiology, 297*, H859–H865.
Zhou, W., Tjen-A-Looi, S. C., & Longhurst, J. C. (2005). Brain stem mechanisms underlying acupuncture modality-related modulation of cardiovascular responses in rats. *Journal of Applied Physiology, 99*, 851–860.

CHAPTER FOURTEEN

Effect and Mechanism of Acupuncture on Gastrointestinal Diseases

Toku Takahashi[1]

Department of Neurology and Department of Surgery, Medical College of Wisconsin and Zablocki VA Medical Center, Milwaukee, Wisconsin, USA

[1]Corresponding author: e-mail address: ttakahashi@mcw.edu

Contents

Abstract

Acupuncture modulates various biomechanical responses, such as prokinetic, antiemetic, and antinociceptive effects. Acupuncture treatment involves the insertion of thin needles into the skin and underlying muscle and the needles are stimulated manually or electrically. Thus, acupuncture stimulates the somatic afferent nerves of the skin and muscles. The somatic sensory information from the body is carried to the cortex area of the brain. Somatic sensory fibers also project to the various nuclei, including the brain stem, periaqueductal gray (PAG), and paraventricular nucleus (PVN) of the hypothalamus. Somatosensory pathways stimulated by acupuncture activate these nuclei. Activation of the brain stem modulates the imbalance between sympathetic activity and parasympathetic activity. Opioid released from the PAG is involved in mediating antiemetic and antinociceptive effects of acupuncture. Oxytocin release from the PVN mediates antistress and antinociceptive effects of acupuncture. Acupuncture may

International Review of Neurobiology, Volume 111
ISSN 0074-7742
http://dx.doi.org/10.1016/B978-0-12-411545-3.00014-6

be effective in patients with functional gastrointestinal (GI) disorders because of its effects on GI motility and visceral pain. It is expected that acupuncture is used in the treatment of patients with functional GI disorders.

1. INTRODUCTION

Acupuncture has been used for treating various gastrointestinal (GI) diseases, including functional dyspepsia (Diehl, 1999; Ma et al., 2012; Xu et al., 2006), irritable bowel syndrome (IBS) (Chan, Carr, & Maberry, 1997; Chey, Maneerattaporn, & Saad, 2011; Fireman, Segal, Kopelman, Sternberg, & Carasso, 2001; Schneider et al., 2006), diabetic gastroparesis (Chang, Ko, Wu, & Chen, 2001; Wang, 2004; Yin, Chen, & Chen, 2010), constipation (Broide et al., 2001; van Wunnik, Baeten, & Southwell, 2011; Zhu, Li, Chen, Wang, & Kan, 2003), diarrhea (Anastasi & McMahon, 2003; Burgmann, Rawsthorne, & Bernstein, 2004), postoperative ileus (Balestrini, Tsuchida, Fukuda, Pappas, & Takahashi, 2005; Deng et al., 2013; Ng et al., 2013), and visceral hypersensitivity (Chu et al., 2012; Tian et al., 2006; Xu, Winston, & Chen, 2009).

Despite the beneficial effects of acupuncture, acupuncture has not been well accepted by the Western physicians. This may be due, at least in part, to the mysterious and unexplainable mechanisms of acupuncture. The beneficial effects of acupuncture should be proven by our Western medicine terminology and methodologies in order to become widely accepted by both patients and physicians (Takahashi, 2006, 2011). In this chapter, the possible mechanisms of acupuncture treatment for GI diseases are discussed from the viewpoint of Western medicine, rather than through the concepts of traditional Chinese medicine.

2. ENERGY FLOW VERSUS SOMATOAUTONOMIC REFLEX

2.1. Energy flow

According to traditional Chinese medicine concepts, the energy force (*Qi*) runs through the body. *Qi* energy enters the body through specific acupuncture points and flows to deeper organ structures, bringing life-giving nourishment of a subtle energetic nature. A person's health is influenced by the flow of *Qi* in the body, in combination with the universal forces of *yin* and *yang*. If the flow of *Qi* is insufficient, unbalanced, or interrupted, *yin* and *yang* become unbalanced and illness may occur (Helms, 1998). However, to date, scientists have been unable to find evidence that supports their existence. Acupuncture is believed to restore the balance of *yin* and *yang*. Acupuncture points (acupoints) are the skin needling points used for acupuncture treatment. In humans, more than 300 acupoints are located along the meridians (Melzack, Stillwell, & Fox, 1977). Despite the fact that the specific acupoints are used for treating specific symptoms and/or diseases, it is not fully understood how their specificity applies and how the needling at acupoints works. Present evidence does not conclusively support that acupuncture points or meridians are electrically distinguishable (Ahn et al., 2008).

2.2. Somatoautonomic reflex

Acupuncture treatment involves the insertion of thin needles into the skin and underlying muscle layer. These acupuncture needles are stimulated manually (manual acupuncture) or electrically (electroacupuncture, EA). Thus, the procedure may stimulate the somatic afferent nerves of the skin and muscles. It has been demonstrated that somatic afferents from the skin and muscle are involved in the control of various autonomic functions. Electrical stimulation of the sciatic nerve inhibits intestinal motility and the splanchnic nerve is responsible for mediating the inhibitory response in dogs (Lehman, 1913). Electrical stimulations of the hind limb stimulate gastric motility via somatic afferents in cats (Jansson, 1969). Sensory stimulation of the abdominal skin by pinching inhibits gastric motility by increasing gastric sympathetic efferent nerve activity in rats (Koizumi, Sato, & Terui, 1980; Sato, Sato, Shimada, & Torigata, 1975). In contrast, pinching the hind paw enhances gastric motility by exciting gastric vagal efferent nerve activity (Kametani, Sato, Sato, & Simpson, 1979). In humans, transcutaneous electrical nerve stimulation applied to the hand and abdomen reduces antral motility (Camilleri, Malagelada, Kao, & Zinsmeister, 1984). These studies indicate the importance of cutaneous input in autonomic control of GI functions.

2.3. Spinal–supraspinal pathways for somatosensory stimulation

The spinal–supraspinal pathways responsible for somatosensory stimulation mainly comprise the posterior spinal cord column pathway and the spinothalamic pathway. Most peripheral thick myelinated afferent fibers activated by the discriminative touch and sense of vibration enter the ipsilateral dorsal column–medial lemniscus tract (posterior column pathway) and emerge into the contralateral spinothalamic pathway. In contrast, the thinly myelinated or unmyelinated afferent fibers activated by pain and temperature are carried up by the contralateral spinothalamic tract to supraspinal levels (spinothalamic pathway). These impulses are further relayed to the thalamus and ultimately sent to the primary somatosensory cortex. In addition, these impulses are also relayed to other brain areas, including the brain stem, periaqueductal gray (PAG), and paraventricular nucleus (PVN) of the hypothalamus, via collateral connections (Hendelman, Humphreys, & Skinner, 2010) (Fig. 14.1).

The nucleus tractus solitarius (NTS) is the primary brain stem relay for visceral information from cardiovascular, respiratory, and GI systems. In addition, recent studies indicate that the NTS also receives somatic afferent

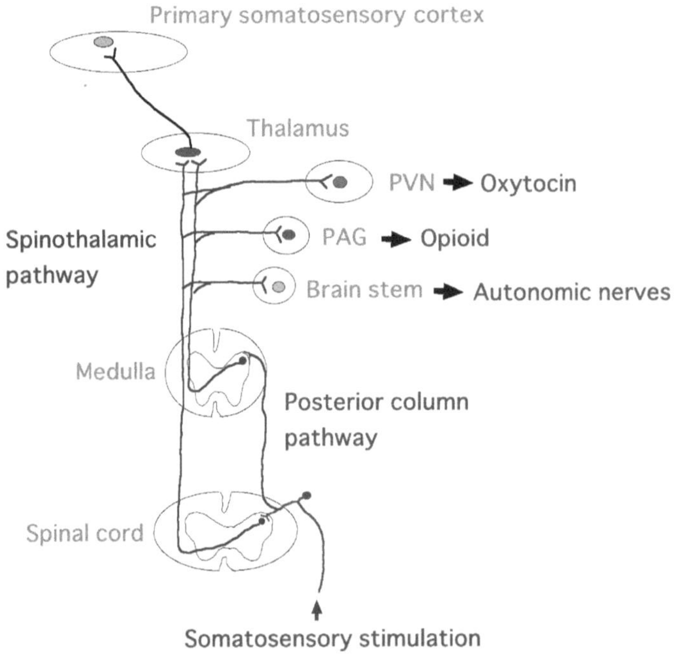

Figure 14.1 Spinal–supraspinal pathways for somatosensory stimulation. The spinal–supraspinal pathways mainly comprise the posterior spinal cord column pathway and the spinothalamic pathway. Spinothalamic pathways are further relayed to the thalamus and ultimately sent to the primary somatosensory cortex. In addition, these impulses are relayed to other brain areas, including the brain stem, periaqueductal gray (PAG), and paraventricular nucleus (PVN) of the hypothalamus, via collateral connections. (For color version of this figure, the reader is referred to the online version of this chapter.)

inputs (Gamboa-Esteves, Lima, & Batten, 2001; Menetrey & Basbaum, 1987). Electrophysiologic study demonstrated that the NTS receives input from cutaneous mechanoreceptors (Toney & Mifflin, 2000). Neurons of the NTS and the spinal cord were labeled by injection of neuroanatomical tracers to the acupoints of the hind limb (ST-36) in rats (Lee et al., 2001).

c-Fos is a cellular proto-oncogene belonging to the immediate early gene family of transcription factors. Once neurons are activated, c-Fos expression is upregulated. By c-Fos immunohistochemistry, the location of activated neurons can easily be detected. Acupuncture at the facial acupoints increases the number of c-Fos immunopositive cells in the NTS in rats (Liu, Li, et al., 2004). Acupuncture at the face increases the number of c-Fos-immunopositive cells in the NTS in rats (Liu, Li, et al., 2004). Therefore, it is highly likely that somatic stimulation induced by acupuncture is conveyed to the NTS.

The NTS is adjacent to the dorsal motor nucleus of vagi (DMV) and both compose the dorsal vagal complex (DVC). The DVC integrates vagovagal reflex, which plays a major role in regulation of GI function (Travagli, Gillis, & Vicini, 1992; Washabau, Fudge, Price, & Barone, 1995). The NTS neurons also project to the rostral ventrolateral medulla (RVLM) of the brain stem (Agarwal & Calaresu, 1991; Kantzides, Owens, De Matteo, & Badoer, 2005; Suzuki, Takayama, & Miura, 1997). The RVLM neurons provide drive to the sympathetic preganglionic neurons in the intermediolateral nucleus of the spinal cord (Laskey & Polosa, 1988). Thus, once NTS neurons are stimulated by acupuncture, autonomic nerve function is altered via the activation of the DMV and/or RVLM (Fig. 14.1).

In addition to the NTS, DMV, and RVLM, acupuncture may activate some other nuclei in the brain. EA at PC-6 (wrist) activates neurons in the arcuate nucleus (ARC) and PAG and inhibits the activity of RVLM, suggesting the existence of ARC–PAG–RVLM neuronal pathway. This pathway may mediate EA inhibition on visceral excitatory cardiovascular reflexes via activation of opioid pathway (Li, Tjen, Guo, Fu, & Longhurst, 2009) (Fig. 14.1). It is conceivable that similar mechanisms might be utilized to produce the effects of EA on GI motor function. It has been shown that the PAG is responsible for mu receptor agonist-induced inhibition of intestinal transit (Tache, Garrick, & Raybould, 1990).

Stimulation of the uterine afferent nerves (the hypogastric and pelvic nerves) excites the neuronal activity at the PVN of the hypothalamus. Stimulation of a somatic nerve of the hind limbs (the sciatic nerve) also activates the PVN neurons. These indicate that specific sensory afferents arrive at the PVN from the uterus and lower legs (Akaishi, Robbins, Sakuma, & Sato, 1988). Oxytocin (OXT) is mainly produced at the PVN and supraoptic nucleus of the hypothalamus. Previous studies showed that OXT levels in plasma and cerebrospinal fluid (CSF) are increased in response to somatosensory nerve stimulation in humans (Stock & Uvnas-Moberg, 1988; Uvnas-Moberg, Bruzelius, Alster, & Lundeberg, 1993). It is conceivable that increased OXT levels in plasma and CSF are associated with OXT expression at the hypothalamus (Fig. 14.1).

3. EFFECTS OF ACUPUNCTURE ON GI MOTILITY

3.1. Region-specific effects of acupuncture

Among more than 300 acupoints, several special points have been used for treating GI symptoms. Acupoint of ST-36 is one of the most effective points, with a wide range of effects for analgesia, immunity, homeostasis, and GI disorders. ST-36 is located near the knee joint approximately 1 cm lateral to the anterior tubercle of the tibia in humans. In contrast, acupoint of ST-25 is used for treating chronic colitis and diarrhea (Lin et al., 1993; Yang & Yan, 1999). ST-25 is located 5 cm lateral to the umbilicus on the abdomen in humans. It is well known that acupuncture has region-specific effects. Acupuncture of the hind limbs stimulates gastric motility in rats (Fukumoto et al., 2003; Li, Zhu, Rong, Ben, & Li, 2007; Sato, Sato, Suzuki, & Uchida, 1993), while acupuncture of the abdominal skin inhibits gastric motility (Li et al., 2007; Sato et al., 1993; Tada et al., 2003) in rats. The excitatory gastric responses to the hind limb are abolished by vagotomy (dissection of vagus nerve) in rats (Li et al., 2007; Sato et al., 1993; Tatewaki et al., 2003). In contrast, the inhibitory gastric responses to abdominal skin stimulation are abolished by sympathetic nerve dissection in rats (Sato et al., 1993; Tada et al., 2003). Acupuncture of the hind limbs increases vagus discharges, while acupuncture on the abdominal surface increases sympathetic discharges in rats (Li et al., 2007; Sato et al., 1993).

Sympathetic and parasympathetic activity can be evaluated by heart rate variability (HRV) analysis in conscious animals noninvasively. The power in the high-frequency (HF) band of HRV represents parasympathetic activities in rats (Imai, Ariga, & Takahashi, 2009; Kuwahara et al., 1994) and humans (Pomeranz et al., 1985). In contrast, the power in the low-frequency (LF) band of HRV represents sympathetic activities in rats (Imai et al., 2009; Kuwahara et al., 1994) and a combination of sympathetic and parasympathetic activities in humans (Pomeranz et al., 1985), respectively. EA at ST-36 decreases, while EA at ST-25 increases the ratio of LF/HF in normal rats (Imai et al., 2008), suggesting that inhibitory gastric responses to the abdomen are mediated via sympathetic nerves. In contrast, the excitatory gastric responses to the hind limb are mediated via parasympathetic nerves.

EA at ST-36 increases the number of c-Fos-immunopositive cells at the DMV and the caudal part of the NTS, while EA at ST-25 increases the number of c-Fos-immunopositive cells at the RVLM and mediocaudal part of the NTS in rats. This suggests that somatic afferents activated by EA at ST-36 are conveyed to the caudal NTS and stimulate the DMV neurons. In contrast, somatic afferents activated by EA at ST-25 are conveyed to the mediocaudal NTS and stimulate the RVLM neurons (Iwa, Tateiwa, Sakita, Fujimiya, & Takahashi, 2007) (Fig. 14.2). Thus, EA at ST-25 inhibits gastric motility via the somatosensory–NTS–RVLM–sympathetic efferent pathway. In contrast, EA at ST-36 stimulates gastric motility via the somatosensory–NTS–DMV–parasympathetic efferent pathway.

The verification of neural pathway may contribute to the mechanisms of the regional difference of acupuncture in regulating GI motor function.

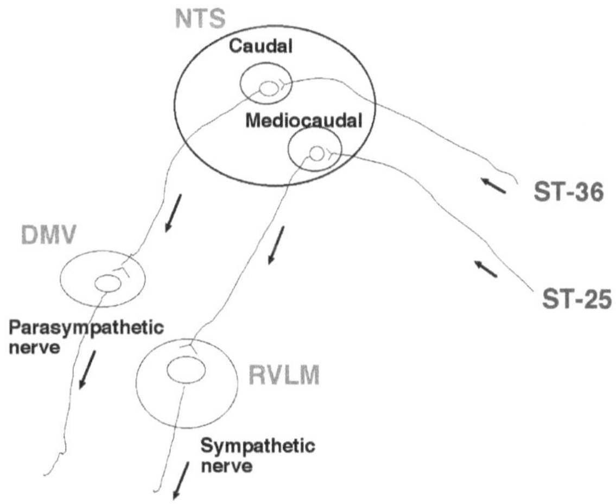

Figure 14.2 Region-specific effects of acupuncture on GI motility. Somatic afferents activated by EA at ST-36 are conveyed to the caudal NTS and stimulate the DMV neurons. In contrast, somatic afferents activated by EA at ST-25 are conveyed to the mediocaudal NTS and stimulate the RVLM neurons. Thus, EA at ST-25 inhibits gastric motility via the somatosensory–NTS–RVLM–sympathetic efferent pathway. In contrast, EA at ST-36 stimulates gastric motility via the somatosensory–NTS–DMV–parasympathetic efferent pathway. (For color version of this figure, the reader is referred to the online version of this chapter.)

Migrating motor complex (MMC) is well characterized by the appearance of GI contractions in the interdigestive state. In a canine study, combined EA at ST-36 and PC-6 enhances the interdigestive gastric MMC (Qian, Peters, & Chen, 1999). EA at ST-36 and PC-6 accelerates gastric emptying and increases the regularity of gastric slow waves in the stomach (Ouyang, Yin, Wang, Pasricha, & Chen, 2002).

The stimulatory effect of EA on gastric motility is associated with increased vagal activity in dogs (Ouyang et al., 2002). Although there is no direct evidence that EA stimulates vagal activity in humans, acupuncture at ST-36 increases the plasma pancreatic polypeptide (PP) levels in humans (Chang et al., 2001). As PP release is dependent on vagal activity, this supports the notion that acupuncture at ST-36 may stimulate the vagal activity in humans. EA at ST-36 accelerates colonic transit and stimulates distal colonic motility in rats (Iwa, Matsushima, et al., 2006; Luo, Liu, Xie, & Hou, 2008). The stimulatory effect of EA on distal colonic motility is mediated via the activation of Barrington's nucleus of the pons and parasympathetic pelvic nerves (Iwa, Matsushima, et al., 2006).

3.2. Effects of acupuncture on GI motility in stressful conditions

It is well known that various stressors stimulate the release of corticotropin-releasing factor (CRF) from the hypothalamus. Released CRF stimulates sympathetic activity, resulting in delayed gastric emptying (Nakade et al., 2005). In contrast, released CRF stimulates parasympathetic activity, resulting in acceleration of colonic transit (Nakade et al., 2007). Delayed gastric emptying and accelerated colonic transit induced by acute stress are mediated via central CRF_2 and CRF_1receptors, respectively (Nakade et al., 2007, 2005) (Fig. 14.3).

Animal studies demonstrate that acupuncture or EA alters various stress-induced physiological responses (Han, Yoon, Cho, Kim, & Min, 1999; Li, Yin-Xiang, Hong, Peng, & Da-Nian, 2001; Manni, Aloe, & Fiore, 2009; Yang et al., 2002). Delayed gastric emptying and accelerated colonic transit induced by acute restraint stress are restored by EA at ST-36 in rats (Iwa, Nakade, Pappas, & Takahashi, 2006). The occurrence of gastric interdigestive contractions (MMC) is abolished by acoustic stress in dogs. EA at ST-36 significantly restores the inhibitory effects of acoustic stress on gastric MMC in dogs (Taniguchi, Imai, Ludwig, & Takahashi, 2012).

Heart rate and LF component of HRV analysis are increased in response to restraint stress in rats (Imai et al., 2009) and acoustic stress in dogs

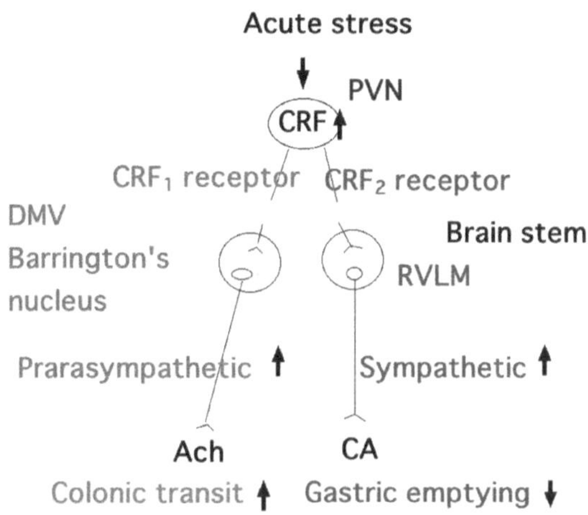

Figure 14.3 Released CRF in response to stress stimulates sympathetic activity, resulting in delayed gastric emptying. In contrast, released CRF stimulates parasympathetic activity, resulting in acceleration of colonic transit. Delayed gastric emptying and accelerated colonic transit induced by acute stress are mediated via central CRF_2 and CRF_1receptors, respectively. (For color version of this figure, the reader is referred to the online version of this chapter.)

(Taniguchi et al., 2012), suggesting increased activity of sympathetic tone during the stress loading. EA at ST-36 attenuates the LF component, while it increases the HF component in response to stress in rats (Imai et al., 2009) and dogs (Taniguchi et al., 2012). These suggest that EA at ST-36 stimulates parasympathetic activity and inhibits sympathetic activity under the stressful conditions, resulting in the restoration of impaired gastric motility. According to traditional Chinese medicine, "Acupuncture restores the balance of *Yin* and *Yang*." This can be translated into the Western medicine terminology as "Acupuncture restores the balance of parasympathetic and sympathetic nerves" (Takahashi, 2011).

GI dysmotility (delayed gastric emptying and accelerated colonic transit) is observed when stressors are chronic and heterotypic (Yoshimoto, Cerjak, et al., 2012; Zheng et al., 2010). Hypothalamic OXT has been shown to have antianxiety effects and antistress effects (Neumann, 2002). Antistress effects of OXT are mediated via its inhibitory effects on hypothalamic CRF expression (Windle et al., 2004; Windle, Shanks, Lightman, & Ingram, 1997). Centrally administered OXT restores GI dysmotility following chronic heterotypic stress in rodents (Yoshimoto, Cerjak, et al., 2012; Zheng et al., 2010). Accelerated colonic transit induced by chronic heterotypic stress is mediated via central CRF_1 receptors (Yoshimoto, Cerjak, et al., 2012). Chronic heterotypic stress fails to increase OXT expression at the PVN and elevated CRF expression is observed following chronic heterotypic stress (Yoshimoto, Cerjak, et al., 2012; Zheng et al., 2010). It is conceivable, therefore, that GI dysmotility induced by chronic heterotypic stress could be treatable once endogenous OXT expression is upregulated.

Yoshimoto, Babygirija, Dobner, Ludwig, and Takahashi (2012) and Yoshimoto, Cerjak, et al. (2012) showed that accelerated colonic transit induced by chronic heterotypic stress is significantly attenuated by EA at ST-36 in rats. The inhibitory effect of EA on accelerated colonic transit is antagonized by icv injection of an OXT antagonist (Yoshimoto, Babygirija, et al., 2012). EA increased the number of OXT-immunopositive cells and decreased CRF-immunopositive cells at the PVN following chronic heterotypic stress (Yoshimoto, Babygirija, et al., 2012). These indicate that EA may act on OXT neurons at the PVN, resulting in reduced CRF expression and restoration of GI dysmotility following chronic heterotypic stress. Others also showed that EA reduced CRF expression at the hypothalamus in rat IBS model (Ma et al., 2009).

3.3. How can we help patients suffering from stress-associated GI disorders?

Animal studies demonstrated that stress-related changes in colonic motility are blocked by selective CRF_1 receptor antagonists (Tache, Martinez, Wang, & Million, 2004 Tache & Perdue, 2004). To treat IBS patients, selective CRF_1 receptor antagonists have been developed. The group at Mayo Clinic assessed whether modulation of central and CRF_1 receptors affects colonic transit and bowel function in female patients with diarrhea-predominant IBS (D-IBS). This randomized, double-blind, placebo-controlled, 2-week study evaluated the effects of oral pexacerfont, a selective CRF_1 receptor antagonist, on colonic transit. Pexacerfont failed to show any significant effects relative to placebo on colonic transit, stool frequency, consistency, and ease of passage (Sweetser et al., 2009). Although preclinical studies suggest that pexacerfont crosses the blood–brain barrier (BBB), there are no human data available whether pexacerfont crosses the BBB and effectively acts on central CRF_1 receptors (Sweetser et al., 2009). Thus, the efficacy of CRF_1 receptor antagonist for IBS patients remains unclear.

Posttraumatic stress disorders (PTSD) are marked by deficits in anxiety, stress regulation, and social functioning. Although cognitive-behavioral therapy is an effective treatment for PTSD, many patients fail to attain remission with cognitive-behavioral therapy alone. Olff et al. showed an augmentation of cognitive-behavioral therapy with OXT in the treatment of PTSD and suggested a dual explanatory mechanism of how OXT is effective: through a reduction of fear response (decreasing amygdala activation, inhibiting fear response, and enhancing learning) and through an increase of social interaction (activating social reward-related brain regions and increasing engagement in the therapeutic alliance). OXT is implicated in both of these areas (Olff, Langeland, Witteveen, & Denys, 2010). Thus, stress-associated diseases such as functional GI disorders may be treatable by OXT.

A clinical trial demonstrated that acupuncture is effective for treating the patients with PTSD. People diagnosed with PTSD were randomized to either an empirically developed acupuncture treatment or a group cognitive-behavioral therapy. Compared with the cognitive-behavioral therapy, acupuncture provided significant effects on PTSD. Symptom reductions, observed at the time that acupuncture treatments ended, were maintained at 3-month follow-up (Hollifield, Sinclair-Lian, Warner, & Hammerschlag, 2007).

As acupuncture upregulates endogenous OXT expression (Yoshimoto, Babygirija, et al., 2012), acupuncture may be effective in reducing daily stress and ameliorating stress-related GI dysmotility.

For anxiety disorders, patients are often given supportive medications such as selective serotonin reuptake inhibitors (SSRIs) and diazepam (a minor tranquilizer). One of the actions of SSRIs is through increased OXT release (Uvnas-Moberg, 1996; Uvnas-Moberg, Bjokstrand, Hillegaart, & Ahlenius, 1999). Administration of SSRIs increased plasma OXT levels in rats (Uvnas-Moberg et al., 1999). The median raphe nucleus is a main site of action for SSRI antidepressants. OXT infusion facilitates serotonin (5-HT) release within the median raphe nucleus and reduces anxiety-related behavior in mice. Infusion of a 5-$HT_{2A/2C}$ receptor antagonist blocks the anxiolytic effect of OXT, suggesting that OXT receptor activation in 5-HT neurons mediates the anxiolytic effects of OXT. Thus, OXT may regulate 5-HT release and exert anxiolytic effects via direct activation of OXT receptor expressed in 5-HT neurons of the raphe nuclei (Yoshida et al., 2009).

The inhibitory effect of OXT on CRF expression is mediated by $GABA_A$ receptors at the PVN (Bulbul et al., 2011). The anxiolytic agent, diazepam, is a $GABA_A$ receptor agonist. Similar to acupuncture, delayed gastric emptying and accelerated colonic transit induced by chronic heterotypic stress were restored by diazepam (2 mg/kg) in rats (Babygirija & Takahashi, unpublished observations). However, anxiolytic agents such as SSRIs or diazepam may not be the first choice to treat GI dysmotility associated with stress, as far as acupuncture is able to activate endogenous OXT pathway and restores GI dysmotility. Acupuncture is promising treatment for GI motility disorders in any population of patients.

4. ACUPUNCTURE FOR VISCERAL PAIN

4.1. Opioid pathway

The analgesic effects of acupuncture are mediated by an endogenous opioid pathway (Diehl, 1999; Stux & Pomeranz, 1995). The painful sensation induced by rectal distension is attenuated by EA at ST-36 in dogs. Naloxone hydrochloride is a nonselective antagonist of opioid receptors. The antinociceptive effect of EA at ST-36 is abolished by the pretreatment with naloxone hydrochloride, a drug that can cross the BBB, but not naloxone methiodide, which does not cross the BBB. This suggests that EA at ST-36 reduces visceral pain via a central opioid pathway (Iwa, Strickland, Nakade, Pappas, & Takahashi, 2005).

Opioid-induced antinociception is mediated by a pathway from the PAG to the various brain nuclei. Recent research has documented that opioids are likely to exert direct effects on PAG projection neurons through both delta- and mu-opioid receptors (Wang & Wessendorf, 2002). EA can induce the expression of opioid peptides and opioid receptors in the CNS (Gao, Wang, Li, & He, 1997; Guo, Wang, Tian, Huo, & Han, 1997).

Using functional MRI (fMRI), acupuncture modulation in CNS structures can be studied noninvasively in humans, allowing for comparison to animal studies. The PAG and cortical areas respond to manual acupuncture at LI-4 (hand), while sham acupuncture results in reduced levels of the PAG and cortical activity (Liu, Feldman, et al., 2004). EA at ST-36 modulates activity in the substantia nigra, nucleus raphe magnus, locus coeruleus, nucleus cuneiformis, and the PAG. Activation in the ventrolateral PAG is greater for EA at ST-36, compared to sham EA (Napadow et al., 2009).

Visceral hypersensitivity develops following chronic heterotypic stress in rats (Bradesi et al., 2005; Winston, Xu, & Sarna, 2010). Increased cortisol and epinephrine in the periphery has been shown to mediate visceral hypersensitivity following chronic heterotypic stress (Hong et al., 2001; Sengupta, 2009). As mentioned earlier, acupuncture has an antinociceptive effect via central opioid release. Visceromotor response to colorectal distension was

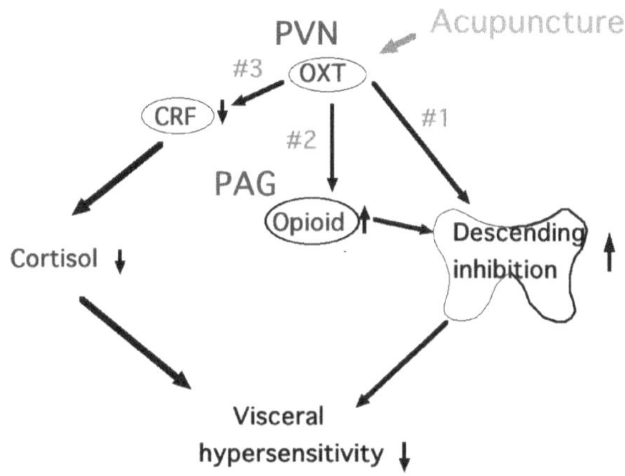

Figure 14.4 The possible mechanisms of antinociceptive effects of hypothalamic OXT. #1: Spinal pathway, a direct descending projection of OXTergic neurons from the PVN to spinal cord and modulation of spinal dorsal horn neurons. #2: Pontomedullary pathway, a descending projection of OXTergic neurons from the PVN to PAG to enhance opioid release. #3: CRF–HPA axis pathway, OXT-induced inhibition on CRF release and reduction of systemic cortisol. (For color version of this figure, the reader is referred to the online version of this chapter.)

attenuated by acupuncture in rats (Gao et al., 1997; Guo et al., 1997). As acupuncture and EA activate PAG neurons (Guo et al., 2004, 2007), it is conceivable that acupuncture impulses are relayed via collateral connections to the PAG, which contains the endogenous opioids. PAG sends inhibitory information to the spinal dorsal horn and blocks noxious signal inputs from the periphery (descending inhibition) (Figs. 14.2 and 14.4).

4.2. OXT pathway

In addition to opioid system, recent studies suggested that antinociceptive effect of somatosensory stimulation is mediated via OXT system (Agren, Lundeberg, Uvnas-Moberg, & Sato, 1995; Stock & Uvnas-Moberg, 1988; Uvnas-Moberg et al., 1993; Yang et al., 2007). Greater OXT levels were correlated with greater tolerance to ischemic pain or cold pressor pain in healthy women. Reduced OXTergic function may be one of multiple biological factors contributing to the greater sensitivity to experimental ischemic pain (Grewen, Light, Mechlin, & Girdler, 2008).

The descending inhibitory system also receives a wide array of afferent modulations from the supra-brain stem regions, including the hypothalamus. It has been documented that there is direct descending OXT neuron projection from the PVN to the spinal cord. This is the way by which stimulation of hypothalamic OXT neurons modulates nociceptive responses (Breton, Poisbeau, & Darbon, 2009; Rojas-Piloni, Mejia-Rodriguez, Martinez-Lorenzana, & Condes-Lara, 2010). The possible mechanisms of antinociceptive effects of hypothalamic OXT are (1) a direct descending projection of OXTergic neurons from the PVN to the spinal cord and modulation of spinal dorsal horn neurons (#1, spinal pathway), (2) a descending projection of OXTergic neurons from the PVN to PAG to enhance opioid release (#2, pontomedullary pathway), and (3) OXT-induced inhibition on CRF release and reduction of systemic cortisol (#3, CRF–HPA axis pathway) (Fig. 14.4).

Intravenous infusion of OXT significantly increases thresholds for visceral perception in IBS patients (Louvel et al., 1996). It has been demonstrated that OXT system is involved in mediating the antinociceptive effects of EA (Yang et al., 2007). Thus, it is conceivable that acupuncture stimulates opioid neurons and OXT neurons, giving pain relief.

5. ACUPUNCTURE FOR EMESIS

Acupoint of PC-6 has been used for nausea and vomiting. PC-6 is located in the groove caudal to the flexor carpi radialis and cranial to the superficial digital flexor muscles. There is clear evidence that acupuncture treatment is effective on postoperative and chemotherapy-induced nausea and vomiting. Acupuncture at PC-6 decreases cisplatin-associated nausea and vomiting (Dundee, Ghaly, Fitzpatrick, Lynch, & Abram, 1987). EA at PC-6 significantly inhibits the postoperative nausea and vomiting after the surgery (Dundee et al., 1989; Gan, Jiao, Zenn, & Georgiade, 2004). EA at PC-6 reduces gastric tachyarrhythmia in vection-induced motion sickness in healthy volunteers (Hu, Stern, & Koch, 1992). Combined acupuncture at ST-36 and PC-6 increases the percentage of regular slow waves, resulting in the normalization of arrhythmia in healthy humans (Lin et al., 1997).

The mechanisms of antiemetic effect of acupuncture need to be elucidated. Level of plasma arginine vasopressin is rapidly elevated in response to diverse emetic stimuli such as motion stimuli (Kim, Chey, Owyang, & Hasler, 1997) and anticancer drugs (Cubeddu, Lindley, Wetsel, Carl, & Negro-Vilar, 1990; Edwards, Carmichael, Baylis, & Harris, 1989). Intravenous injection of vasopressin results in vomiting and retching in humans (Kim et al., 1997; Rowe, Shelton, Helderman, Vestal, & Robertson,

1979) and dogs (Carpenter, Briggs, & Strominger, 1984; Chen, Qian, Ouyang, & Yin, 2003). Acupuncture at ST-36 and PC-6 attenuates symptom scores of emesis induced by vasopressin infusion in dogs (Chen et al., 2003). Retrograde propulsive movement is highly associated with the frequency of retching and vomiting in dogs. EA at PC-6 reduces the frequency of retrograde propulsive movement induced by vasopressin (Tatewaki et al., 2005).

Opioids have dual effects: an antiemetic effect and an emetic effect. The emetic effect of opioids is mediated via the chemoreceptor trigger zone (CTZ), whereas the antiemetic effect of opioids is mediated via the vomiting center (Blancquaert, Lefebvre, & Willems, 1986; Costello & Borison, 1977; Foss, Yuan, Roizen, & Goldberg, 1998). The CTZ is contained in the area postrema on the caudal margin of the fourth ventricle. As area postrema has no BBB, naloxone methiodide (a peripheral opioid antagonist) can antagonize the emetic effect of opioid mediated via the CTZ.

In contrast, the vomiting center is deeply located beneath the solitary tract of the caudal brain stem. Both the emetic and antiemetic effects of opioid can be blocked by naloxone hydrochloride, because naloxone hydrochloride can cross the BBB (Foss et al., 1998).

Naloxone hydrochloride, but not naloxone methiodide, abolishes the antiemetic effect of EA at PC-6 in dogs (Tatewaki et al., 2005). This suggests that the antiemetic effect of EA at PC-6 is mediated via the central opioid pathway. The most frequently experienced side effects of chemotherapy are nausea and vomiting. The recent review article concluded that acupuncture may alleviate chemotherapy-induced nausea and vomiting (Spencer, Beaumont, Del Carmen, Growdon, & Goodman, 2012).

6. CONCLUSION

The mechanisms behind the beneficial effects of acupuncture can be explained by Western medicine terminology. Acupuncture stimulation of somatosensory neurons activates various nuclei at the CNS, including the NTS, DMV, RVLM, PAG, and PVN. Activation of the NTS, DMV, and RVLM modulates the imbalance between sympathetic activity and parasympathetic activity. Opioid released from the PAG is involved in mediating antiemetic and antinociceptive effects of acupuncture. OXT release from the PVN is involved in mediating antistress and antinociceptive effects of acupuncture.

Functional GI diseases are common in the general population with a reported prevalence of 25–40%. Functional GI disorders are the multifactorial disorders in which the pathophysiological mechanisms are variably combined in different patients. Stress is highly associated with the pathogenesis of functional GI disorders. GI dysmotility and visceral hypersensitivity are especially important factors. Acupuncture may be effective in patients with functional GI disorders because acupuncture has been shown to alter GI motility and visceral pain.

REFERENCES

Agarwal, S. K., & Calaresu, F. R. (1991). Monosynaptic connection from caudal to rostral ventrolateral medulla in the baroreceptor reflex pathway. *Brain Research*, *555*, 70–74.

Agren, G., Lundeberg, T., Uvnas-Moberg, K., & Sato, A. (1995). The oxytocin antagonist 1-deamino-2-D-Tyr-(Oet)-4-Thr-8-Orn-oxytocin reverses the increase in the withdrawal response latency to thermal, but not mechanical nociceptive stimuli following oxytocin administration or massage-like stroking in rats. *Neuroscience Letters*, *187*, 49–52.

Ahn, A. C., Colbert, A. P., Anderson, B. J., Martinsen, O. G., Hammerschlag, R., Cina, S., et al. (2008). Electrical properties of acupuncture points and meridians: A systematic review. *Bioelectromagnetics*, *29*, 245–256.

Akaishi, T., Robbins, A., Sakuma, Y., & Sato, Y. (1988). Neural inputs from the uterus to the paraventricular magnocellular neurons in the rat. *Neuroscience Letters*, *84*, 57–62.

Anastasi, J. K., & McMahon, D. J. (2003). Testing strategies to reduce diarrhea in persons with HIV using traditional Chinese medicine: Acupuncture and moxibustion. *The Journal of the Association of Nurses in AIDS Care*, *14*, 28–40.

Balestrini, J. L., Tsuchida, D., Fukuda, H., Pappas, T. N., & Takahashi, T. (2005). Acupuncture accelerates delayed gastrointestinal transit after abdominal surgery in conscious rats. *Scandinavian Journal of Gastroenterology*, *40*, 734–735.

Blancquaert, J. P., Lefebvre, R. A., & Willems, J. L. (1986). Emetic and antiemetic effects of opioids in the dog. *European Journal of Pharmacology*, *128*, 143–150.

Bradesi, S., Schwetz, I., Ennes, H. S., Lamy, C. M., Ohning, G., Fanselow, M., et al. (2005). Repeated exposure to water avoidance stress in rats: A new model for sustained visceral hyperalgesia. *American Journal of Physiology Gastrointestinal and Liver Physiology*, *289*, G42–G53.

Breton, J. D., Poisbeau, P., & Darbon, P. (2009). Antinociceptive action of oxytocin involves inhibition of potassium channel currents in lamina II neurons of the rat spinal cord. *Molecular Pain*, *5*, 63.

Broide, E., Pintov, S., Portnoy, S., Barg, J., Klinowski, E., & Scapa, E. (2001). Effectiveness of acupuncture for treatment of childhood constipation. *Digestive Diseases and Sciences*, *46*, 1270–1275.

Bulbul, M., Babygirija, R., Cerjak, D., Yoshimoto, S., Ludwig, K., & Takahashi, T. (2011). Hypothalamic oxytocin attenuates CRF expression via GABA(A) receptors in rats. *Brain Research*, *1387*, 39–45.

Burgmann, T., Rawsthorne, P., & Bernstein, C. N. (2004). Predictors of alternative and complementary medicine use in inflammatory bowel disease: Do measures of conventional health care utilization relate to use? *The American Journal of Gastroenterology*, *99*, 889–893.

Camilleri, M., Malagelada, J. R., Kao, P. C., & Zinsmeister, A. R. (1984). Effect of somatovisceral reflexes and selective dermatomal stimulation on postcibal antral pressure activity. *The American Journal of Physiology*, *247*, G703–G708.

Carpenter, D. O., Briggs, D. B., & Strominger, N. (1984). Behavioral and electrophysiological studies of peptide-induced emesis in dogs. *Federation Proceedings*, *43*, 2952–2954.
Chan, J., Carr, I., & Maberry, J. (1997). The role of acupuncture in the treatment of irritable bowel syndrome: A pilot study. *Hepato-Gastroenterology*, *44*, 1328–1330.
Chang, C. S., Ko, C. W., Wu, C. Y., & Chen, G. H. (2001). Effect of electrical stimulation on acupuncture points in diabetic patients with gastric dysrhythmia: A pilot study. *Digestion*, *64*, 184–190.
Chen, J. D., Qian, L., Ouyang, H., & Yin, J. (2003). Gastric electrical stimulation with short pulses reduces vomiting but not dysrhythmias in dogs. *Gastroenterology*, *124*, 401–409.
Chey, W. D., Maneerattaporn, M., & Saad, R. (2011). Pharmacologic and complementary and alternative medicine therapies for irritable bowel syndrome. *Gut Liver*, *5*, 253–266.
Chu, W. C., Wu, J. C., Yew, D. T., Zhang, L., Shi, L., Yeung, D. K., et al. (2012). Does acupuncture therapy alter activation of neural pathway for pain perception in irritable bowel syndrome? A comparative study of true and sham acupuncture using functional magnetic resonance imaging. *Journal of Neurogastroenterology and Motility*, *18*, 305–316.
Costello, D. J., & Borison, H. L. (1977). Naloxone antagonizes narcotic self blockade of emesis in the cat. *The Journal of Pharmacology and Experimental Therapeutics*, *203*, 222–230.
Cubeddu, L. X., Lindley, C. M., Wetsel, W., Carl, P. L., & Negro-Vilar, A. (1990). Role of angiotensin II and vasopressin in cisplatin-induced emesis. *Life Sciences*, *46*, 699–705.
Deng, G., Wong, W. D., Guillem, J., Chan, Y., Affuso, T., Yeung, K. S., et al. (2013). A phase II, randomized, controlled trial of acupuncture for reduction of Postcolectomy Ileus. *Annals of Surgical Oncology*, *20*, 1164–1169.
Diehl, D. L. (1999). Acupuncture for gastrointestinal and hepatobiliary disorders. *Journal of Alternative and Complementary Medicine*, *5*, 27–45.
Dundee, J. W., Ghaly, R. G., Bill, K. M., Chestnutt, W. N., Fitzpatrick, K. T., & Lynas, A. G. (1989). Effect of stimulation of the P6 antiemetic point on postoperative nausea and vomiting. *British Journal of Anaesthesia*, *63*, 612–618.
Dundee, J., Ghaly, R., Fitzpatrick, K., Lynch, G. A., & Abram, W. P. (1987). Acupuncture to prevent cisplatin-associated vomiting. *Lancet*, *1*, 1083.
Edwards, C. M., Carmichael, J., Baylis, P. H., & Harris, A. L. (1989). Arginine vasopressin—A mediator of chemotherapy induced emesis? *British Journal of Cancer*, *59*, 467–470.
Fireman, Z., Segal, A., Kopelman, Y., Sternberg, A., & Carasso, R. (2001). Acupuncture treatment for irritable bowel syndrome. A double-blind controlled study. *Digestion*, *64*, 100–103.
Foss, J. F., Yuan, C. S., Roizen, M. F., & Goldberg, L. I. (1998). Prevention of apomorphine- or cisplatin-induced emesis in the dog by a combination of methylnaltrexone and morphine. *Cancer Chemotherapy and Pharmacology*, *42*, 287–291.
Fukumoto, S., Tatewaki, M., Yamada, T., Fujimiya, M., Mantyh, C., Voss, M., et al. (2003). Short-chain fatty acids stimulate colonic transit via intraluminal 5-HT release in rats. *American Journal of Physiology Regulatory, Integrative and Comparative Physiology*, *284*, R1269–R1276.
Gamboa-Esteves, F. O., Lima, D., & Batten, T. F. (2001). Neurochemistry of superficial spinal neurones projecting to nucleus of the solitary tract that express c-fos on chemical somatic and visceral nociceptive input in the rat. *Metabolic Brain Disease*, *16*, 151–164.
Gan, T. J., Jiao, K. R., Zenn, M., & Georgiade, G. (2004). A randomized controlled comparison of electro-acupoint stimulation or ondansetron versus placebo for the prevention of postoperative nausea and vomiting. *Anesthesia & Analgesia*, *99*, 1070–1075, table of contents.
Gao, M., Wang, M., Li, K., & He, L. (1997). Changes of mu opioid receptor binding sites in rat brain following electroacupuncture. *Acupuncture & Electro-Therapeutics Research*, *22*, 161–166.

Grewen, K. M., Light, K. C., Mechlin, B., & Girdler, S. S. (2008). Ethnicity is associated with alterations in oxytocin relationships to pain sensitivity in women. *Ethnicity & Health, 13*, 219–241.

Guo, H. F., Wang, X. M., Tian, J. H., Huo, Y. P., & Han, J. S. (1997). 2 Hz and 100 Hz electroacupuncture accelerate the expression of genes encoding three opioid peptides in the rat brain. *Sheng Li Xue Bao, 49*, 121–127.

Guo, Z. L., & Longhurst, J. C. (2007). Expression of c-Fos in arcuate nucleus induced by electroacupuncture: relations to neurons containing opioids and glutamate. *Brain Research, 1166*, 65–76.

Guo, Z. L., Moazzami, A. R., & Longhurst, J. C. (2004). Electroacupuncture induces c-Fos expression in the rostral ventrolateral medulla and periaqueductal gray in cats: relation to opioid containing neurons. *Brain Research, 1030*, 103–115.

Han, S. H., Yoon, S. H., Cho, Y. W., Kim, C. J., & Min, B. I. (1999). Inhibitory effects of electroacupuncture on stress responses evoked by tooth-pulp stimulation in rats. *Physiology & Behavior, 66*, 217–222.

Helms, J. M. (1998). An overview of medical acupuncture. *Alternative Therapies in Health and Medicine, 4*, 35–45.

Hendelman, W., Humphreys, P., & Skinner, C. (2010). *The integrated nervous system*. Boca Raton, FL: CRC Press.

Hollifield, M., Sinclair-Lian, N., Warner, T. D., & Hammerschlag, R. (2007). Acupuncture for posttraumatic stress disorder: A randomized controlled pilot trial. *The Journal of Nervous and Mental Disease, 195*, 504–513.

Hong, S., Zheng, G., Wu, X., Snider, N. T., Owyang, C., & Wiley, J. W. (2001). Corticosterone mediates reciprocal changes in CB 1 and TRPV1 receptors in primary sensory neurons in the chronically stressed rat. *Gastroenterology, 140*, 627.e4–637.e4.

Hu, S., Stern, R. M., & Koch, K. L. (1992). Electrical acustimulation relieves vection-induced motion sickness. *Gastroenterology, 102*, 1854–1858.

Imai, K., Ariga, H., Chen, C., Mantyh, C., Pappas, T. N., & Takahashi, T. (2008). Effects of electroacupuncture on gastric motility and heart rate variability in conscious rats. *Autonomic Neuroscience, 138*, 91–98.

Imai, K., Ariga, H., & Takahashi, T. (2009). Electroacupuncture improves imbalance of autonomic function under restraint stress in conscious rats. *The American Journal of Chinese Medicine, 37*, 45–55.

Iwa, M., Matsushima, M., Nakade, Y., Pappas, T. N., Fujimiya, M., & Takahashi, T. (2006). Electroacupuncture at ST-36 accelerates colonic motility and transit in freely moving conscious rats. *American Journal of Physiology Gastrointestinal and Liver Physiology, 290*, G285–G292.

Iwa, M., Nakade, Y., Pappas, T. N., & Takahashi, T. (2006). Electroacupuncture elicits dual effects: Stimulation of delayed gastric emptying and inhibition of accelerated colonic transit induced by restraint stress in rats. *Digestive Diseases and Sciences, 51*, 1493–1500.

Iwa, M., Strickland, C., Nakade, Y., Pappas, T. N., & Takahashi, T. (2005). Electroacupuncture reduces rectal distension-induced blood pressure changes in conscious dogs. *Digestive Diseases and Sciences, 50*, 1264–1270.

Iwa, M., Tateiwa, M., Sakita, M., Fujimiya, M., & Takahashi, T. (2007). Anatomical evidence of regional specific effects of acupuncture on gastric motor function in rats. *Autonomic Neuroscience, 137*, 67–76.

Jansson, G. (1969). Extrinsic nervous control of gastric motility. *Acta Physiologica Scandinavica Supplementum, 326*, 1–42.

Kametani, H., Sato, A., Sato, Y., & Simpson, A. (1979). Neural mechanisms of reflex facilitation and inhibition of gastric motility to stimulation of various skin areas in rats. *The Journal of Physiology (London), 294*, 407–418.

Kantzides, A., Owens, N. C., De Matteo, R., & Badoer, E. (2005). Right atrial stretch activates neurons in autonomic brain regions that project to the rostral ventrolateral medulla in the rat. *Neuroscience*, *133*, 775–786.
Kim, M. S., Chey, W. D., Owyang, C., & Hasler, W. L. (1997). Role of plasma vasopressin as a mediator of nausea and gastric slow wave dysrhythmias in motion sickness. *The American Journal of Physiology*, *272*, G853–G862.
Koizumi, K., Sato, A., & Terui, N. (1980). Role of somatic afferents in autonomic system control of the intestinal motility. *Brain Research*, *182*, 85–97.
Kuwahara, M., Yayou, K., Ishii, K., Hashimoto, S., Tsubone, H., & Sugano, S. (1994). Power spectral analysis of heart rate variability as a new method for assessing autonomic activity in the rat. *Journal of Electrocardiology*, *27*, 333–337.
Laskey, W., & Polosa, C. (1988). Characteristics of the sympathetic preganglionic neuron and its synaptic input. *Progress in Neurobiology*, *31*, 47–84.
Lee, C. H., Jung, H. S., Lee, T. Y., Lee, S. R., Yuk, S. W., Lee, K. G., et al. (2001). Studies of the central neural pathways to the stomach and Zusanli (ST36). *The American Journal of Chinese Medicine*, *29*, 211–220.
Lehman, A. V. (1913). Studien uber reflektorische Darmbewegungen beim Hunde. *Pflugers Archiv für die Gesamte Physiologie*, *149*, 413–433.
Li, P., Tjen, A. L. S. C., Guo, Z. L., Fu, L. W., & Longhurst, J. C. (2009). Long-loop pathways in cardiovascular electroacupuncture responses. *Journal of Applied Physiology*, *106*, 620–630.
Li, L., Yin-Xiang, C., Hong, X., Peng, L., & Da-Nian, Z. (2001). Nitric oxide in vPAG mediates the depressor response to acupuncture in stress-induced hypertensive rats. *Acupuncture & Electro-Therapeutics Research*, *26*, 165–170.
Li, Y., Zhu, B., Rong, P. J., Ben, H., & Li, Y. H. (2007). Neural mechanism of acupuncture-modulated gastric motility. *World Journal of Gastroenterology*, *13*, 709–716.
Lin, X., Liang, J., Ren, J., Mu, F., Zhang, M., & Chen, J. D. (1997). Electrical stimulation of acupuncture points enhances gastric myoelectrical activity in humans. *The American Journal of Gastroenterology*, *92*, 1527–1530.
Lin, Y., Zhou, Z., Shen, W., Shen, J., Hu, M., Zhang, F., et al. (1993). Clinical and experimental studies on shallow needling technique for treating childhood diarrhea. *Journal of Traditional Chinese Medicine*, *13*, 107–114.
Liu, W. C., Feldman, S. C., Cook, D. B., Hung, D. L., Xu, T., Kalnin, A. J., et al. (2004). fMRI study of acupuncture-induced periaqueductal gray activity in humans. *Neuroreport*, *15*, 1937–1940.
Liu, J. H., Li, J., Yan, J., Chang, X. R., Cui, R. F., He, J. F., et al. (2004). Expression of c-fos in the nucleus of the solitary tract following electroacupuncture at facial acupoints and gastric distension in rats. *Neuroscience Letters*, *366*, 215–219.
Louvel, D., Delvaux, M., Felez, A., Fioramonti, J., Bueno, L., Lazorthes, Y., et al. (1996). Oxytocin increases thresholds of colonic visceral perception in patients with irritable bowel syndrome. *Gut*, *39*, 741–747.
Luo, D., Liu, S., Xie, X., & Hou, X. (2008). Electroacupuncture at acupoint ST-36 promotes contractility of distal colon via a cholinergic pathway in conscious rats. *Digestive Diseases and Sciences*, *53*, 689–693.
Ma, X. P., Tan, L. Y., Yang, Y., Wu, H. G., Jiang, B., Liu, H. R., et al. (2009). Effect of electro-acupuncture on substance P, its receptor and corticotropin-releasing hormone in rats with irritable bowel syndrome. *World Journal of Gastroenterology*, *15*, 5211–5217.
Ma, T. T., Yu, S. Y., Li, Y., Liang, F. R., Tian, X. P., Zheng, H., et al. (2012). Randomised clinical trial: An assessment of acupuncture on specific meridian or specific acupoint vs. sham acupuncture for treating functional dyspepsia. *Alimentary Pharmacology & Therapeutics*, *35*, 552–561.

Manni, L., Aloe, L., & Fiore, M. (2009). Changes in cognition induced by social isolation in the mouse are restored by electro-acupuncture. *Physiology & Behavior, 98*, 537–542.

Melzack, R., Stillwell, D. M., & Fox, E. J. (1977). Trigger points and acupuncture points for pain: Correlations and implications. *Pain, 3*, 3–23.

Menetrey, D., & Basbaum, A. I. (1987). Spinal and trigeminal projections to the nucleus of the solitary tract: A possible substrate for somatovisceral and viscerovisceral reflex activation. *The Journal of Comparative Neurology, 255*, 439–450.

Nakade, Y., Fukuda, H., Iwa, M., Tsukamoto, K., Yanagi, H., Yamamura, T., et al. (2007). Restraint stress stimulates colonic motility via central corticotropin-releasing factor and peripheral 5-HT3 receptors in conscious rats. *American Journal of Physiology Gastrointestinal and Liver Physiology, 292*, G1037–G1044.

Nakade, Y., Tsuchida, D., Fukuda, H., Iwa, M., Pappas, T. N., & Takahashi, T. (2005). Restraint stress delays solid gastric emptying via a central CRF and peripheral sympathetic neuron in rats. *American Journal of Physiology Regulatory, Integrative and Comparative Physiology, 288*, R427–R432.

Napadow, V., Dhond, R., Park, K., Kim, J., Makris, N., Kwong, K. K., et al. (2009). Time-variant fMRI activity in the brainstem and higher structures in response to acupuncture. *NeuroImage, 47*, 289–301.

Neumann, I. D. (2002). Involvement of the brain oxytocin system in stress coping: Interactions with the hypothalamo-pituitary-adrenal axis. *Progress in Brain Research, 139*, 147–162.

Ng, S. S., Leung, W. W., Mak, T. W., Hon, S. S., Li, J. C., Wong, C. Y., et al. (2013). Electroacupuncture reduces duration of postoperative ileus after laparoscopic surgery for colorectal cancer. *Gastroenterology, 144*, 307.e1–313.e1.

Olff, M., Langeland, W., Witteveen, A., & Denys, D. (2010). A psychobiological rationale for oxytocin in the treatment of posttraumatic stress disorder. *CNS Spectrums, 15*, 522–530.

Ouyang, H., Yin, J., Wang, Z., Pasricha, P. J., & Chen, J. D. (2002). Electroacupuncture accelerates gastric emptying in association with changes in vagal activity. *American Journal of Physiology Gastrointestinal and Liver Physiology, 282*, G390–G396.

Pomeranz, B., Macaulay, R. J., Caudill, M. A., Kutz, I., Adam, D., Gordon, D., et al. (1985). Assessment of autonomic function in humans by heart rate spectral analysis. *The American Journal of Physiology, 248*, H151–H153.

Qian, L., Peters, L. J., & Chen, J. D. (1999). Effects of electroacupuncture on gastric migrating myoelectrical complex in dogs. *Digestive Diseases and Sciences, 44*, 56–62.

Rojas-Piloni, G., Mejia-Rodriguez, R., Martinez-Lorenzana, G., & Condes-Lara, M. (2010). Oxytocin, but not vassopressin, modulates nociceptive responses in dorsal horn neurons. *Neuroscience Letters, 476*, 32–35.

Rowe, J. W., Shelton, R. L., Helderman, J. H., Vestal, R. E., & Robertson, G. L. (1979). Influence of the emetic reflex on vasopressin release in man. *Kidney International, 16*, 729–735.

Sato, A., Sato, Y., Shimada, F., & Torigata, Y. (1975). Changes in gastric motility produced by nociceptive stimulation of the skin rats. *Brain Research, 87*, 151–159.

Sato, A., Sato, Y., Suzuki, A., & Uchida, S. (1993). Neural mechanisms of the reflex inhibition and excitation of gastric motility elicited by acupuncture-like stimulation in anesthetized rats. *Neuroscience Research, 18*, 53–62.

Schneider, A., Enck, P., Streitberger, K., Weiland, C., Bagheri, S., Witte, S., et al. (2006). Acupuncture treatment in irritable bowel syndrome. *Gut, 55*, 649–654.

Sengupta, J. N. (2009). Visceral pain: The neurophysiological mechanism. *Handbook of Experimental Pharmacology, 194*, 31–74.

Spencer, R., Beaumont, E., Del Carmen, M., Growdon, W., & Goodman, A. (2012). Acupuncture for chemotherapy-induced nausea and vomiting. *Medical Acupuncture, 24*, 104–112.

Stock, S., & Uvnas-Moberg, K. (1988). Increased plasma levels of oxytocin in response to afferent electrical stimulation of the sciatic and vagal nerves and in response to touch and pinch in anaesthetized rats. *Acta Physiologica Scandinavica, 132*, 29–34.

Stux, G., & Pomeranz, B. (1995). *Basics of acupuncture* (4th ed.). New York: Springer.

Suzuki, T., Takayama, K., & Miura, M. (1997). Distribution and projection of the medullary cardiovascular control neurons containing glutamate, glutamic acid decarboxylase, tyrosine hydroxylase and phenylethanolamine N-methyltransferase in rats. *Neuroscience Research, 27*, 9–19.

Sweetser, S., Camilleri, M., Linker Nord, S. J., Burton, D. D., Castenada, L., Croop, R., et al. (2009). Do corticotropin releasing factor-1 receptors influence colonic transit and bowel function in women with irritable bowel syndrome? *American Journal of Physiology Gastrointestinal and Liver Physiology, 296*, G1299–G1306.

Tache, Y., Garrick, T., & Raybould, H. (1990). Central nervous system action of peptides to influence gastrointestinal motor function. *Gastroenterology, 98*, 517–528.

Tache, Y., Martinez, V., Wang, L., & Million, M. (2004). CRF1 receptor signaling pathways are involved in stress-related alterations of colonic function and viscerosensitivity: Implications for irritable bowel syndrome. *British Journal of Pharmacology, 141*, 1321–1330.

Tache, Y., & Perdue, M. H. (2004). Role of peripheral CRF signalling pathways in stress-related alterations of gut motility and mucosal function. *Neurogastroenterology and Motility, 16*(Suppl. 1), 137–142.

Tada, H., Fujita, M., Harris, M., Tatewaki, M., Nakagawa, K., Yamamura, T., et al. (2003). Neural mechanism of acupuncture-induced gastric relaxations in rats. *Digestive Diseases and Sciences, 48*, 59–68.

Takahashi, T. (2006). Acupuncture for functional gastrointestinal disorders. *Journal of Gastroenterology, 41*, 408–417.

Takahashi, T. (2011). Mechanism of acupuncture on neuromodulation in the gut: A review. *Neuromodulation, 14*, 8–12.

Taniguchi, H., Imai, K., Ludwig, K., & Takahashi, T. (2012). Effects of acupuncture on stress-induced gastrointestinal dysmotility in conscious dogs. *Medical Acupuncture, 24*, 43–49.

Tatewaki, M., Harris, M., Uemura, K., Ueno, T., Hoshino, E., Shiotani, A., et al. (2003). Dual effects of acupuncture on gastric motility in conscious rats. *American Journal of Physiology Regulatory, Integrative and Comparative Physiology, 285*, R862–R872.

Tatewaki, M., Strickland, C., Fukuda, H., Tsuchida, D., Hoshino, E., Pappas, T. N., et al. (2005). Effects of acupuncture on vasopressin-induced emesis in conscious dogs. *American Journal of Physiology Regulatory, Integrative and Comparative Physiology, 288*, R401–R408.

Tian, X. Y., Bian, Z. X., Hu, X. G., Zhang, X. J., Liu, L., & Zhang, H. (2006). Electroacupuncture attenuates stress-induced defecation in rats with chronic visceral hypersensitivity via serotonergic pathway. *Brain Research, 1088*, 101–108.

Toney, G. M., & Mifflin, S. W. (2000). Sensory modalities conveyed in the hindlimb somatic afferent input to nucleus tractus solitarius. *Journal of Applied Physiology, 88*, 2062–2073.

Travagli, R. A., Gillis, R. A., & Vicini, S. (1992). Effects of thyrotropin-releasing hormone on neurons in rat dorsal motor nucleus of the vagus, in vitro. *The American Journal of Physiology, 263*, G508–G517.

Uvnas-Moberg, K. (1996). Neuroendocrinology of the mother-child interaction. *Trends in Endocrinology and Metabolism, 7*, 126–131.

Uvnas-Moberg, K., Bjokstrand, E., Hillegaart, V., & Ahlenius, S. (1999). Oxytocin as a possible mediator of SSRI-induced antidepressant effects. *Psychopharmacology (Berlin), 142*, 95–101.

Uvnas-Moberg, K., Bruzelius, G., Alster, P., & Lundeberg, T. (1993). The antinociceptive effect of non-noxious sensory stimulation is mediated partly through oxytocinergic mechanisms. *Acta Physiologica Scandinavica, 149*, 199–204.

van Wunnik, B. P., Baeten, C. G., & Southwell, B. R. (2011). Neuromodulation for constipation: Sacral and transcutaneous stimulation. *Best Practice & Research Clinical Gastroenterology, 25,* 181–191.

Wang, L. (2004). Clinical observation on acupuncture treatment in 35 cases of diabetic gastroparesis. *Journal of Traditional Chinese Medicine, 24,* 163–165.

Wang, H., & Wessendorf, M. W. (2002). Mu- and delta-opioid receptor mRNAs are expressed in periaqueductal gray neurons projecting to the rostral ventromedial medulla. *Neuroscience, 109,* 619–634.

Washabau, R. J., Fudge, M., Price, W. J., & Barone, F. C. (1995). GABA receptors in the dorsal motor nucleus of the vagus influence feline lower esophageal sphincter and gastric function. *Brain Research Bulletin, 38,* 587–594.

Windle, R. J., Kershaw, Y. M., Shanks, N., Wood, S. A., Lightman, S. L., & Ingram, C. D. (2004). Oxytocin attenuates stress-induced c-fos mRNA expression in specific forebrain regions associated with modulation of hypothalamo-pituitary-adrenal activity. *The Journal of Neuroscience, 24,* 2974–2982.

Windle, R. J., Shanks, N., Lightman, S. L., & Ingram, C. D. (1997). Central oxytocin administration reduces stress-induced corticosterone release and anxiety behavior in rats. *Endocrinology, 138,* 2829–2834.

Winston, J. H., Xu, G. Y., & Sarna, S. K. (2010). Adrenergic stimulation mediates visceral hypersensitivity to colorectal distension following heterotypic chronic stress. *Gastroenterology, 138,* 294–304.

Xu, S., Hou, X., Zha, H., Gao, Z., Zhang, Y., & Chen, J. D. (2006). Electroacupuncture accelerates solid gastric emptying and improves dyspeptic symptoms in patients with functional dyspepsia. *Digestive Diseases and Sciences, 51,* 2154–2159.

Xu, G. Y., Winston, J. H., & Chen, J. D. (2009). Electroacupuncture attenuates visceral hyperalgesia and inhibits the enhanced excitability of colon specific sensory neurons in a rat model of irritable bowel syndrome. *Neurogastroenterology and Motility, 21,* 1302–e1125.

Yang, C. H., Lee, B. B., Jung, H. S., Shim, I., Roh, P. U., & Golden, G. T. (2002). Effect of electroacupuncture on response to immobilization stress. *Pharmacology, Biochemistry, and Behavior, 72,* 847–855.

Yang, C., & Yan, H. (1999). Observation of the efficacy of acupuncture and moxibustion in 62 cases of chronic colitis. *Journal of Traditional Chinese Medicine, 19,* 111–114.

Yang, J., Yang, Y., Chen, J. M., Liu, W. Y., Wang, C. H., & Lin, B. C. (2007). Effect of oxytocin on acupuncture analgesia in the rat. *Neuropeptides, 41,* 285–292.

Yin, J., Chen, J., & Chen, J. D. (2010). Ameliorating effects and mechanisms of electroacupuncture on gastric dysrhythmia, delayed emptying, and impaired accommodation in diabetic rats. *American Journal of Physiology Gastrointestinal and Liver Physiology, 298,* G563–G570.

Yoshida, M., Takayanagi, Y., Inoue, K., Kimura, T., Young, L. J., Onaka, T., et al. (2009). Evidence that oxytocin exerts anxiolytic effects via oxytocin receptor expressed in serotonergic neurons in mice. *The Journal of Neuroscience, 29,* 2259–2271.

Yoshimoto, S., Babygirija, R., Dobner, A., Ludwig, K., & Takahashi, T. (2012). Anti-stress effects of transcutaneous electrical nerve stimulation (TENS) on colonic motility in rats. *Digestive Diseases and Sciences, 57,* 1213–1221.

Yoshimoto, S., Cerjak, D., Babygirija, R., Bulbul, M., Ludwig, K., & Takahashi, T. (2012). Hypothalamic circuit regulating colonic transit following chronic stress in rats. *Stress, 15,* 227–236.

Zheng, J., Babygirija, R., Bulbul, M., Cerjak, D., Ludwig, K., & Takahashi, T. (2010). Hypothalamic oxytocin mediates adaptation mechanism against chronic stress in rats. *American Journal of Physiology Gastrointestinal and Liver Physiology, 299,* G946–G953.

Zhu, Z., Li, H., Chen, L., Wang, G., & Kan, C. (2003). Acupuncture treatment of habitual constipation. *Journal of Traditional Chinese Medicine, 23,* 133.

INDEX

Note: Page numbers followed by "*f*" indicate figures and "*t*" indicate tables.

O

P

R

S

CONTENTS OF RECENT VOLUMES

Volume 39

Volume 40

Volume 41

Volume 42

Volume 43

Volume 44

Volume 45

Volume 46

Volume 47

Volume 48

Volume 49

Volume 50

Volume 51

Volume 52

Volume 53

Volume 54

Volume 55

Volume 56

Volume 57

Volume 58

Volume 59

Volume 60

Volume 61

Volume 62

Volume 63

Volume 64

Volume 65

Volume 66

Volume 67

Volume 68

Volume 69

Volume 70

Volume 71

Volume 72

Volume 73

Volume 74

Volume 75

Volume 76

Volume 77

Volume 78

Volume 79

Volume 80

Volume 81

Volume 82

Volume 83

Volume 84

Volume 85

Volume 86

Volume 87

Volume 88

Volume 89

Volume 90

Volume 91

Volume 92

Volume 93

Volume 94

Volume 95

Volume 96

Volume 97

Volume 98

Volume 99

Volume 100

Volume 101

Volume 102

Volume 103

Volume 104

Volume 105

Volume 106

Volume 107

Volume 108

Volume 109

Volume 110

FSC
www.fsc.org
MIX
Papier | Fördert gute Waldnutzung
FSC® C083411

Kolibri 360 GmbH
Ferdinand-Jühlke-Straße 7
99095 Erfurt, Deutschland
produktsicherheit@kolibri360.de